Fretting Fatigue

CONFERENCE COMMITTEES

Organizing Committee

R. B. Waterhouse
T. C. Lindley
K. J. Miller

Review Committee

R. Akid	E. R. Leheup
J. Beard	T. C. Lindley
J. G. P. Binner	J. R. Moon
M. W. Brown	B. Noble
R. Cook	D. Nowell
I. Greaves	M. Raoof
G. W. Greenwood	D. P. Rooke
D. A. Hills	D. E. Taylor
A. Kfouri	R. B. Waterhouse

Fretting Fatigue

Edited by

R. B. Waterhouse

and

T. C. Lindley

ESIS Publication 18

Papers presented at the International Symposium on Fretting Fatigue held at the University of Sheffield.
Sponsored by the European Structural Integrity Society (ESIS).

Mechanical Engineering Publications Limited
LONDON

First published 1994

© 1994 European Structural Integrity Society

ISBN 0 85298 940 7

A CIP catalogue record for this book is available from the British Library.

Printed in Great Britain by Antony Rowe Ltd, Chippenham, Wiltshire

Typeset by Santype International Limited, Salisbury

Contents

Other titles in the ESIS Series

Introduction

The present volume contains the vast majority of the papers presented at the International Conference on Fretting Fatigue held in the University of Sheffield between 19 and 22 April 1993. The conference was sponsored by ESIS and this volume forms the latest in a series based on various aspects of fatigue including the behaviour of short cracks and the effects of environment, both aqueous and high temperature, on the development and propagation of fatigue cracks.

The contribution of fretting to the onset of fatigue failure is well established in the fields of aviation and general engineering. The vulnerable situations include press-fitted and shrink-fitted joints, riveted joints, turbine blade/disc fixings (dovetail and firtree), generator rotors, as well as a wide variety of wire rope applications such as hawsers, mooring ropes, cable railways, bridge suspensions, and overhead powerlines. Fretting fatigue is a form of contact fatigue and much recent work has concentrated on the analysis of the contact situation. Other researches have involved the susceptibility of certain materials and metallurgical structures to fretting damage as well as the influence of a number of variables such as surface roughness, residual stress and surface hardness on the process. Finally, much time and effort has been devoted to eliminating or minimizing the problem by improved design or modification of the surface by treatments, including shot peening or applications of surface coatings; all these aspects are dealt with extensively in the volume.

The delegates attending numbered forty-nine and came from fourteen countries with the UK, France, and Japan dominating the membership.

The editors particularly wish to express their thanks and appreciation for the considerable effort contributed by the local organisers from SIRIUS, namely Professor Keith Miller and Mrs June Devlin. Finally the editors' thanks are also due to the authors and referees who have made this timely publication possible.

R. B. Waterhouse
T. C. Lindley

Delegates to the Conference

Editors' Summary

The meeting was opened with a brief review by Professor Miller on short crack micromechanics and barriers to the growth of small fretting cracks. This introduction was opportune since many presentations subsequently referred to the rapid introduction of small fretting cracks at a very early stage in fatigue life and at stresses well below the fretting fatigue limit. As a consequence, much research has often been centred on the modelling of the growth of small fretting cracks with a view to understanding why, on many occasions, the cracks become non-propagating. On the other hand, it was recognised that we need to develop a better understanding of the factors controlling fretting crack formation. Moreover, as emphasized by both Hoeppner and Hattori, the 'fitness for purpose' assessment procedures in certain industries do not permit the known presence of cracks in certain components. Indeed, it became apparent during the meeting that different industries adopt quite different strategies in assessing the importance of fretting fatigue defects and the subject of 'Assessment Methodologies' was identified as a key topic for the new ESIS TC3 sub-committee on 'Contact Fatigue'.

Professor Hoeppner gave an interesting commentary on the milestones in fretting research, highlighting key contributions in the field. A valuable presentation was made by the University of Utah Group regarding the formation of fretting cracks using an innovative in-situ fretting system in a scanning electron microscope. On the same theme, optical and quantitative electron microscopy was used by Lundh and Norden to study the inter-relation between wear, deformation, and microstructure in nickel. A good understanding of mechanisms (including time-dependent fretting processes either at high temperature (Mutoh/Satoh) or in an aggressive environment (Taylor)) is vital if, in making assessments of fretting life, we are required to extrapolate with confidence beyond our experimental database.

The Oxford Contacts Mechanics Group (Hills/Nowell) presented a critical analysis of fretting fatigue experiments discussing, for example, the pros and cons of the various contact geometries previously used by earlier researchers. It is widely recognized that there is a need for standardization of fretting fatigue experiments which would allow better comparison of inter-laboratory results. The analytical papers were often concerned with the derivation of stress intensity factors to be used in fracture mechanics assessments of the growth of small fretting cracks. The pros and cons of using Green's functions, boundary element, or distributed dislocation methods were intensely debated.

Multiple crack initiation invariably occurs at fretting contacts and the paper by Dubourg and Lamacq modelled the mutual interaction between neighbouring cracks with prediction of crack retardation, and arrest in terms of crack length, distance between cracks, interfacial coefficient of friction, crack position in the contact zone, and type of loading. Regarding the inter-relation between fretting wear and fretting fatigue, Vincent demonstrated that cracks up to 2–3 mm long can develop at fretting contacts in the absence of fatigue body stresses. Professor Vincent was also involved in an interesting debate regarding the nature and role played by 'white etching layers' sometimes developed during fretting.

The practical importance of fretting fatigue was demonstrated by the wide selection of papers from many industries, and fretting wear, corrosion, and fatigue, and their interactions, continue to cause problems.

With such widespread concern regarding fretting fatigue in industry, the use of palliatives and studies to understand their behaviour was a major theme of the conference. Some palliatives reduce friction and hence surface stresses whilst others improve the material fatigue properties, e.g., peening which introduces beneficial compressive residual stresses. Perhaps not particularly innovative, but peening was universally identified as a proven method in combatting fretting fatigue in many practical situations. With all palliatives, caution is needed in the presence of excessive fretting wear.

Finally, design engineers hoping to make use of advanced materials such as metal matrix composites and ceramics should note that fretting fatigue can have an important influence on their performance.

HISTORICAL REVIEW

*D. W. Hoeppner**

Mechanisms of Fretting Fatigue

REFERENCE Hoeppner, D. W., **Mechanisms of fretting fatigue**, *Fretting Fatigue*, ESIS 18 (Edited by R. B. Waterhouse and T. C. Lindley) 1994, Mechanical Engineering Publications, London, pp. 3–19.

ABSTRACT From the time period since fretting was first identified to the present day, a great deal of progress has been made in understanding the mechanisms of fretting fatigue. After a brief review of why it is being attempted to study mechanisms of fretting fatigue, this paper reviews the developments related to increasing our understanding of fretting fatigue.

This paper emphasizes that, depending on parameters that influence fretting fatigue, a significant reduction in fatigue life may occur as a result of fretting. This reduction of life occurs due to stages of fretting fatigue that are now recognized. These are:

- cohesion of surfaces;
- breaking of cohesion;
- slip of surfaces ;
- production of fretting debris;
- corrosion of 'fresh surface' and/or debris;
- third body production;
- generation of surface damage such as pits, fretting scars, subsurface fractures, and (eventually) cracks;
- nucleation of cracks that may propagate at various angles to the surface, depending on loading conditions and the material and its microstructure and texture;
- propagation of cracks that may be influenced by either the friction forces or debris, or both, in their early stages of propagation but may become independent of contact conditions as they become longer in length;
- fatigue crack propagation that is independent of the conditions that produce fretting;
- instability.

The historical events that led to the understanding of these stages are reviewed. Major meetings and books that have aided the development of focus on fretting fatigue also are presented. The paper also discusses the effects that fretting is known to have on surfaces and the host (parent) materials.

Methods of fretting fatigue component life prediction that have led to an improvement of our ability to estimate fretting fatigue occurrences also are presented. The paper concludes with recommendations for further research in the area of fretting fatigue.

Introduction

Intensive research has been conducted to develop an understanding of fretting fatigue since Eden, Rose, and Cunningham (**1**) first reported the observation of fretting in relation to fatigue tests in 1911. The progress made in developing an understanding of fretting, including fretting corrosion and fretting wear, has been significant (**2**)–(**187**). Fretting also acts conjointly with fatigue (cyclic loading) on components and specimens both in use and in laboratory evaluation programs.

The purpose of this paper is to discuss numerous significant developments that have led to an increased understanding of the mechanism(s) of fretting

* Quality and Integrity Design Engineering Center, 3209 MEB University of Utah, Salt Lake City, Utah 84112, USA.

3

fatigue. This paper attempts to include as much material as feasible without making it excessively long. Regrettably, however, some important material and works may have been overlooked and this is by omission.

In preparation of this paper, an attempt was made to 'sift and winnow' in order to provide a comprehensive summary of the works that have been carried out on fretting fatigue mechanisms. At the outset, it is important to indicate why we wish to study and understand the challenge of fretting fatigue. Of course, for many of us the question is unimportant since we do it because we love it, and our curiosity about fretting fatigue got the better of us. There are many reasons to study fretting fatigue mechanisms besides the derivation of 'existential pleasure.' Some of these are:

(1) physical/chemical understanding drives the models of fretting fatigue;
(2) by understanding IT, the science and engineering community is guided in developing alleviation and prevention schemes;
(3) maintenance, inspection, and component replacement intervals are guided by our understanding of fretting mechanisms;
(4) the application of holistic damage tolerance concepts (134)–(175) requires models of fretting fatigue that permit more accurate estimation of the various stages of fretting fatigue;
(5) development of standard test practices for fretting fatigue require continued evolution of our understanding.

The research that has been conducted from 1911 to 1993 has assisted a great deal in all of these issues.

The following sections show the evolution of our knowledge of the fretting mechanisms from 1911 to the present. Subsequently, a discussion of future work on developing an increased understanding of fretting fatigue is presented that will assist in the five tasks cited above.

Epistemology of fretting fatigue mechanisms

Early times

Shortly after the research of Eden et al. (1), Gillet and Mack (2) performed research that showed a significant reduction in fatigue of machine grips. Shortly thereafter, Tomlinson (3) performed the first systematic investigation of fretting; he clearly recognized that corrosion is a secondary factor and that surface damage could be caused by movements of very small magnitude. He also described slip and proposed a mechanism for fretting. Numerous investigators (4)–(9) performed additional studies that aided the development of a phenomenological recognition and understanding of fretting.

In 1941, Warlow–Davies (10) conducted an extensive investigation of the effect of fretting on fatigue 'properties' of material. He was an early proponent of the idea that fretting accelerates fatigue degradation. Other investigators

(12)–(17) advanced the state of knowledge on fretting and fretting fatigue. Godfrey (14) performed extensive microscopic evaluation of fretting corrosion to attempt to characterize the nature of fretting and clearly delineate fretting mechanisms. He concluded that adhesion resulted from contact, and extremely fine particles of debris were noted to have broken loose and oxidized. By 1950 Bowden and Tabor (13) had published the first part of their classic work on friction and lubrication of solids. In 1952 Feng and Rightmire proposed a theory on fretting mechanisms (15). Also, Wright and Mann (16)(17)(27) performed extensive studies on the role of oxidation in fretting and recognized that ferric oxide was separating the two original surfaces. They observed that the formation of oxides accelerated the development of fretting damage.

By 1952 enough progress had been made to organize an ASTM symposium on fretting corrosion (18). This was a landmark event because it brought together investigators in the field and provided focus and a stimulus for additional activity. Five papers were presented at this symposium. It is interesting to note the terms which were used to describe fretting at this meeting:

– friction oxidation;
– wear oxidation;
– false brinelling;
– bleeding;
– cocoa.

Subsequently, Horger (21)(28), Uhlig (22)–(24), Waterhouse (25), and Corten (26) made contributions. Several papers on fretting fatigue were presented at the International Conference on Fatigue held in 1956 (27)–(29). Halliday and Hirst (30) and Liu, Corten, and Sinclair (31) also made important contributions. The research of Waterhouse started to make its mark in the 1950s and 1960s (32)(33).

By 1963 the field of fretting corrosion had reached such concern that the US Army (34) issued a major literature review. Shortly thereafter Bowden and Tabor (35) published the second part of their work on friction and lubrication of solids. Parts I and II of this classic work (13)(35) have had a great influence on the evolution of knowledge on fretting and fretting fatigue.

In 1964 the author's activity on fretting and fretting fatigue accelerated with work on wire rope, orthopaedic implants, fixed wing and rotary wing aircraft components including reciprocating and gas turbine engines, compressor station shafts, and energy generating equipment. It is interesting to reflect that very little knowledge existed at that time (circa 1965) in the engineering community related to the importance of fretting. The idea of a life reduction factor for fretting fatigue was emerging and alleviation and prevention practices were rapidly developed. Numerous investigators (36)–(43) developed additional knowledge on the subject. In 1968/69 the significant work of Nishioka, Nishimura, and Hirakawa (44) emerged followed by Nishioka and Hirakawa's publications (45)–(48)(56). These investigations aided our research a great deal.

They also proposed one of the early models of fretting fatigue that allowed a prediction of fretting fatigue life from knowledge of the slip amplitude, contact pressure, and materials. Endo, Goto, and Nakamura published a work in 1969 that was an early study of environment **(49)** followed by the work of Waterhouse and Taylor **(50)** related to the relative effects of fretting and corrosion.

In 1970, the first part of a NATO/AGARD manual by Barrois **(51)** emerged. This set the stage for his treatise on fretting corrosion **(82)** which was truly a significant contribution to the literature. In 1970, Hurricks **(53)** provided an extensive review of the mechanisms of fretting and noted that fretting mechanisms involved the following three stages:

- initial adhesion and metal transfer;
- production of debris in a normally oxidized state;
- steady state wear condition.

Further research by Waterhouse *et al.* **(54)(55)** and Hurricks **(57)** added additional insight to the knowledge related to the effect of environment on fretting.

In 1971 the first International Conference on corrosion fatigue was held **(58)** and numerous papers on fretting fatigue were presented **(59)–(64)**. The author summarized the major fretting mechanisms of the day **(61)** and the concept of a fretting fatigue damage threshold was presented. At about the same time that the conference proceedings were published the first book **(65)** of Waterhouse emerged. This seminal book has become a classic in the field and is used extensively, even today.

In the 1970s and 80s the author of the present paper, and colleagues, published numerous works on fretting fatigue **(59)(61)(69)(74)–(77) (90)(91a)(91b)(93)(94)(96)(100)(110)–(112)(120)**. During this period these researches concluded that fretting is predominantly influenced by mechanical surface damage. The concept of the fretting fatigue damage envelope and damage threshold was introduced. Extensive research was done on aluminium and titanium alloys during this period. Also, principles and concepts that could be used in engineering design that could either present or alleviate fretting fatigue were presented **(111)(112)**. The relative role of environment on fretting was also extensively studied.

In 1974 a specialists' meeting on fretting in aircraft systems was held in Munich, Germany **(73)** and additional impetus was provided to fretting fatigue studies related to aircraft problems.

The systems view of fretting began to emerge during this period **(61)(74)(81)**. Views of this are shown in Fig. 1 **(61)**, Fig. 2 **(81)**, and Fig. 3 **(111)(112)**. The systems view is extremely helpful in approaching both research in fretting fatigue and engineering design challenges in fretting fatigue. The later work by Czichos **(99)** is an extensive treatise on the systems approach to the science and technological challenges of tribology.

During the 1970s the number of investigations in fretting and fretting fatigue

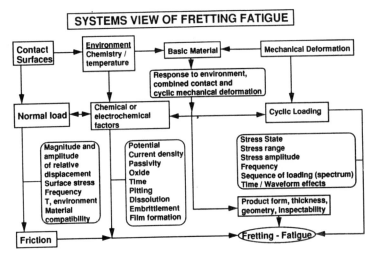

Fig 1   Systems view of fretting fatigue (61)(74)

increased markedly. Noticeable works continued to emerge from Waterhouse and colleagues (50)(54)(60)(66)(67)(70)–(72)(86)(91b)(92). Taylor's work with Waterhouse (67) on surface treatments was an important study. Furthermore, the microscopic studies conducted (67) clearly showed that the origin of fretting cracks was in the boundary between the slip and non-slip region of the contact area.

In 1975 Barrois (82) published another section of his extensive treatise on fatigue with emphasis on fretting. This too proposed a systems view of fretting fatigue.

Additional research began to emerge from Japan during this period (78)(80)(85)(108) and provided significant contributions to the field. The

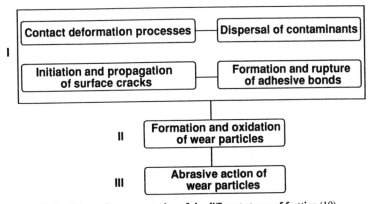

Fig 2   Schematic representation of the different stages of fretting (19)

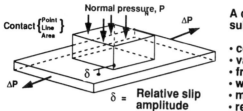

A component is
subjected to cyclic loads

• constant load amplitude
• variable load amplitude
• frequency
• waveform
• mean stress
• residual stress

FRETTING FATIGUE OF A COMPONENT
Obtain fatigue response for conditions of interest (coi)

• δ, P_N - magnitude, frequency

• material compatibility
• friction
• temperature
• environment

• stress state
• geometric detail
• material
• surface condition
• failure criteria

Fig 3   Schematic of minimum parameters to be considered in fretting fatigue (74)(111)(112)

research of Endo and Goto **(85)** is important because they applied the concepts of fracture mechanics to fretting fatigue (see also **(89)**). In 1977 an extensive review of fretting initiated fatigue was prepared by a group working on behalf of the National Research Council – National Materials Advisory Board **(88)**. The work of that group was extensive and intense and the report was an extensive review of the state of information to that time. I was privileged to be part of that group and can report that it resulted from significant fretting fatigue failures in the field. Although one of the goals of that report **(88)** was to stimulate managers and strategists, as well as researchers in fretting fatigue, it had almost the reverse effect – at least in the US. This can be noted in the relative decline of activity in the US fretting community activity from 1978 to the present day.

Additional researches in the 1970s and early 80s were conducted by numerous investigators **(75)–(109)** and by 1981 a major book on fretting fatigue emerged **(109)** and another ASTM symposium was held **(122)**. Various papers of importance on environmental effects in fretting fatigue emerged during this period **(110)–(112)(115)–(120)**. Edwards **(114)** published an important and extensive paper on the application of fracture mechanics to fretting. This research, along with Endo and Goto's earlier paper and the work of the present author **(88)(110)** and other workers has formed the basis for extensive application of fracture mechanics up to the present time.

In 1981 Bill published a noteworthy paper on a comparison between fretting wear and fretting fatigue. In the early 1980s significant developments on research on fretting began to emerge **(117)–(141)**. Some of these were more extensive study related to orthopaedic implants **(118)(125)(141)(161)(162)**. Even though fretting and fretting fatigue of orthopaedic implants was recognized by 1980, the concern was accelerated around this time since the orthopaedic surgeons and implant companies were interested in modular implants. Because these implants possessed more mechanical joints the concern for fretting damage increased markedly, especially in certain titanium alloy materials. This is due to the concern for infection in the body that may result from the debris.

A major book on contact mechanics (128) appeared and the concepts of a third body was reinforced (127)(145)(151)(157)–(159). This concept, and the analysis, were significant contributions related to improving our understanding of fretting fatigue. Nix and Lindley (129)–(131) performed extensive researches on fretting damage formation and fracture mechanics applied to fretting fatigue crack propagation. Also during this period the NATO/AGARD/SMP (Structures and Materials Panel) stimulated the production of a book on aircraft corrosion (133). This was another attempt to focus attention on the important role of corrosion, including fretting corrosion and fretting wear, on structural integrity of aircraft components. This book has been referred to extensively during the concern about 'ageing' military and commercial aircraft.

By mid 1980 Attia and colleagues were performing extensive research at Ontario Hydro in Canada (138)(142)(167). They were attempting to apply thermal evaluation and modelling techniques to the challenge of fretting. Attia also continued the work of many in ASTM (Horger, Grover, Hyler, Hoeppner, Niefert, Marble et al.) on fretting fatigue. This culminated in the ASTM symposium held in 1990 and planned by Attia and Waterhouse (179). Sato et al. also studied fretting fatigue damage formation and crack propagation during this period (147)–(150). Hattori also began his extensive studies on applying fracture mechanics to fretting fatigue (156). The concept of fretting maps emerged during this period as well (152)(167).

In 1985 in a keynote paper which the present author presented at a NATO/AGARD/SMP a plea was made for holistic damage tolerance design concepts for all critical components of gas turbine aeroengines. Also, it was emphasized that extraneous influences such as creep, corrosion, fretting, and mechanical damage could all play a role in nucleating (forming) damage that could result in fatigue crack propagation from these damage sites (134). This was, in part, based on the extensive research in the field of fretting and fretting fatigue performed up to 1985 (1)–(133). These concepts have taken hold in many companies but much remains to be done. In 1992 (175) these concepts were in part reiterated during a NATO/AGARD/SMP meeting on impact of materials defects on engine structural integrity. The paper presented by Domas (GE – USA) at this conference is of special interest.

During the 1980s other books on wear and surfaces began to emerge (155)(163)(168). Recently, ASM published a new volume of the handbook which includes an important and extensive contribution by Waterhouse (178). From 1985 to the present time the contributions on fretting fatigue mechanisms have been extensive (135)–(187). During the past eighty years significant progress has been made in understanding fretting fatigue mechanisms.

Current state of knowledge

From all of the previous research, and undoubtedly some has inadvertently been overlooked, it is clear, that depending on the conditions of interest, fretting results in the following forms of damage:

- pits;
- oxide and debris (third body);
- scratches – fretting and/or wear tracks;
- material transfer;
- surface plasticity;
- subsurface cracking and/or voids;
- fretting craters;
- cracks at various angles to the surface.

The effects of this damage on fretting fatigue are known to cause the following events related to component/system integrity:

- the surface may become unfunctional;
- fretting damage may lead to other corrosion mechanisms becoming operative;
- the debris results in contamination of the lubricant;
- cracks may form from the conjoint action of fretting fatigue that may lead to component integrity problems;
- the host body may become infected by the debris.

The knowledge that has been gained in understanding the mechanisms of fretting fatigue has done much to assist the science, engineering, and business communities to deal with this exciting phenomenon. It is now also known that the mechanisms are extremely complex, that damage formation leads to surface and near surface degradation that causes cracks to become nucleated and, under appropriate conditions, propagate. We have learned to apply concepts of linear elastic fracture mechanics, elastic–plastic fracture mechanics, and fully plastic fracture mechanics to models for fretting fatigue life estimation. Even though much progress has been made much remains to be done.

The future

In conclusion, the damage diagram developed by Jeal and Hoeppner (see e.g. (134)) will be utilized (Fig. 4). The upper portion of the figure is a conceptual view of four stages of damage nucleation and growth. As indicated in the introduction the knowledge of mechanisms relates to these phases (1)–(4) of damage nucleation (formation), structurally dependent crack propagation, stress and strain dominated crack propagation and final instability, respectively. Since these stages are interrelated, as Freudenthal and the present author have discussed on numerous occasions, it is imperative to reach greater understanding of the role of fretting in each stage and also the transition between stages.

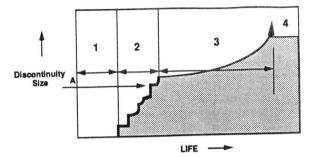

A : "FIRST" detectable crack
1 : Nucleation phase, "NO CRACK"
2 : "SMALL CRACK" phase - steps related to local structure (Anisotropy)
3 : Stress dominated crack growth, LEFM, EPFM
4 : Crack at length to produce instability

METHODS FOR EACH LIFE PHASE

Nucleation	"SMALL CRACK" Growth	Stress Dominated Crack Growth	Failure (Fracture)
Material failure Mechanism with appropriate stress / strain life data	Crack Prop. threshold related to structure (micro)	Fracture mechanics • similitude • boundary condition Data base **	K_{Ic} etc. C.O.D.
Nucleated discontinuity (not inherent) type, size, location	Structure dominated crack growth Mechanisms, rate	Appropriate stress intensity factor	Tensile / compressive buckling
Presence of malignant D*, H*	Onset of stress dominated crack growth	Initial D*, H* Size, location, type	
Possibility of extraneous effects • Corrosion	Effects of	Effects of • R ratio	
• Fretting	• R ratio	• Stress state	
• Creep	• Stress state ╱chem	• Environment ╱chem	
• Mechanical damage	• Environment ╱T	• Spectrum - waveform T	
	• Spectrum - waveform		

Fig 4   Damage nucleation and growth (134)

In the future increased emphasis needs to be provided on the following topics:

– development of standardized terminology on fretting fatigue;
– evolution of a standardized fretting fatigue experimental practice;
– planning of regular conferences and symposia on this topic to provide focus, interaction, and stimulation;
– more quantitative information on stage 1 Fig. 4, i.e., on the nucleation stage of fretting fatigue and transition to cracking;
– expanded insight into both chemical and thermal environmental effects on fretting fatigue;
– increased knowledge related to the role of multiaxial stresses and variable amplitude loading on fretting fatigue.

Additional quantification of fretting fatigue damage nucleation and growth is necessary to guide the evolution of models! It is also needed, however, to provide additional insight to the non-destructive evaluation community in order to be able to accurately assess the extent of damage both in the field and the laboratory.

The past and present activities have assisted the community a great deal to bring us to a much more knowledgeable state about fretting fatigue than when Eden *et al.* made the first report (1). Much, however, remains to be done!

Acknowledgement

My thanks go to many colleagues and students who have worked with me on fretting fatigue over the years. I am particularly indebted to Dr. Saeed Adibnazari, V. Chandrasekaran, and Thomas Mills for assistance in preparation of this manuscript and presentation. I owe much to my wife, Sue, and daughters (Laura, Lynne, and Amy) for tolerating a fretter for many years.

References

(1) EDEN, E. M., ROSE, W. N., and CUNNINGHAM, F. L. (1911) The endurance of metals, *Proc. Instn mech. Engrs*, 875.
(2) GILLET, H. W. and MACK, E. L. (1924) Notes on some endurance tests of metals, *Proc. Am. Soc. Testing Mater.*, **24**, 476.
(3) TOMLINSON, G. A. (1927) The rusting of steel surfaces in contact, *Proc. R. Soc. Lond. Ser. A.*, **115**, 472–483.
(4) PETERSON, R. E. and WAHL, A. M. (1935) Fatigue of shafts at fitted members, with a related photoelastic analysis, *Trans. Am. Soc. mech. Engrs*, **57**, A-1.
(5) EVANS, U. R. and MILEY, H. A. (1937) Measurements of oxide films on copper and iron, *Nature*, **139**, 283.
(6) ALMEN, J. O. (1937) Lubricants and false brinelling of ball and roller bearings, *Mech. Engng*, **59**, 415.
(7) CAMPBELL, W. E. and THOMAS, U. B. (1938) Films on freshly abraded copper surfaces, *Nature*, **142**, 253.
(8) CAMPBELL, W. E. (1939) Studies in boundary lubrication, *Trans. Am. Soc. mech. Engrs*, **61**, 633.
(9) TOMLINSON, G. A., THORPE, P. L., and GOUGH, H. J. (1939) An investigation of fretting corrosion of closely fitting surfaces, *Proc. Inst. mech. Engrs*, **141**, 223.
(10) WARLOW-DAVIES, E. J. (1941) Fretting corrosion and fatigue strength: Brief results of preliminary experiments, *Proc. Instn mech. Engrs*, **146**, 32.
(11) HORGER, O. J. and NEIFERT, H. R. (1941) Effect of surface conditions on fatigue properties, *Surface treatment of metals*, American Society of Metals, pp. 279–298.
(12) ALMEN, O. J. (1948) Fretting corrosion, *Corrosion handbook* (Edited by H. H. Uhlig), John Wiley, pp. 590–597.
(13) BOWDEN, F. P. and TABOR, D. (1950) The friction and lubrication of solids – Part I, Oxford University Press, Oxford.
(14) GODFREY, D. (1950) Investigation of fretting corrosion by microscopic observation, NACA Technical Note 2039.
(15) FENG, I-MING and RIGHTMIRE, B. G. (1952) The mechanism of fretting, Mass. Inst. of Tech., Cambridge, Mass., AD 4463.
(16) WRIGHT, K. H. R. (1952)–(1953) An investigation of fretting corrosion, *Proc. Instn mech. Engrs*, **1B**, 556–571.
(17) WRIGHT, K. H. R. (1952)–(1953) Discussion and communication on an investigation of fretting corrosion, *Proc. Instn mech. Engr.*, **1B**, 571–574.
(18) *Symposium on fretting corrosion, ASTM STP 144*, (1952) ASTM, Philadelphia, USA.

(19) FENG, I-MING and RIGHTMIRE, B. G. (1953) The mechanism of fretting, *Lubric. Engng*, **9**, 134–136, 158–161.
(20) McDOWELL, J. R. (1953) Fretting corrosion tendencies of several combinations of materials, *Symposium on fretting corrosion, ASTM STP 144*, ASTM, Philadelphia, USA, pp. 24–39.
(21) HORGER, O. J. (1953) Influence of fretting corrosion on the fatigue strength of fitted members. *Symposium on fretting corrosion, ASTM STP 144*, ASTM, Philadelphia, USA, pp. 40–53.
(22) UHLIG, H. H., TIERNEY, W. D., and McCLELLAN, A. (1953) Test equipment for evaluating fretting corrosion, *Symposium on fretting corrosion, ASTM STP 144*, ASTM, Philadelphia, USA, pp. 71–81.
(23) UHLIG, H. H., FENG, I-MING, *et al.* (1953) Fundamental investigation of fretting corrosion, NACA Technical Note 3029.
(24) UHLIG, H. H. (1954) Mechanism of fretting corrosion, *J. appl. mech.*, **21**, 401–407.
(25) WATERHOUSE, R. B. (1955) Fretting corrosion, *Proc. Instn mech. Engrs*, **169**, 1159–1172.
(26) CORTEN, H. T. (1955) Factors influencing fretting fatigue strength, Department of Theoretical and Applied Mechanics, University of Illinois, Report No. 88.
(27) FENNER, A. J., WRIGHT, K. H. R., and MANN, J. Y. (1956) Fretting corrosion and its influence on fatigue failure, *Proceedings of the international conference on fatigue of metals*, ASME, pp. 386–393.
(28) HORGER, O. J. (1956) Fatigue of large shafts by fretting corrosion, *Proceedings of the international conference on fatigue of metals*, ASME, pp. 352–360.
(29) ODING, I. A. and IVANOVA, V. S. (1956) Fatigue of metals under contact friction, *Proceedings of the international conference on fatigue of metals*, ASME, pp. 408–413.
(30) HALLIDAY, J. S. and HIRST, W. (1956) The fretting corrosion of mild steel, *Proc. R. Soc., Lond.*, Ser. A, **236**, 411–425.
(31) LIU, H. W., CORTEN, H. T., and SINCLAIR, G. M. (1957) Fretting fatigue strength of titanium alloy RC 130B, *ASTM*, **57**, 623.
(32) WATERHOUSE, R. B. (1961) Influence of local temperature increases on the fretting corrosion of mild steel, *J. Iron Steel Inst.*, **197**, 301–305.
(33) WATERHOUSE, R. B., BROOK, P. A., and LEE, G. M. (1962) The effect of electro deposited metals on the fatigue behavior of mild steel under conditions of fretting corrosion, *Wear*, **5**, 235.
(34) COMYN, R. H. and FURLANI, C. W. (1963) Fretting corrosion – a literature survey, United States Army Material Command, Harry Diamond Laboratories, Washington, USA.
(35) BOWDEN, F. P. and TABOR, D. (1964) The friction and lubrication of solids – Part II (Oxford University Press, London).
(36) COLLINS, J. A. and MARCO, S. M. (1964) The effect of stress direction during fretting on subsequent fatigue life, *Proceedings, ASTM*, **64**, 547.
(37) COLLINS, J. A. (1965) Fretting fatigue damage factor determination, *J. Engng Ind.*, **87**, 298.
(38) BETHUNE, B. and WATERHOUSE, R. B. (1965) Adhesion between fretting steel surfaces, *Wear*, **8**, 22–29.
(39) WATERHOUSE, R. B. and ALLERY, M. (1965) The effect of non-metallic coatings on the fretting corrosion of mild steel, *Wear*, **8**, 112–120.
(40) HARRIS, W. J. (1967) The influence of fretting on fatigue, NATO, AGARD Symposium, Advisory Report No. 8.
(41) BETHUNE, B. (1968) Adhesion of metal surfaces under fretting conditions I – like metals in contact, *Wear*, **12**, 289–296.
(42) BETHUNE, B. (1968) Adhesion of metal surfaces under fretting conditions I – unlike metals in contact, *Wear*, **12**, 369–374.
(43) BETHUNE, B. (1968) Electro chemical studies of fretting corrosion. *Wear*, **12**, 27–34.
(44) NISHIOKA, K., NISHIMURA, S., and HIRAKAWA, K. (1968) Fundamental investigation of fretting fatigue – part 1. On the relative slip amplitude of press-fitted axle assemblies, *Bull. JSME*, **11**, 437–445.
(45) NISHIOKA, K. and HIRAKAWA, K. (1969) Fundamental investigation of fretting fatigue – part 2. Fretting fatigue testing machine and some test results, *Bull. JSME*, **12**, 180–187.
(46) NISHIOKA, K. and HIRAKAWA, K. (1969) Fundamental investigation of fretting fatigue – part 3. Some phenomena and mechanisms of surface cracks, *Bull. JSME*, **12**, 397–407.

(47) NISHIOKA, K. and HIRAKAWA, K. (1969) Fundamental investigation of fretting fatigue – part 4. The effect of mean stress, *Bull. JSME*, **12**, 408–414.
(48) NISHIOKA, K. and HIRAKAWA, K. (1969) Fundamental investigation of fretting fatigue – part 5, The effect of relative slip amplitude, *Bull. JSME*, **12**, 692–697.
(49) ENDO, K., GOTO, H., and NAKAMURA, T. (1969) Effects of cycle frequency on fretting fatigue life of carbon steel, *Bull. JSME*, **12**, 1300–1308.
(50) WATERHOUSE, R. B. and TAYLOR, D. E. (1970) The relative effects of fretting and corrosion on the fatigue strength of a eutectoid steel, *Wear*, **15**, 449–451.
(51) BARROIS, W. G. (1970) Manual on the fatigue of structures – fundamental and physical aspects, NATO–AGARD Manual No. 8, NATO, AGARD Symposium.
(52) HARRIS, W. J. (1970) The influence of fretting on fatigue, NATO, AGARD Symposium, Advisory Report No. 21.
(53) HURRICKS, P. L. (1970) The mechanisms of fretting – a review, *Wear*, **15**, 389–409.
(54) WATERHOUSE, R. B., DUTTA, M. K., and SWALLOW, P. J. (1971) Fretting fatigue in corrosion environments, *Proceedings of the conference on the mechanical behavior of materials*, 3, The Society of Materials Science, Japan, pp. 294–300.
(55) WATERHOUSE, R. B. and TAYLOR, D. E. (1971) The initiation of fatigue cracks in a 0.7% carbon steel by fretting, *Wear*, **17**, 139–147.
(56) NISHIOKA, K. and HIRAKAWA, K. (1972) Fundamental investigation of fretting fatigue – part 6. Effects of contact pressure and hardness of materials, *Bull. JSME*, **15**, 135–144.
(57) HURRICKS, P. L. (1972) The fretting wear of mild steel from room temperature to 200 deg C, *Wear*, **30**, 189–212.
(58) DEVEREAUX, O. F., McEVILY, A. J., and STAEHLE, R. W. (Editors), *Corrosion fatigue: chemistry, mechanics, and microstructure*, National Association of Corrosion Engineers, USA.
(59) HOEPPNER, D. W. and UHLIG, H. H. (1972) Fretting, cavitation and rolling contact fatigue – critical introduction. *Corrosion fatigue: chemistry, mechanics, and microstructure*, National Association of Corrosion Engineers, USA, p. 607.
(60) WATERHOUSE, R. B. (1972) The effect of fretting corrosion in fatigue crack initiation, *Corrosion fatigue: chemistry, mechanics, and microstructure*, National Association of Corrosion Engineers, USA, pp. 608–616.
(61) HOEPPNER, D. W. and GOSS, G. L. (1972) Research on the mechanism of fretting fatigue, *Corrosion fatigue: chemistry, mechanics, and microstructure*, National Association of Corrosion Engineers, USA, pp. 617–626.
(62) SALKIND, M J. and LUCAS, J. J. (1972) Fretting fatigue in titanium helicopter components, *Corrosion fatigue: chemistry, mechanics, and microstructure*, National Association of Corrosion Engineers, USA, pp. 627–630.
(63) LUM, D. W. and CROSBY, J. J. (1972) Fretting resistant coatings for titanium alloys, *Corrosion fatigue: chemistry, mechanics, and microstructure*, National Association of Corrosion Engineers, USA, pp. 631–641.
(64) STARKEY, W. L. (1972) A new fretting fatigue testing machine, *Corrosion fatigue: chemistry, mechanics, and microstructure*, National Association of Corrosion Engineers, USA, pp. 642–645.
(65) WATERHOUSE, R. B. (1972) *Fretting corrosion*, Pergamon Press, USA.
(66) WATERHOUSE, R. B. and TAYLOR, D. E. (1972) Fretting debris and delamination theory of wear, *Wear*, **29**, 337–344.
(67) TAYLOR, D. E. and WATERHOUSE, R. B. (1972) Sprayed molybdenum coatings as a protection against fretting fatigue, *Wear*, **20**, 401.
(68) COLLINS, J. A. and TOVEY, F. M. (1972) Fretting fatigue mechanisms and the effect of direction of fretting motion on fretting strength, *J. Maters*, **7**, 460.
(69) GOSS, G. L. and HOEPPNER, D. W. (1973) Characterization of fretting fatigue damage by SEM analysis, *Wear*, **24**, 77–95.
(70) WATERHOUSE, R. B. and DUTTA, M. K. (1973) The fretting fatigue of titanium and some titanium alloys in a corrosive environment, *Wear*, **25**, 171–175.
(71) WHARTON, M. H., WATERHOUSE, R. B., HIRAKAWA, K., and NISHIOKA, K. (1973) The effect of different contact materials on the fretting fatigue strength of an aluminium alloy, *Wear*, **26**, 253–260.
(72) WHARTON, M. H., TAYLOR, D. E., and WATERHOUSE, R. B. (1973) Metallurgical factors in the fretting fatigue behavior of 70/30 brass and 0.7 carbon steel, *Wear*, **26**, 251.

(73) Specialists meeting on fretting in aircraft systems (1974), NATO–AGARD conference proceeding No. 161. *AGARD*.

(74) HOEPPNER, D. W. (1974) Fretting of aircraft control surfaces, Specialists meeting on fretting in aircraft systems, NATO–AGARD conference proceeding No. 161, pp. 9–13.

(75) GOSS, G. L. and HOEPPNER, D. W. (1974) Normal load effects in fretting fatigue of titanium and aluminium alloys, *Wear*, 27, 153–159.

(76) HOEPPNER, D. W. and GOSS, G. L. (1974) A fretting fatigue damage threshold concept, *Wear*, 27, 61–70.

(77) HOEPPNER, D. W. and GOSS, G. L. (1974) Metallographic analysis of fretting fatigue damage in Ti-6Al-4V MA and 7075-T6 aluminum, *Wear*, 27, 175–187.

(78) OHMAE, N., KOBAYASHI, K., and TSUKIZOE, T. (1974) Characteristics of fretting of carbon fibre reinforced plastics, *Wear*, 29, 345–353.

(79) OHMAE, N. and TSUKIZOE, T. (1974) The effect of slip amplitude on fretting, *Wear*, 27, 281–294.

(80) OHMAE, N. and TSUKIZOE, T. (1974) Prevention of fretting by ion plated film, *Wear*, 30, 299–309.

(81) CZICHOS, H. (1974) Discussion of paper 13, *op-cit* reference (74), 13-16–13-18.

(82) BARROIS, W. G. (1975) Manual on the Fatigue of Structures – II. Causes and prevention of structural damage 6. Fretting – corrosion damage in aluminum alloys, NATO, AGARD Symposium, Manual No. 9.

(83) WATERHOUSE, R. B. (1975) Fretting in hostile environment, *Wear*, 34, 301–309.

(84) MUELLER, K. (1975) How to reduce fretting corrosion – influence of lubricants, Tribology International, 57.

(85) ENDO, K. and GOTO, H. (1976) Initiation and propagation of fretting fatigue cracks, *Wear*, 38, 311–324.

(86) WATERHOUSE, R. B. and WHARTON, M. H. (1976) The behavior of three high strength titanium alloys in fretting fatigue in a corrosive environment, *Lubric. Engng*, 32, 294.

(87) Selection and use of wear tests for metals, *ASTM STP* 615, 1977, ASTM, Philadelphia.

(88) Control of fretting fatigue. (1977) Report of The Committee On Control Of Fretting-Initiated Fatigue, National Materials Advisory Board, Commission on Sociotechnical Systems, National Research Council, Publication NMAB-33, National Academy of Sciences, Washington, D.C.

(89) HOEPPNER, D. W. (1977) Comments on initiation and propagation of fretting fatigue cracks (letter to the editor), *Wear*, 43, 267–270.

(90) REEVES, R. K. and HOEPPNER, D. W. (1977) An apparatus for investigating fretting fatigue in vacuum, *Wear*, 45, 127–134.

(91a) HOEPPNER, D. W. and SALIVAR, G. C. (1977) The effect of crystallographic orientation on fatigue and fretting initiated fatigue of copper single crystals, *Wear*, 43, 227–237.

(91b) WATERHOUSE, R. B. (1977) The role of adhesion and delamination in the fretting wear of metallic materials, *Wear of Materials 1977*, (Edited by W. A. Glaeser, K. C. Ludema, and S. K. Rhee), ASME, p. 55.

(92) WATERHOUSE, R. B. (1978) Effect of environment in wear processes and the mechanisms of fretting wear, *Fundamentals Tribol.*, 567–584.

(93) REEVES, R. K. and HOEPPNER, D. W. (1978) The effect of fretting on fatigue (short communication), *Wear*, 40, 395–397.

(94) REEVES, R. K. and HOEPPNER, D. W. (1978) Microstructural and environmental effects on fretting fatigue, *Wear*, 47, 221–229.

(95) ENDO, K. and GOTO, H. (1978) Effect of environment on fretting fatigue, *Wear*, 48, 347.

(96) REEVES, R. K. and HOEPPNER, D. W. (1978) Scanning electron microscope analysis of fretting fatigue damage, *Wear*, 48, 87–92.

(97) ALYAB'EV, A. YA., SHEVELYA, V. V., *et al*. (1978) The effect of plasma spraying on the fatigue strength of 30KhGSA steel under fretting corrosion conditions, Plenum Publishing, USA.

(98) SPROLES, Jr., E. S. and DUQUETTE, D. J. (1978) The mechanism of material removal in fretting, *Wear*, 49, 339–352.

(99) CZICHOS, H. (1978) *Tribology – a systems approach to the science and technology of friction, lubrication and wear*, Elsevier, Amsterdam.

(100) POON, C. and HOEPPNER, D. W. (1979) The effect of environment on the mechanism of fretting fatigue, *Wear*, 52, 175–191.

(101) WHARTON, M. H. and WATERHOUSE, R. B. (1990) Environmental effects in fretting fatigue of Ti-6Al-4V, *Wear*, **62**, 287–297.

(102) ALIC, J. A. and KANTIMATHI, A. (1979) Fretting fatigue, with reference to aircraft structures, SAE paper No. 790612, Business Aircraft Meeting and Exposition, Century II.

(103) ALIC, J. A., HAWLEY, A. L., and UREY, J. M. (1979) Formation of fretting fatigue cracks in 7075-T7351 aluminum alloy, *Wear*, **56**, 351–361.

(104) ALIC, J. A. and HAWLEY, A. L. (1979) On the early growth of fretting fatigue crack, *Wear*, **56**, 377–389.

(105) BARROIS, W. (1979) Repeated plastic deformation as a cause of mechanical surface damage in fatigue, wear, fretting-fatigue, and rolling fatigue: a review, French Air Force, Report No. ICAF–1116.

(106) WATERHOUSE, R. B. and SAUNDERS, D. A. (1979) The effect of shot peening on the fretting fatigue behavior of an austenitic stainless steel and a mild steel, *Wear*, **53**, 381–386. 1

(107) ALYAB'EV, A. Y. A., SHEVELYA, V. V., *et al.* (1980) *The resistance of thick layer chromium coatings to fretting corrosion* (Planum Publishing, USA).

(108) SATO, J., IGARASHI, J., and SHIMA, M. (1980) Fretting of glass, *Wear*, **65**, 55–65.

(109) WATERHOUSE, R. B. (1981) *Fretting fatigue*, (Edited by R. B. Waterhouse), Applied Science Publishers, UK.

(110) HOEPPNER, D. W. (1981) Environmental effects in fretting fatigue, *Fretting fatigue*, (Edited by R. B. Waterhouse), Applied Science Publishers, UK, pp. 143–158.

(111) HOEPPNER, D. W. (1981) Material/Structure degradation due to fretting and fretting-initiated fatigue, *Canadian Aeronautics Space J.*, **27**, 213–221.

(112) HOEPPNER, D. W. and GATES, F. L. (1981) Fretting fatigue considerations in engineering design, *Wear*, **70**, 155–164.

(113) Specialist meeting on corrosion fatigue NATO–AGARD. (1981), Fifty-second meeting of the structures and materials panel.

(114) EDWARDS, P. R. (1981) The application of fracture mechanics to predicting fretting fatigue, *Fretting fatigue*, (Edited by R. B. Waterhouse), Applied Science Publishers, UK, pp. 67–99.

(115) BILL, R. C. (1981) The role of oxidation in the fretting wear process, International Conference on wear of materials.

(116) BILL, R. C. (1981) Fretting wear and fretting fatigue: how are they related, National Aeronautics and Space Administration, Lewis Research Center, Report No. NASA–TM–82633.

(117) LEADBEATER, G., KOVALEVSKII, V. V., NOBLE, B., and WATERHOUSE, R. B. (1981) Fractographic investigation of fretting-wear and fretting fatigue in aluminium alloys, *Fatigue Engng Mater. Structures*, **3**, 237–246.

(118) BROWN, S. A. and MERRITT, K. (1981) Fretting corrosion in saline and serum, *J. Biomedical Mater. Res.*, **15**, 479–488.

(119) *Proceedings of the SAE fatigue conference* (1982) Society of Automotive Engineers, PA, USA.

(120) HOEPPNER, D. W. (1982) Corrosion and fretting effects on fatigue, *Proceedings of the SAE fatigue conference*, Society of Automotive Engineers, PA, USA, pp. 111–119.

(121) MANN, D. S. (1982) *The design and development of an experimental apparatus for fracture mechanics based fretting fatigue studies with electrohydraulic closed loop servo-control of axial load, normal load and slip amplitude*, MA Thesis, University of Toronto, Canada.

(122) Materials evaluation under fretting conditions, *ASTM STP 780*, 1982, ASTM, Philadelphia.

(123) KUSNER, D., POON, C., and HOEPPNER, D. W. (1982) A new machine for studying surface damage due to wear and fretting, *Materials evaluation under fretting conditions*, *ASTM STP 780*, ASTM, Philadelphia, USA, PP. 17–29.

(124) BILL, R. C. (1982) Review of factors that influence fretting wear, *Materials evaluation under fretting conditions*, *ASTM STP 780*, ASTM, Philadelphia, USA, pp. 165–182.

(125) COOK, S. D., GIANOLI, G. J., CLEMOW, A. J. T., and HADDAD, JR. (1983–84), Fretting corrosion in orthopaedic alloys, 1 *Biomat., Med., Dev., Art. Org.*, **11**, 281–292.

(126) WATERHOUSE, R. B. (1984) Fretting wear, *Wear*, **100**, 107–118.

(127) COLOMBIE, C., BERTHIER, Y. FLOQUET, A. VINCENT, L., and GODET, M. (1984) Fretting: Load carrying capacity of wear debris, *Trans ASME*, **106**, 194.

(128) JOHNSON, K. L. (1985) *Contact mechanics*, Cambridge University Press, UK.

(129) LINDLEY, T. C. and NIX, K. J. (1985) The role of fretting in the initiation and early growth of fatigue cracks in turbo-generator materials, *Conference: multiaxial fatigue*, ASTM, Philadelphia, USA, pp. 340–360.

(130) NIX, K. J. and LINDLEY, T. C. (1985) The application of fracture mechanics to fretting fatigue, *Fatigue Fracture Engng Mater. Structures*, **8**, 143–160.

(131) NIX, K. J. and LINDLEY, T. C. (1985) The initiation and propagation of small defects in fretting fatigue, *Conference: life assessment of dynamically loaded materials and structures*, Engineering Materials Advisory Services Limited, UK, pp. 89–98.

(132) HARRIS, S. J., OVERS, M. P., and GOULD, A. J. (1985) The use of coatings to control fretting wear at ambient and elevated temperatures, *Wear*, **106**, 35–52.

(133) WALLACE, W. and HOEPPNER, D. W. (1985) AGARD Corrosion Handbook, Vol. 1, Aircraft corrosion: cause and case histories, AGARDograph No. 278, VI, NATO, AGARD Symposium.

(134) HOEPPNER, D. W. (1985) Parameters that input to application of damage tolerance concepts to critical engine components, in NATO–AGARD CP – 393, Conference on damage tolerance concepts for critical engine components, NATO–AGARD.

(135) TANAKA, K., MUTOH, Y., SAKODA, S., and LEADBEATER, G. (1985) Fretting fatigue in 0.55C spring steel and 0.45C carbon steel, *Fatigue Fracture Engng Mater. Structures*, **8**, 129–142.

(136) IWABUCHI, A. (1985) Fretting wear of Inconel 625 at high temperature in high vacuum, *Wear*, **106**, 163–175.

(137) WATERHOUSE, R. B. and IWABUCHI, A. (1985) The composition and properties of surface films formed during the high temperature fretting of titanium alloy, *Proceedings of the international tribology conference*, Japan Society of Lubrication Engineers, Tokyo, Japan, pp. 53–58.

(138) ATTIA, M. H. and D'SILVA, N. S. (1985) Effect of mode of motion and process parameters on the prediction of temperature rise in fretting wear, *Wear*, **106**, 206–224.

(139) SATO, J. (1985) Fundamental problems of fretting wear, *Proceedings of the JSLE international tribology conference*, Elsevier Science, NY, USA, pp. 635–640.

(140) SATO, J., SHIMA, M., and TAKEUCHI, M. (1985) Fretting wear in sea water, *Proceedings of the JSLE international tribology conference*, Japan Society of Lubrication Engineers, Tokyo, Japan, pp. 47–52.

(141) BROWN, S.A. and MERRIT, K. (1985) Fretting corrosion of plates and screws: an in vitro test method, *Corrosion and degradation of implant materials: second symposium, ASTM STP 859*, (edited by A. C. Fraker and C. D. Griffin), ASTM, Philadelphia, USA, pp. 105–116.

(142) ATTIA, M. H. and KO, P. L. (1986) On the thermal aspects of fretting wear-temperature measurements in the subsurface layer, *Wear*, **111**, 363–376.

(143) WATERHOUSE, R. B. (1986) Fretting wear of nitrogen bearing austenitic stainless steel at temperatures to 600 deg, *J. Tribology*, **108**, 359–370.

(144) BRYGGMAN, U. and SODERBERG, S. (1986) Contact conditions in fretting, *Wear*, **110**, 1–17.

(145) COLOMBIE, C., BERTHIER, Y., VINCENT, L., and GODET, M. (1986) How to choose coatings in fretting, *Conference: advances in surface treatments: technology, applications and effects*, Pergamon Press, UK, pp. 321–334.

(146) WATERHOUSE, R. B. (1986) Residual stresses and fretting, crack initiation and propagation, *Conference: advances in surface treatments, technology applications, effects*, Pergamon Press, UK, pp. 511–525.

(147) SATO, K., FUJII, H., and KODAMA, S. (1986) Crack propagation behavior in fretting fatigue, *Wear*, **110**, 19–34.

(148) SATO, K., FUJII, H., and KODAMA, S. (1986) Crack propagation behavior in fretting fatigue of S45C Carbon steel, *Bull. JSME*, **29**, 3253–3258.

(149) SATO, K., FUJII, H., and KODAMA, S. (1986) Fretting fatigue crack propagation behavior of A2024–T3 aluminium alloy, *J. Soc. Mater. Sci. Jpn*, **34**, 1076–1081.

(150) SATO, K., FUJII, H., and KODAMA, S. (1986) Effects of stress ratio fretting fatigue cycles on the accumulation of fretting fatigue damage to carbon steel S45C, *Bull. JSME*, **29**, 2759.

(151) BERTHIER, Y., COLOMBIE, CH., VINCENT, L., and GODET, M. (1987) Fretting wear mechanisms and their effects on fretting fatigue, *ASLE/ASME Tribology Conference*, ASME, USA.

(152) VINGSBO, O. B. and SODERBERG, S. (1987) On fretting maps, *Conference: wear of materials 1987*, ASME, NY, 1987, 885–894.

(153) DOBROMIRSKI, J. and SMITH, I. O. (1987) Fretting fatigue failure, 1987 Symposium of the Australian Fracture Group; Publ: University of Sydney, N.S.W., Australia, 1987, 179–196.

(154) NOWELL, D., HILLS, D. A., and O'CONNOR, J. J. (1987) An analysis of fretting fatigue, Tribology – friction, lubrication and wear fifty years on, Mechanical Engineering Publications, London, UK, pp. 965–973.

(155) BUDINSKI, K. G. (1988) *Surface engineering*, Prentice Hall, New Jersey, USA.

(156) HATTORI, T., NAKAMURA, M., SAKATA, H., and WATANABE, T. (1988) Fretting fatigue analysis using fracture mechanics. *JSME Int. J. I.*, **31**, 100–107.

(157) BERTHIER, Y., FLAMAND, L., GODET, M., SCHMUCK, J. and VINCENT, L. (1988) Tribological behavior of titanium alloy Ti–6Al–4V, Sixth World Conference on Titanium.

(158) BERTHIER, Y., COLOMBIE, CH., VINCENT, L., and GODET, M. (1988) Fretting wear mechanisms and their effects on fretting fatigue, *J. Tribology*, **110**, 517–524.

(159) BERTHIER, Y., GODET, M., and VINCENT, L. (1988) Fretting wear and fatigue: initiations, mechanisms, and prevention, *Mec. Matel. Elect.*, **428**, 20–26.

(160) COLLIN, G., GUERIN, J. J., BARATTO, G., and MONGIS, J. (1988) Fretting Corrosion, *Cetim Inf*, **107**, 47–52.

(161) BROWN, S. A., HUGHES, P. J., and MERRITT, K. (1988) In vitro studies of fretting corrosion of orthopaedic materials, *J. Orthopaedic Res.*, **6**, 572–579.

(162) MERRITT, K. and BROWN, S. A. (1988) Effect of proteins and pH on fretting corrosion and metal ion release, *J. Biomedical Mater. Res.*, **22**, 111–120.

(163) Metal Behavior and Surface Engineering (1989) (Edited by S. Curioni, R. B. Waterhouse, and D. Kirk). IITT – International, Technology Transfer Series, France.

(164) TAYLOR, D. E. and WATERHOUSE, R. B. (1989) Wear, fretting and fretting fatigue, metal behavior and surface engineering, (Edited by S. Curioni, R. B. Waterhouse, and D. Kirk), IITT – International, Technology Transfer Series, France, pp. 13–36.

(165) TROSHCHENKO, V. T., TSYBANEV, G. V., and KHOTSYANOVSKII, A. O. (1989) Life of steels in fretting fatigue, **20**, 703–709.

(166) ATTIA, M. H. (1989) A thermally controlled fretting wear tribometer – a step towards standardization of test equipment and methods. International Conference on Wear of Materials, p. 709.

(167) ATTIA, M. H. (1989) Fretting fatigue testing: Current practice and future prospects for standardization, *ASTM Standard News*, **17**, 26–31.

(168) *Surface engineering*, (1990) (Edited by S. A. Meguid), Elsevier Science Publishing, NY, USA.

(169) WATERHOUSE, R. B. (1990) The effect of surface treatments on the fretting wear of an aluminum alloy (RR58)/steel (BS 970 080 M40) couple, *Surface engineering*, (Edited by S. A. Meguid), Elsevier Science, NY, USA, pp. 325–334.

(170) SINNOTT, M. M. and HOEPPNER, D. W. (1990) Degradation of a biomedical polyurethane: Surface integrity and fatigue effects, *Surface engineering* (Edited by S. A. Meguid), Elsevier Science Publishing, NY, USA), pp. 335–343.

(171) NAKAZAWA, K., SUMITA, M., and MARUYAMA, N. (1990) Saturation of damage in fretting fatigue of high strength steels in sea water. *J. Iron Steel Inst. Jpn*, **76**, 917–923.

(172) NOWELL, D. and HILLS, D. A. (1990) Crack initiation criteria in fretting fatigue, *Wear*, **136**, 329–343.

(173) NOWELL, D. (1990) An analysis of fretting fatigue, *Dissertation Abs Int.*, **50**, 245.

(174) ADIBNAZARI, S. (1991) *Investigation of fretting fatigue mechanisms on 7075–T6 aluminum alloy and Ti–6Al–4V titanium alloy*, PhD Thesis, University of Utah, USA.

(175) HOEPPNER, D. W. (1992) History and prognosis of material discontinuity effects on engine components structural integrity, Impact of Materials Defects on Engine Structural Integrity, to be published.

(176) ASM Handbook. (1992), Vol. 18, *Friction, Lubrication, and Wear Technology*, ASM International, USA.

(177) ADIBNAZARI, S. and HOEPPNER, D. W. (1992) Characteristics of the fretting fatigue damage threshold, *Wear*, **159**, 43–46.

(178) WATERHOUSE, R. B. (1992) Fretting wear, ASM Handbook, Vol. 18, *Friction, Lubrication, and Wear Technology*, ASM International, USA.

(179) *Standardization of fretting fatigue test methods and equipment, ASTM STP 1159,* (Edited by M. H. Attia and R. B. Waterhouse), ASTM, Philadelphia, USA, p. 1992.

(180) WATERHOUSE, R. B. (1992) A historical introduction of fretting fatigue, *Standardization of fretting fatigue test methods and equipment, ASTM STP 1159,* (Edited by M. H. Attia and R. B. Waterhouse), ASTM, Philadelphia, USA, p. 8.

(181) WATERHOUSE, R. B. (1992) The problems of fretting fatigue testing, *Standardization of fretting fatigue test methods and equipment, ASTM STP 1159,* (Edited by M. H. Attia and R. B. Waterhouse), ASTM, Philadelphia, USA, pp. 13–22.

(182) HOEPPNER, D. W. (1992) Mechanisms of fretting fatigue and their impact on test methods development. *Standardization of fretting fatigue test methods and equipment, ASTM STP 1159,* (Edited by M. H. Attia and R. B. Waterhouse), ASTM, Philadelphia, USA. pp. 23–32.

(183) ADIBNAZARI, S. and HOEPPNER, D. W. (1992) A fretting fatigue normal pressure threshold concept. *Wear,* **160**, 33–35.

(184) HOEPPNER, D. W., ADIBNAZARI, S., and MOESSER, M. W. Literature review and preliminary studies of fretting and fretting fatigue including special applications to aircraft joints, QIDEC–U. of UTAH report to US Department of Transportation to be issued as a NIST document.

(185) MOESSER, M., ADIBNAZARI, S., AND HOEPPNER, D. W. (1994) Finite element model of fretting fatigue with variable coefficient of friction over time and space, *Fretting Fatigue,* Mechanical Engineering Publications, London, pp. 103–109. *This volume.*

(186) ADIBNAZARI, S. and HOEPPNER, D. W. (1993) The role of normal pressure in modelling fretting fatigue, *Fretting Fatigue,* Mechanical Engineering Publications, London, pp. 125–133. *This volume.*

(187) ELLIOTT, C. B. and HOEPPNER, D. W. (1993) A fretting fatigue system usable in a scanning microscope, *Fretting Fatigue,* Mechanical Engineering Publications, London, pp. 211–217. *This volume.*

ANALYTICAL METHODS

*D. P. Rooke**

The Development of Stress Intensity Factors

REFERENCE Rooke, D. P., **The development of stress intensity factors,** *Fretting Fatigue,* ESIS 18 (Edited by R. B. Waterhouse and T. C. Lindley) 1994. Mechanical Engineering Publications, London, pp. 23–58.

ABSTRACT Cracks are costly: their presence increases the time and effort spent on maintenance and repair and they may ultimately lead to component fracture and subsequent structural failure, which in extreme cases can endanger human life. A US National Committee estimated that the cost of fracture amounted to as much as 4 percent of the Gross National Product: it is, therefore, imperative to try and reduce these effects. Since cracks cannot be eliminated, procedures must be devised to quantify and predict the behaviour of cracked structures under service conditions. The cracks themselves may be present as small flaws in the material manufacturing stage; they may arise during fabrication, or they may be the result of damage (fatigue, impact, corrosion etc) to the completed structure. Systematic scientific rules must be devised to characterize cracks and their effects and to predict if and when they may become unsafe during the structures' operational service life. This science has come to be termed 'fracture mechanics' and the prime characterization and prediction parameter is the stress intensity factor. This paper outlines the sources and methods for obtaining stress intensity factors for a wide range of structural configurations in both two and three dimensions.

Introduction

The first systematic investigation of fracture phenomena was carried out by A. A. Griffith (1) seventy years ago. His experimental studies were carried out on a brittle material, glass. In order to explain the measured strengths of different glass rods, Griffith postulated the existence of crack-like flaws which could increase in size under the action of external loads. Griffith further postulated that the growth of cracks was controlled by the balance between the available strain energy and the energy required to form new crack surfaces. This theory allows the strength of brittle solids to be estimated, and provides a relationship between fracture strength and defect size.

Irwin (2) and Orowan (3) independently pointed out that for less brittle materials Griffith's energy balance must be between stored strain energy and the surface energy plus the work done in plastic deformation. Irwin also recognized that for relatively ductile materials the energy required to form new crack surfaces is generally insignificant compared to the work done in plastic deformation. Thus if the strain energy release rate is greater than the 'total energy absorbed' during crack extension, further crack growth will occur. Another important contribution by Irwin (4) was the recognition of the universality of the crack tip stress field; it always has the same functional dependence on the spatial coordinates, and the stress magnitudes are determined by a single parameter, the stress intensity factor K. Furthermore, Irwin showed

* DRA, Farnborough, Hampshire, GU14 6TD, UK.

that the 'energy-balance' approach to the characterization of fracture is equivalent to the 'critical stress intensity factor' approach.

The fundamental postulate of Linear Elastic Fracture Mechanics (LEFM) is
that the behaviour of cracks (i.e., whether they grow or not and how fast they
grow) is determined solely by the value of the stress intensity factor. The above
fundamental postulate implies that it is necessary to evaluate the stress intensity factor in order to be able to predict the behaviour of cracked solids. This
factor, which is a function of the applied loading and the geometry of the
cracked component, has been evaluated for many hundreds of structural configurations. The methods available for obtaining stress intensity factor solutions to crack problems form the main part of this paper. Many of the
available solutions for simple structural configurations, obtained by using
these methods, have been collected together in handbooks **(5)–(8)**.

The stress intensity factor once determined is used in two main areas:

(1) the determination of the static strength of a crack structure (residual
    strength);
(2) the determination of the rate of growth of a crack in a structure subjected
    to variable loading (fatigue).

The solution of the equations of linear elasticity for a cracked body always
(4) exhibits singularities in the stress field at the crack tips. In general the
singularity is proportional to $r^{-1/2}$ where r is the distance from the crack tip.
The constant of proportionality is the stress intensity factor. The presence of
this theoretical singularity imposes limitations on the procedures (analytical
and numerical) that may be used to solve crack problems. The extent to which
the stress singularity is modelled largely determines the accuracy of the resultant stress intensity factor calculation. Several analytical solutions have been
obtained for the stress field in the vicinity of the tip of a crack in an infinite
domain. In two dimensions the work of Westergaard **(9)** and Williams **(10)** are
particularly useful, and in three dimensions that of Sneddon **(11)**.

It has been established that a knowledge of the stress intensity factor K is
necessary for the calculation of the residual strength of cracked structural
components, the evaluation of critical crack lengths and the deduction of rates
of crack growth in fatigue, etc. Over the last forty years many methods of
obtaining K solutions have been developed. Several methods are set out in
Table 1, where they are divided into three categories depending on their
degree of sophistication and the times required to obtain a solution. For
simple geometrical configurations, or where a complex structure can be simply
modelled, it may be possible to use one of the reference books **(5)–(8)**. Where
a solution cannot be obtained directly from a reference book then one of the
relatively simple methods in stage 2 may be adequate; they will seldom require
more than a few man-hours to obtain K values.

Although for most practical crack problems in real engineering structures
stage 2 methods cannot produce very accurate solutions, the worth of these

Table 1   Methods of determining stress intensity factors

Stage 1	Stage 2	Stage 3
Handbooks	Superposition	Collocation (mapping)
	Stress concentration	Integral transform/continuous dislocation
	Stress distribution	Body force method
	Compounding	Edge function method
	Green's function	Method of lines
	Weight function	Finite element method
		Boundary element method
		Alternating technique

methods should not be underestimated. Many of the simple analytical and semi-empirical methods clearly illustrate the underlying physical principles that govern crack problems in general. An understanding of these principles is necessary in order to model all the necessary features of a particular crack problem before attempting a numerical solution. These simpler methods can also be used in enabling rough, approximate K solutions to be obtained relatively quickly. The use of more than one model and/or more than one method may enable upper and lower bounds to be placed on the K solutions to the real problem.

When a particular stress intensity factor is required repeatedly, say for a standard test-piece, and high accuracy is important, then numerical methods in stage 3 would be used. These methods will also be needed for complex structural configurations. In general, methods based on boundary elements or finite elements will be the most widely applicable.

This paper includes some of the methods listed in stages 2 and 3. The list is by no means exhaustive, but ample references are given; a more comprehensive list is available in reference **(12)**. The chronological development of methods largely parallels the ability to solve more complex (and realistic) structural problems. In particular, important structural aspects such as the proximity of boundaries and the presence of stress concentrations are emphasized, and various techniques which are used in different methods are highlighted and their universality demonstrated. Although most of the examples shown are for static problems, many of the numerical techniques can be extended to consider dynamic problems, such as impact loading, or fast-moving cracks.

Many of various methods described in this paper have been used for determining the stress intensity factors required in the fracture mechanics analysis of fretting fatigue. Early work in the realm of fretting analysis assumed a linear response; this is not always adequate in practice. In general, the contact area changes with load and there is friction between the rubbing surfaces, and so fretting problems are often non-linear. Techniques of analysis, in particular the boundary element method and methods based on continuous dislocations are currently being developed.

Historical development

Crack growth phenomena in glass were first described in the 1920s by Griffith (1) in terms of an energy-balance concept. This concept was extended to less brittle materials by Irwin (2) and Orowan (3). The theoretical stress analysis of crack and notch problems was simultaneously being developed by Westergaard (9), Williams (10), Sneddon (11), Neuber (13), Muskhelishvili (14), and many others. In 1957 Irwin (4) showed that crack-tip stress fields could always be characterized by a single parameter, the stress intensity factor K, which was uniquely related to the strain-energy release rate introduced by Griffith (1), some thirty-five years previously.

During the late 1950s there were further important developments. Irwin (15) and Dugdale (16) studied ways of incorporating the plastic-zone phenomena into linear elastic fracture mechanics. Bueckner (17) published an important principle which permitted stress intensity factors to be calculated from a knowledge of the stress-field in the uncracked body and crack Green's functions or influence functions. In 1961 the first Symposium on Crack Propagation was held at Cranfield, UK; only three of the papers in the proceedings were concerned with stress intensity factors. Three years later Paris (18) demonstrated that fatigue crack growth was largely determined by the range of the stress intensity factor during the operational load cycle. This established fracture mechanics, based on the stress intensity factor, as a valuable engineering design tool for both strength and life calculations. There are now statutory requirements in many fields of engineering which depend on the principles of fracture mechanics; a fact which demonstrates its effectiveness and reliability.

Energy balance

The first systematic study of fracture phenomena was carried out by Griffith (1) who measured the tensile strengths of glass rods. He observed that freshly drawn glass rods fractured at a higher stress than old rods of the same diameter and that thin rods fractured at a higher stress than thick rods. To explain these phenomena he postulated the existence of crack-like flaws which weakened the glass rods; fracture occurred when these flaws spread across the section. How failure initiated at sharp cracks could not be explained by extrapolating what was known about notches. Griffith suggested that the criterion for failure due to crack growth was determined by the balance of strain energy and surface energy. He postulated that if the strain energy released by the strain field when the crack advanced a small distance was greater than the energy required to form the new surfaces then unstable crack growth would occur, that is if

$$G \geqslant 2\Gamma \tag{1}$$

where G is the strain energy release rate (per unit area of crack growth) and Γ is the work required to form unit area of new crack surface (two surfaces). The strain energy release rate G is a function of the loading and the crack size.

By considering a crack as the limiting case of a thin elliptical cavity Griffith showed (1) that the strain energy release rate was given by

$$G = \frac{\pi\sigma^2 a}{E} \tag{2}$$

where σ is the tensile stress remote from the crack in a direction perpendicular to the crack, $2a$ is the crack length, and E is the Young's modulus of the material. From equation (2) it can be seen that at the onset of instability $(G = 2\Gamma)$ the failure stress σ_c is related to the critical crack length a_c as follows

$$\sigma_c\sqrt{a_c} = \text{constant} \tag{3}$$

This functional dependence was verified experimentally.

Griffith's experiments were conducted on glass, a 'brittle' material, which fractures with little or no permanent deformation. Most structural materials, for example metals, are ductile which means fracture is accompanied by permanent deformation. Irwin (2) and Orowan (3) independently suggested that Griffith's energy-balance criterion could be extended to ductile materials. They suggested that failure by unstable crack-growth would occur if

$$G \geqslant 2\Gamma + \Delta \tag{4}$$

where Δ is the non-recoverable work associated with the permanent deformation at the crack tip. For ductile materials such as metals $\Delta \gg \Gamma$, it therefore follows that the energy-balance criterion becomes

$$G \geqslant \Delta \tag{5}$$

This relationship illustrates why more work must be done to fracture a ductile material than to fracture a brittle material. Materials with a large work of fracture are said to be 'tough' – the parameter used to measure the ability of a material to withstand cracks is called the 'fracture toughness'.

Stress analysis

Westergaard (9) developed stress functions in 1939 which automatically satisfy traction-free conditions at the crack faces; his functions $Z(z)$ are frequently used to solve two-dimensional problems in cracked structures. The stress fields for opening-mode (mode I) deformation are given (9) by

$$\sigma_{xx} = Re\ \{Z(z)\} - yIm\ \{Z'(z)\} \tag{6}$$

$$\sigma_{yy} = Re\ \{Z(z)\} + yIm\ \{Z'(z)\}$$

and

$$\sigma_{xy} = -yRe\ \{Z'(z)\}$$

and the displacement fields from

$$2\mu u_x = \tfrac{1}{2}(\kappa - 1)\ Re\ \{\tilde{Z}(z)\} - yIm\{Z(z)\} \tag{7}$$

and

$$2\mu u_y = \tfrac{1}{2}(\kappa + 1)\ Im\ \{\tilde{Z}(z)\} - yRe\ \{Z(z)\}$$

where z is the coordinate $(x + iy)$, $\tilde{Z} = \int Z(z)\ dz$ and μ and κ are elastic constants. Similar expressions exist for the other modes of deformation (II and III). Westergaard stress functions are a special type of stress function which is specific to cracks. More general complex stress functions have been developed by Muskhelishvili (14), and have been used by Eftis, Subramonian, and Liebowitz (19), among others, to solve crack problems. The use of Westergaards stress functions has been generalized to finite regions by Thein Wah (20).

In 1946 Sneddon studied a three-dimensional crack problem. He obtained (11) the asymptotic behaviour of the stress field near the tip of a penny-shaped crack in an infinite solid and showed that the field for the two-dimensional crack was essentially the same as Westergaard's (for a one-dimensional crack) apart from a multiplicative constant.

Neuber's (13) work on notch stresses has been very valuable in the development of stress intensity factors. Although he did not study cracks per se the stress concentrations he derived for many narrow notch configurations have been used, coupled with limiting procedures, to obtain approximate stress intensity factors for several crack configurations. The stress and displacement fields for a sharp notch were obtained in a series form by Williams (10). In the limit as the notch angle approaches zero, a notch becomes a crack, and Williams' analysis is applicable to cracks in infinite sheets. His fields are the basis of a singularity subtraction technique, much used in the numerical analysis of crack problems.

Stress intensity factors

Irwin (4) solved several two-dimensional crack problems in linear elasticity and showed that the stress field in the vicinity of the crack tip was always of the same form. He showed that the stress field component σ_{ij} at the point (r, θ) near the crack tip is given by

$$\sigma_{ij}(r, \theta) = \frac{K}{\sqrt{(2\pi r)}}\ f_{ij}(\theta) + \text{other terms} \tag{8}$$

where the origin of the polar coordinates (r, θ) is at the crack tip and $f_{ij}(\theta)$ contains trigonometric functions. As the coordinate r tends to zero the first term in equation (8) dominates; the other terms are constant or tend to zero. The constant K in the first term is known as the stress intensity factor. It therefore follows that the stress field in the vicinity of the crack tip is characterized by the stress intensity factor.

By considering the elastic work to close up the tip of a crack Irwin **(4)** derived a relationship between the strain energy release rate and the stress intensity factor; it was

$$G \propto K^2 \tag{9}$$

The constant of proportionality in equation (9) is a function of the elastic constants of the material. This relationship provides a link between the crack tip stress field and the energy-balance criterion for crack growth which can now be interpreted in terms of a critical K value that is required for crack growth.

Since the basic assumption of linear elastic fracture mechanics is that the growth of a crack (stable or unstable) is controlled by the stress field at the crack tip it follows that the crack growth will be characterized by the parameter K. This implies that two different cracks which have the same value of K will behave in the same manner. Thus, in order to predict the growth of a crack in a structural component it is necessary to know the value of K. In general K will be a function of the crack size and shape, the type of loading and the geometrical configuration of the structure. The stress intensity factor is often written as

$$K = Y\sigma\sqrt{(\pi a)} \tag{10}$$

where σ is a stress, a is a measure of the crack-length and Y is a non-dimensional function of the geometry.

Since Irwin **(4)** demonstrated the importance of the stress intensity factor in determining crack tip stress fields many different methods have been devised for obtaining K and many K solutions now exist **(5)–(8)**. These methods will be discussed in detail later.

Residual strength

The criterion for failure due to the unstable growth of a crack can be expressed in the following way: failure occurs if

$$K \geqslant K_{\mathrm{Ic}} \quad \text{(plane strain condition)}$$

or

$$K \geqslant K_{\mathrm{c}} \quad \text{(plane stress condition)} \tag{11}$$

where K_{Ic} and K_{c} are considered to be constants, called the fracture toughness of the material. From equations (10) and (11) the failure criterion can be written as

$$Y\sigma_{\mathrm{c}}\sqrt{(\pi a_{\mathrm{c}})} = K_{\mathrm{c}} \tag{12}$$

where σ_{c} and a_{c} are, respectively, the critical stress and the critical crack length at failure. The functional relationship between σ_{c} and a_{c} given in equation (12) is a generalization of that derived by Griffith, see equation (3).

It follows from equation (12) that for a given crack length the critical stress depends on both the toughness and the crack length; in fact.

$$\sigma_c \propto \frac{K_c}{\sqrt{a_c}}.$$ (13)

Thus the critical stress increases as the toughness increases and decreases as the crack length increases. It also follows from equation (12) that, for a given stress level σ_c, the critical crack length for failure depends on the toughness K_c and the stress σ_c; in particular

$$a_c \propto \left[\frac{K_c}{\sigma_c}\right]^2$$ (14)

Thus the critical crack length increases as the toughness increases, and decreases as the stress level increases.

Fatigue crack growth

Paris (18) suggested that the growth of cracks which results when the applied stress is varied (fatigue) may also be described by the stress intensity factor, even though the maximum stresses may be much less than the critical stress. He postulated that the rate of growth per cycle of stress (da/dN) was a function of the stress intensity factor range ΔK; which is defined as $K_{max} - K_{min}$ that is

$$\frac{da}{dN} = Cf(\Delta K)$$ (15)

where C is a constant to be determined experimentally. Many experimental data are now available to confirm that, for many stress conditions, fatigue crack growth is largely controlled by ΔK. The simplest explicit form of equation (15) suggested was

$$\frac{da}{dN} = C(\Delta K)^m$$ (16)

where C and m are constants. Later work suggested that C was not strictly constant but depended on such parameters as $R(= K_{min}/K_{max})$, K_c and the threshold stress intensity factor ΔK_{th}. At high K values C must depend on K_{max}, since if $K_{max} = K_c$ static failure $(da/dN = \infty)$ must occur; and at low values of ΔK, C must depend on ΔK_{th}, since if $\Delta K \leqslant \Delta K_{th}$ no crack-growth occurs $(da/dN = 0)$.

Crack tip fields

It is not usually possible to obtain an analytic expression for the full elastic fields in the vicinity of a crack tip. Only for some simple configurations can the elasticity equations be solved exactly; they include isolated cracks in an infinite body and some periodic arrays of cracks. If the full field is known, then the stress intensity factor can be obtained directly. Most of the analytical solutions available have been derived using Westergaard stress functions.

In more general structures, the fields can only be obtained approximately. Although the near-tip fields can still be expressed as a Williams (10) series, the coefficients are all unknown and depend on the positions of and the conditions on the boundaries of the body. The coefficient of the leading term in the Williams expansion is proportional to K. Numerical methods can obtain K by indirect means, using limiting procedures on the fields near the tip; these methods may not be very accurate or reliable. Some modelling procedures incorporate information about the fields (e.g., traction singularity) into the modelling and obtain an accurate K in a more direct way. Alternatively energy methods which exploit the relationship between strain energy and stress intensity factor are often used to obtain K without detailed knowledge of the total field.

Westergaard stress functions

Westergaard stress functions (9) are frequently used to solve two-dimensional problems in cracked structures. For instance the stress function for an infinite sheet containing a crack of length $2a$ subjected to a remote biaxial stress is given by

$$Z(z) = \frac{\sigma z}{\sqrt{(z^2 - a^2)}} \tag{17}$$

for traction-free crack surfaces. The explicit stress and displacement fields can be obtained from equations (6) and (7), respectively. In the vicinity of the crack-tip ($r \ll a$), the stresses are given by

$$\sigma_{xx} = \frac{K_1}{\sqrt{(2\pi r)}} \cos\frac{\theta}{2}\left(1 - \sin\frac{\theta}{2}\sin\frac{3}{2}\theta\right) \tag{18}$$

$$\sigma_{yy} = \frac{K_1}{\sqrt{(2\pi r)}} \cos\frac{\theta}{2}\left(1 + \sin\frac{\theta}{2}\sin\frac{3}{2}\theta\right)$$

and

$$\sigma_{xy} = \frac{K_1}{\sqrt{(2\pi r)}} \sin\frac{\theta}{2}\cos\frac{\theta}{2}\cos\frac{3}{2}\theta$$

where (r, θ) are polar coordinates centred at the crack tip at $z = a$, and $K_1 = \sigma\sqrt{(\pi a)}$. The displacements are given by

$$u_x = \frac{K_1}{\mu}\sqrt{\left(\frac{r}{2\pi}\right)}\cos\frac{\theta}{2}\left\{\frac{1}{2}(\kappa - 1) + \sin^2\frac{\theta}{2}\right\} \tag{19}$$

and

$$u_y = \frac{K_1}{\mu} \sqrt{\left(\frac{r}{2\pi}\right)} \sin \frac{\theta}{2} \left\{ \frac{1}{2}(\kappa + 1) - \cos^2 \frac{\theta}{2} \right\}$$

The functional forms in equations (18) and (19) are used to define two general limiting expressions for K_1; namely

$$K_1 = \lim_{r \to 0} \left\{ \sqrt{(2\pi r)} \sigma_{yy}(r, 0) \right\} \tag{20}$$

and

$$K_1 = \lim_{r \to 0} \left\{ \frac{2\mu}{\kappa + 1} \sqrt{\frac{2\pi}{r}} u_y(r, \pi) \right\}$$

A similar analysis can be done for mode II and mode III deformation and general definitions of K_{II} and K_{III} obtained.

Boundary effects

In the previous section, there were no boundaries present other than the cracks. If there are boundaries close enough to the cracks, the full fields can no longer be obtained analytically, and the stress intensity factors become a function of the position and type of boundary. The effect of boundaries and their proximity to the cracks is very important and must be taken into account when calculating stress intensity factors. It was this realization that led to the development of the compounding method which systematically (albeit approximately) accounts for the presence of boundaries in the vicinity of the crack. Some of the success of the more powerful boundary element method for solving crack problems may be attributed to its ability to model accurately and efficiently the interactions between boundaries (which include the cracks).

Energy relationships

There are two main energy methods which enable the crack tip parameter to be obtained directly from global parameters, which have the advantage of not requiring detailed modelling in the vicinity of the crack tip. The first involves directly calculating the strain energy release rate from two analyses at two slightly different crack lengths (a and $a + \delta a$ say). If U is the strain energy in the body, under the given loading at crack length a, and $U + \delta U$ is the strain energy at crack length $a + \delta a$, then

$$\frac{\delta U}{\delta a} = G = K^2/E' \tag{21}$$

where $E' = E$ (Young's modulus) for plane stress and $E' = E/(1 - v^2)$ for plane strain; thus K can be determined.

The stress intensity factor can also be related to a path-independent integral, termed the 'J' integral, described by Rice (21). This integral is independent of the actual path chosen, provided that the contour Γ starts and finishes on opposite faces of the crack and that the contour includes the crack tip. If the crack is parallel to the x axis the 'J' integral can be defined as

$$J = \int_{\Gamma} \left(U'\, dy - t_i \frac{\partial u_i}{\partial x}\, dS \right) \tag{22}$$

where U' is the strain energy density, t_i are components of the traction vector, u_i are components of the displacement vector (repeated suffix summation is assumed) and dS is an element of arc along the integration contour Γ. For a linear elastic material it can be shown (21) that

$$J = G = K^2/E' \tag{23}$$

Determination of K: simple methods

The earliest methods for determining stress intensity factors tended to be based on the simplest ideas, although this is not exclusively the case. Nevertheless, this section will outline some of the simpler methods which do not, in general, require large-scale computing power. The methods described here are those listed in stage 2 of Table 1.

Superposition

Superposition is probably the most common and simplest technique in use for obtaining stress intensity factors. Configurations with complex boundary conditions are considered to be a combination of a number of separate simpler boundary conditions on the same configuration, each having a known stress intensity factor. The stress intensity factors for the simple configurations are then added together to obtain the required solution. Superposition is exact but errors can arise from using superposition when the complex conditions being analysed cannot be precisely built up from simpler conditions with known stress intensity factors. An important application of superposition is in the analysis of pin-loaded lugs; opening mode stress intensity factors for non-symmetric loadings can be found by adding the more easily obtainable results for simpler symmetrical loadings. However, it is worth noting that no information about K_{II} can be obtained in this manner.

Bueckner (17) has derived an important result which is related to the principle of superposition. He demonstrated the equivalence of stress intensity factors resulting from external loading on a body and those resulting from internal tractions on the crack face. The stress intensity factor for a crack in a loaded body may be determined by considering the crack to be in an unloaded body with applied tractions on the crack surface only. These surface tractions are equal in magnitude but opposite in sign to those evaluated along the line of the crack site in the uncracked configuration. This method of determining the stress intensity factor is important in the use of the Green's function and the weight function methods which will be discussed later.

The application of Bueckner's principle (17) is subject to the same restrictions as the more usual superposition of stresses and displacements. In addition, the crack surfaces in the final configuration must always be separated along their entire length, although there may be some overlap of the crack surfaces (K_I may even be negative) in some of the ancillary configurations. If overlap does occur in an ancillary configuration it must be ignored in evaluating the ancillary stress intensity factor, otherwise the results of the superposition will be invalid. The limitations of superposition of stress intensity factors have been considered by Aamodt and Bergan (22).

Certain solutions for partially loaded cracks may be superimposed to obtain approximate solutions for cracks in arbitrary stress fields. An approximate solution to a distribution of pressure can be obtained by superimposing results given by Emery et al. (23)(24), for a band of pressure of variable width acting over part of the crack surface. This method can be used for any arbitrary distribution of stress including thermal stresses (for example (24)), providing that the stress intensity factor for the band of pressure in the configuration is known. The method is a special form of the Green's function technique which will be described later. Lachenbruch (25) has obtained a step function loading on the crack surface which has been expressed in algebraic form by Emery (26) and applied to the solution of crack problems in thermally stressed cylinders. More recently Ball (27) used superposition techniques to obtain mode I stress intensity factors for a number of mechanically-fastened joints.

Stress concentrations

Irwin (15) proposed that Neuber's results (13) for the stress concentrations of notches of very small flank angle and very small root radius ρ may be used to obtain theoretical expressions for stress intensity factors. Consider a notch which, in the limit of zero root radius (ρ), tends to a crack along the $y = 0$ axis: if σ_{max} is the maximum value of σ_{yy} at the tip, then

$$K_I = \frac{\sqrt{\pi}}{2} \lim_{\rho \to 0} [\sigma_{max}\sqrt{\rho}]. \qquad (24)$$

As an example of this approach consider a semi-elliptical edge notch of depth c in a semi-infinite sheet subjected to a remote uniaxial tensile stress σ. Equation (24) can be written in terms of K_t, the stress concentration factor (ratio of maximum stress to applied stress), as follows

$$\frac{K_I}{\sigma\sqrt{(\pi c)}} = \frac{K_I}{\sigma\sqrt{(\pi l)}} = \lim_{\rho \to 0} \left\{ \tfrac{1}{2}K_t\sqrt{\frac{\rho}{c}} \right\} \qquad (25)$$

where l the crack length (i.e. $c = l$ at $\rho = 0$). The stress concentration factor K_t has been obtained for this configuration as a function of ρ/c by Bowie (28). From this result for K_t and a plot of $K_t(\rho/c)^{1/2}/2$ versus ρ/c, the value of K_I was determined with an accuracy of better than 1 percent.

For the mode III stress intensity factor Harris (29) suggested that if σ_{max} is the maximum value of σ_{yz} at the tip, then

$$K_{III} = \sqrt{\pi} \lim_{\rho \to 0} \{\sigma_{max}\sqrt{\rho}\} \qquad (26)$$

The mode II stress intensity factors cannot be obtained as in mode I and mode III by considering the limiting forms of the maximum stresses ahead of the notch tip as $\rho \to 0$, since the shear stress $\sigma_{xy} \to 0$ as $\rho \to 0$. Sih and Liebowitz (30) suggested that K_{II} could be obtained from a consideration of the maximum hoop stress on the notch surface, by the following relationship

$$K_{II} = \sqrt{\pi} \lim_{\rho \to 0} \{\sigma_{hoop}^{max}\sqrt{\rho}\} \qquad (27)$$

Further discussions on methods of obtaining K_{II} from stress concentration factors can be found in Chen (31) and Chiang (32).

Harris (29) has made considerable use of equations (24) and (26) and Neuber's work on stress concentration factors K_t in deriving expressions for K_I and K_{III} in circumferentially cracked round bars subjected to bending, transverse shear, torsion, and longitudinal tension. Pook and Dixon (33) have analysed a finite rectangular sheet with an edge crack, loaded such that there is combined tension and bending at the crack tip.

Hasebe and Katanda (34) have suggested a systematic method of determining stress intensity factors which involves expressing the stress concentration factor in a series form. The unknown coefficients in the series are determined by fitting the expression to available data for the stress concentration factor at various values of the root radius ρ. In this way it is possible to make better use of the stress concentration factors determined for values of ρ too large for equations (24) and (26) to be applied directly. The stress concentration factors required may be determined analytically, numerically, or experimentally; many have been determined and are collected together and published by Peterson (35) and Hardy and Malik (36).

Local stress distributions

Some simple methods have been proposed for determining stress intensity factors from local stress distributions. The methods are all based on the use of the stress intensity factor for an edge crack, subjected to a uniform internal pressure p in a semi-infinite plane. This is given, for a crack of length l, by

$$K_I = 1.12p\sqrt{(\pi l)} \qquad (28)$$

the factor 1.12, accurately obtained by Koiter (37), is often called the 'free edge correction'.

Equation (28) can be used to provide approximate stress intensity factors for cracks at a hole providing that l is small compared to the radius of curvature R at the end of the notch. In the application to short cracks at holes or notches the pressure p is replaced by a stress which is characteristic of that over the crack site in the uncracked solid $\sigma(x)$, where x is measured along the line of the crack from the notch end. In the maximum stress method

$$p = \sigma_{max} = \sigma(0) \tag{29}$$

in the mean stress method

$$p = \sigma_{mean} = \frac{1}{l} \int_0^l \sigma(x) \, dx \tag{30}$$

and in the tip stress method

$$p = \sigma_{tip} = \sigma(l) \tag{31}$$

Results using the maximum stress method are reported in (12) for a crack at both a circular and an elliptical cut-out in a sheet subjected to a uniform stress field. For short crack lengths equation (29) gives a reasonable approximation to the accurate solution of Newman (38). The range of crack lengths over which equation (29) gives a reasonable estimate of K_I depends on the tip radius of the notch; the blunter the notch, the better the approximation. At long crack lengths, that is $l \gg R$, the stress intensity factor for two equal cracks, at opposite sides of the hole, approaches that for an isolated crack of length $2(c + l)$ in a uniform stress field; that is

$$K_I = \sigma \sqrt{\{\pi(c + l)\}} \tag{32}$$

where c is the semi-major axis of the ellipse. For a single crack of length l the stress intensity factor approaches that for an isolated crack of length $2c + l$ given by

$$K_I = \sigma \sqrt{\{\pi(2c + l)/2\}} \tag{33}$$

Hence an approximation can be obtained over the whole range of crack lengths by interpolating between the short crack and the long crack limits. This approach of obtaining approximate stress intensity factors by utilizing asymptotic expansions for both short and long cracks has been developed in some detail by Benthem and Koiter (39). They studied both two- and three-dimensional problems and multiple cracks.

The short crack approximation in (12) can be replaced by equation (30) or (31). This procedure involves more calculations since the distribution of stress over the whole crack site in the uncracked solid must be known, rather than just the maximum stress (i.e., K_t for the notch or hole). Another simple approximation has been suggested by Williams and Isherwood (40) who studied an edge crack in a thermally stressed plate. They proposed a method based on the mean stress and suggested an empirical way of making corrections for finite width effects. This method has been used for determining stress intensity factors for a central radial crack in a rotating disc by Chan, Tuba, and Wilson (41) and for an edge crack in a circular bar subjected to pure bending by Cannon and Allen (42).

Compounding method

The compounding method described here is a versatile and quick way of extending available stress intensity factor solutions of simple configurations to other, more complex configurations. The compounding method was developed by Cartwright and Rooke (43) and its application to complex geometrical configurations is reported by Rooke (44)(45). Although a special case of compounding was used earlier by Figge and Newman (46), Smith (47), and Liu (48), its generality was not realized or investigated.

The principle of the compounding method is presented here: the stress intensity factor for a complex configuration is compounded from the factors for a number of simpler ancillary configurations. Each ancillary configuration will usually contain one boundary only, which interacts with the crack. Initially the contributions to the final stress intensity factor are compounded, neglecting any effects due to other boundary–boundary interactions.

Consider a configuration containing a crack near to two stress-free boundaries B_1 and B_2; the configuration is subjected to an applied stress system S_0 on its boundary B_0 which is remote from the crack. Let the stress intensity factor at one of the crack tips be denoted by K_1 if only the stress-free boundary B_1 is present. If the two boundaries B_1 and B_2 are present together, the resultant stress intensity factor K_r is given by

$$K_r = K_1 + K_2 - \bar{K} + K_e \tag{34}$$

where $\bar{K}$ is the stress intensity factor if both internal boundaries (B_1 and B_2) are absent. The extra term K_e represents the possible effects due to the interactions of the boundaries. Thus the stress intensity factor for a crack in a configuration with multiple internal boundaries can be expressed in terms of stress intensity factors derived from configurations with single internal boundaries, apart from a correction term. In general for N boundaries B_n ($n = 1, 2, \ldots, N$), the resultant stress intensity factor is given by

$$K_r = \bar{K} + \sum_{n=1}^{N} (K_n - \bar{K}) + K_e \tag{35}$$

where K_n is the stress intensity factor in the presence of B_n only. In terms of normalized stress intensity factors Q_r, this becomes

$$Q_r = 1 + \sum_{n=1}^{N} (Q_n - 1) + Q_e \tag{36}$$

where $Q_r = K_r/\bar{K}$, $Q_n = K_n/\bar{K}$, and $Q_e = K_e/\bar{K}$; Q_e is the correction term due to the interactions of all the N boundaries. If Q_e can be estimated or can be shown to be small ($\ll 1$), then equation (36) can be used to build up solutions to complex configurations from known simpler ones. The correction term Q_e can be expressed formally **(43)** using the Schwarz alternating technique which has been described by Sokolnikoff **(49)**. It is an important technique in the determination of stress intensity factors, and has been used by Hartranft and Sih **(50)** and many others, some of which are reviewed in reference **(12)**. The simple application of the compounding technique of adding together the effects of the individual boundaries needs to be modified if the crack crosses one of the boundaries, for example a crack at the edge of a hole, or a crack beneath a stiffener (which is treated as a boundary). Before the effect of the other boundaries can be considered, the crack plus the boundary it crosses must be replaced by an equivalent crack **(44)(45)** which interacts with the other boundaries. If the stress intensity factor is K_0 when only the boundary the crack crosses is present, then the equivalent crack is defined **(44)** in terms of K_0. It is defined as an isolated crack of the same total length $2a$ in a sheet with a remote stress σ', which is determined by the condition that the stress intensity factor is K_0; that is

$$\sigma'\sqrt{(\pi a)} = K_0 \tag{37}$$

Since in this case $\bar{K} = \sigma\sqrt{(\pi a)}$, where σ is the remote applied stress, it follows that

$$\sigma' = \frac{K_0}{\bar{K}}\,\sigma = Q_0\sigma \tag{38}$$

The effects of the other boundaries B_n ($n = 1, \ldots, N$) on the original crack plus the attached boundary are now considered to be the same as the effects on the equivalent crack in a configuration subjected to an applied stress σ'. The general compounding formula **(35)** is then modified **(44)** to

$$K_r = K_0 + \sum_{n=1}^{N} (K'_n - K_0) + K_e \tag{39}$$

where K'_n is the stress intensity factor for the equivalent crack in the presence of the nth boundary only. Equation (39) can now be written in terms of the normalized stress intensity factors and becomes

$$Q_r = Q_0\left\{1 + \sum_{n=1}^{N} (Q_n - 1)\right\} + Q_e \tag{40}$$

where $Q_n = K'_n/K_0$. It is worth noting that Q_n is independent of σ' since K'_n and K_0 are both proportional to σ'.

In some configurations, particularly with localized loading near the crack, the equivalent crack is better described by a pair of point forces P acting on the crack to give a stress intensity factor of K_0. For a more complete description and other important considerations of boundary–boundary interactions see (44)(45)(51).

Green's functions

The crack face Green's function $G(x)$ can be obtained from a solution of point forces acting on the crack faces. These solutions can then be used to obtain stress intensity factors for any problem of the same geometry under an arbitrary loading. The expression for the stress intensity factor for a crack of length a can be written, in terms of the Green's function $G(x)$, as

$$K_I = \frac{1}{\sqrt{(\pi a)}} \int_a p(x)G(x)\,dx \qquad (41)$$

where $p(x)$ is the crack pressure acting on the crack surface. The Green's function $G(x_0)$ can be identified from equation (41) as the normalized stress intensity factor for a point force located at $x = x_0$. If a point force P (per unit thickness) acts at $x = x_0$, then $p(x) = P\delta(x - x_0)$ where $\delta(x)$ is the Dirac delta function. Substitution of $p(x)$ into equation (41) gives

$$K_I = \frac{P}{\sqrt{(\pi a)}} G(x_0) \qquad (42)$$

A Green's function which is particularly useful, when coupled with Bueckner's principle is that for a pair of point forces $\pm P$ acting at $x = x_0$ on the crack face; the Green's function for the stress intensity factor is

$$G(x_0) = \sqrt{\left(\frac{a + x_0}{a - x_0}\right)} \qquad (43)$$

The concept of Green's function is not confined to point forces acting of the crack faces. Equation (41) can be generalized for any boundary, b say. Thus, if a stress field $\sigma_b(x)$ acts on a boundary b of a cracked body, then the stress intensity factor will be given by

$$K_I = \frac{1}{\sqrt{(\pi a)}} \int_b \sigma_b(x)G_b(x)\,dx \qquad (44)$$

where $G_b(x)$ is the Green's function (normalized stress intensity factor) for a force acting at the point denoted by x on the boundary b. Such a Green's function enables the stress intensity factors to be obtained for various distributions of load on the boundary b, without the need to re-solve the elasticity equations.

An example of such a Green's function is that for a radial crack at the edge of a circular hole subjected to a point force (either normal or tangential). There are, in fact, four separate Green's functions, because both the normal force P and the tangential force Q produce both mode I and mode II stress intensity factors. These four Green's functions have been calculated numerically and tabulated by Rooke and Hutchins (52), and used to investigate the effects of simulated pin loading on the edge of the hole.

The use of simple Green's functions to systematically derive Green's functions for more complicated structures has been described in detail in references (12)(53), which also contain many source references.

Weight functions

Generalized forms of Green's functions called weight functions have been introduced by Bueckner (54) and by Rice (55). Rice showed that if the crack face displacement $u^{(2)}$ (a, x) and the mode stress I intensity factor $K_I^{(2)}$ are known for a symmetrical load system on a linearly elastic body, then the stress intensity factor $K_I^{(1)}$ for any other symmetrical load system can be obtained from an integral over the boundary Γ

$$K_I^{(1)} = \int_\Gamma \sigma^{(1)}(x) H(a, x) \, \mathrm{d}x \tag{45}$$

where $H(a, x) = E'(\partial u^{(2)}/\partial a)/K_I^{(2)}$ is a weight function for a symmetrical problem. The stress $\sigma^{(1)}$ is the distribution of stress, on the boundary Γ, which produces $K_I^{(1)}$ in the cracked body. Rice also presented the derivation of three-dimensional weight functions in the Appendix of reference (55). The three-dimensional derivation is based on the displacement field variations associated, to first order, with an arbitrary variation in the position of the crack front.

Although the above weight function technique is potentially more powerful than the Green's function (which is a special case of a weight function), it suffers from the disadvantage that a detailed knowledge of one K solution is required for the cracked body in question. Even if a value of K is known for a given loading, it is rare for details of the displacement field, in particular $\partial u/\partial a$ to be known. However, there are now several approximate techniques developed for overcoming these difficulties; the works of Fett, Mattheck and Munz (56) and Wu and Carlsson (57) are particularly important for enabling this method to be more widely used. Apart from these approximate procedures, the technique is being used in both finite element and boundary element analyses where $\partial u/\partial a$ is obtained explicitly (numerically) for a given loading.

Bueckner's (58) generalization of the concept of weight functions was based on Betti's reciprocal theorem and so-called fundamental fields. The fundamental fields are non-physical fields which can be derived from the stress functions for point forces on the crack face at the tip of the crack. The stress intensity factor in two dimensions, due to an applied traction t, is given by

$$K = \frac{E'}{4\sqrt{2\pi B}} \int_\Gamma t \cdot u \mathrm{d}\Gamma \tag{46}$$

where $u^{(2)}$ is the displacement field on the boundary Γ, which results from a fundamental field of strength B at the crack tip. Similar equations exist (12) for all modes. Bueckner (58)(59) also presented a formulation for weight functions in three dimensions based on fundamental fields. Other theoretical advances in three-dimensional weight functions have been made by Rice (60)(61), Gao and Rice (62)–(64), Bueckner (65)–(68) and Gao (69)(70).

Determination of K: numerical methods

The methods listed in stage 3 of Table 1 generally require large-scale computing facilities, but are usually able to model realistic cracked structures accurately. The order of the methods does not indicate any chronological sequence – developments in the various methods were often contemporaneous. There is not always a clear distinction between some of the various methods, and many of the mathematical techniques used are common to more than one numerical method.

Boundary collocation methods

The boundary collocation methods involve replacing the governing differential equations of elasticity with a set of algebraic equations with unknown coefficients. These equations are formed by expressing certain functions, which satisfy the differential equations, as truncated infinite series having unknown coefficients. The coefficients in these series are determined by matching the known conditions for stresses, displacements, or forces on the boundaries of the configuration and on the crack surface. The boundary conditions may be matched exactly at a number of boundary points, or by minimizing the squares of the residual errors on the boundary. Whilst convergence is not guaranteed the boundary collocation methods have been used to obtain a considerable number of stress intensity factor solutions. The accuracy of these solutions, which is usually reported to be between one and two percent, should be assessed by consideration of the residual errors between the known boundary conditions and those obtained in the numerical solution.

   Application of boundary collocation methods to crack problems starts from a series representation of either real or complex stress functions. The real function approach which is based on the Williams' eigenfunction series expansion was originally proposed by Gross, Srawley, and Brown (71) who analysed the single edge crack problem under uniform tensile stress. Alternatively complex stress functions can be used as for example by Newman (38). The complex function approach may involve using series for the region occupied by the actual configuration or may employ conformal mapping first to transform the region occupied by the actual configuration to a simpler one. The advantage of conformally mapping the configuration of the physical plane onto a parameter plane is that singularities (e.g., crack tips) can be removed by the mapping function; this makes the boundary conditions easier to match.

Known properties of the stress function may sometimes be used to exactly satisfy certain boundary conditions on some chosen boundary (e.g., crack surface, hole boundary, etc) thereby reducing the number of boundary conditions to be matched in the numerical solution. A recent development in boundary-point mapping methods has been to divide the configuration in the physical plane into zones and to use a separate stress-function expansion and possibly a separate mapping in each zone. The unknown coefficients in each expansion are determined from the known boundary conditions and the compatibility of displacements along the common boundaries of each adjacent zone.

Newman (38) carried out a study on the convergence of the different boundary collocation procedures of complex stress functions. In his study the problem of two equal cracks emanating from a circular hole in an infinite sheet subjected to a remote tensile stress σ was considered. The simplest stress functions are valid for regions having a single crack and the non-intersecting external boundary. For configurations in which there are multiple internal boundaries, or if the crack is not traction-free, these stress functions have to be modified. Some examples of the modified functions are given by Newman (38), Isida (72), Cartwright and Parker (73), Pei-Qing (74), and Woo, Wang, and Cheung (75).

Integral transforms/continuous dislocations

Integral transforms and continuous dislocation theory have been used to solve many crack problems in fracture mechanics. Full details of these techniques are given by Sneddon (76) and Bilby and Eshelby (77).

The elastic boundary value problem for a crack of length l can be written as an integral equation of the form

$$\int_0^l \frac{F(r, x)P(x)}{\sqrt{\{x(l - x)\}}} \, dx = -f(r) \tag{47}$$

where r is an arbitrary point on the crack and $f(r)$ is the crack line loading in the uncracked state. The kernel $F(r, x)$ has the form

$$F(r, x) = \frac{1}{x - r} + R(r, x) \tag{48}$$

where $R(r, x)$ is not singular for $0 < x, r < l$. The unknown function $P(x)$ is related to the normal displacement $u_2(r)$ on the crack surfaces by

$$u_2(r) = \frac{-2(1 - v^2)}{E} \int_r^l \frac{P(x)}{\sqrt{\{x(l - x)\}}} \, dx \tag{49}$$

The stress intensity factor may be obtained by equating the displacement field in (49) with the near-tip displacement field given in (19). For the application of integral transform and continuous dislocation techniques to crack problems see (78)–(81).

Body force method

The body force method was originally proposed by Nisitani (82) for solving two-dimensional stress problems, and was later extended to the solution of notch (83) and crack problems (84). The basic concept of the body force method is analogous to the indirect formulation of the boundary element method (BEM). As in the BEM, the body force method uses the stress field due to a point force in an infinite domain as a fundamental solution. The prescribed boundary conditions are satisfied by applying the body force along the imaginary boundaries (i.e., the boundaries of the notch or a crack) in an infinite sheet and adjusting the force density so as to satisfy the boundary conditions. The boundaries of the problem are divided into a finite number of straight and/or curved elements with unknowns defined at the mid-points of the elements. The application of the body force method to three-dimensional crack problems can be found in papers by Nisitani and Murakami (85) and Murakami and Nemat-Nasser (86).

Method of lines

The basis of the method of lines (MOL) is to semi-discretize a set of partial differential equations into a system of ordinary differential equations defined on discrete lines. This is done by the substitution of finite differences for the derivatives with respect to all the independent variables except one for which the derivatives are retained. These equations describe the dependent variable along lines which are parallel to the coordinate-direction in which the derivatives were retained. Although the method was developed several decades ago for the solution of boundary value problems (see Liskovets (87)) it still requires further research to extend it to the solution of complex engineering problems. The application of MOL to crack problems has been reported by Gyekenyesi and Mendelson (88), Alam and Mendelson (89), Mendelson and Alam (90), and Xanthis and Edwards (91).

Edge function method

The edge function method, originally developed by Quinlan (92), may be described as a boundary element method that collects together suitable truncated series expansions of harmonic solutions for several parts of the domain. Since the early application of the technique to the torsion of prismatic bars of polygonal cross-sections (92), it has been successfully applied to thin plates by Quinlan and O'Callaghan (93). Two-dimensional crack problems have been studied by Fleming et al. (94), and three-dimensional embedded elliptical cracks by Quinlan et al. (95).

Finite element method

The finite element method (FEM) has emerged over the past two decades as one of the most powerful numerical tools for the solution of crack problems in fracture mechanics. Hundreds of finite-element solutions to crack problems are now available, for examples see reviews by Gallagher (96) and Liebowitz and Moyer (97).

Early applications of the finite element method to crack problems can be found in papers by Swedlow, Williams, and Yang (98) and Chan, Tuba, and Wilson (41), where it was reported that a reasonable level of accuracy (5–10 percent) could be obtained for simple configurations. Later careful studies of convergence by Wilson (99) and Oglesby and Lamackey (100) showed that the solutions in the vicinity of the crack tip could not be guaranteed to be accurate, regardless of the mesh density around the tip. This limitation led to the development of several special crack tip elements, for example by Tracey (101), Blackburn (102), Akin (103) and Yamada, Ezawa, Nishguchi and Okabe (104), which introduce the $\sqrt{r}$ dependence of the displacement into the shape function representation, and hence a $1/\sqrt{r}$ singularity into the stress and strain. Henshell and Shaw (105) and Barsoum (106) independently observed that by moving the mid-side node of a eight noded quadrilateral element to a quarter point position, the desired $1/\sqrt{r}$ variation in the strains could be directly achieved. The so-called quarter-point element has received much attention in the literature since standard FEM programs can be used. They can also be used in conjunction with the boundary element method. Other classes of elements developed to deal with crack problems are those known as 'enriched elements', see for instance Heyliger and Kriz (107); and those known as 'hybrid elements', described by Atluri (108).

Another approach, which is based on removing the singularities associated with the crack tip stress fields, was developed by Morley (109) and Yamamoto (110) for use with finite elements around the same time as Symm (111) developed it for boundary elements. This technique, which has been referred to as a superposition approach, makes use of the Williams' (10) near-tip field solution to subtract the stress singularities associated with the crack tip. The resulting non-singular field problem can be solved by the finite element method, with the stress intensity factor as one of the unknowns. The main drawback of this approach in the finite element method (unlike the boundary element method) is that the singular stress components are not the primary unknowns in the problem. Nevertheless, the technique has received much attention, for example by Yamamoto, Tokudo, and Sumi (112), Amara, Destuynder, and Djaoua (113), and Sinclair and Mullan (114). These techniques, as noted in (114), differ from one another only in the number of terms used from the Williams series expansion or the extra conditions required for the evaluation of the stress intensity factors.

There are many ways of evaluating the stress intensity factors from finite element solutions: these include extrapolation of stresses and/or displacement fields to the crack tip; Rice's contour integral J; the strain energy approach; the virtual crack extension technique. In the strain energy approach two finite element analyses are carried out for two crack lengths which differ by an incremental amount δa and the strain energy difference δU is evaluated. The strain energy release rate G is then calculated from the relationship.

$$G = \frac{\delta U}{\delta a} \tag{50}$$

The stress intensity factors can be evaluated from the following relationships

$$G_I = K_I^2/E; \; G_{II} = (1 - v^2)K_{II}/E \quad \text{and} \quad G_{III} = (1 + v)K_{III}^2/E$$

If cracking occurs in a combined mode, then the total strain energy release rate is expressed as

$$G = G_I + G_{II} + G_{III} \tag{51}$$

and extra analysis is required to separate the individual modes (see below).

A virtual crack extension technique which is a variant of the strain energy release rate approach has been developed by Parks (115) and Hellen (116). Here, the change in strain energy δU is associated with elements surrounding the crack tip and is evaluated by displacing the nodal points within the crack tip elements by an incremental distance δa. The strain energy release rate G is evaluated from a knowledge of the displacement field before the crack extension and the change of stiffness during the crack extension.

A finite element method for evaluating weight functions based on Rice's derivation was developed by Parks and Kamenetzky (117) and Vanderglas (118); they obtained mode I weight functions using the virtual crack extension techniques of Parks (115) and Hellen (116). However, the procedure has the disadvantage of requiring a numerical differentiation as well as certain limitations regarding the special perturbation technique used to simulate the virtual crack extension. An extension of the method of Parks and Kamenetzky (117) to mixed-mode weight functions has been developed by Vainshtok (119) and Sha and Yang (120). The latter authors use a symmetric mesh in the vicinity of the crack tip which permits the analytical separation of the mode I and mode II components of the weight function, as proposed by Ishikawa (121).

Recently Sham (122) has developed a finite element procedure for evaluating weight functions based on the separation of displacement and stress fields into highly singular terms due to the Bueckners's fundamental field, and the remaining terms associated with a traction-free crack problem. In his procedure the singular fundamental fields were imposed on a few layers of elements surrounding the crack tip and 'quarter-point' cubic elements were used in the immediate vicinity of the tip to model the singular stress behaviour. Sham and Zhou (123)(124) have extended the above procedure to two-dimensional orthotropic and three-dimensional isotropic problems.

Boundary element method

The methods described above have been used to solve many crack problems in the last thirty years or so. Some methods have been found more suitable for particular classes of problems than others, and many of the methods are still under active development. Up to recent times the most versatile method has probably been the finite element method and it has been used to solve many practical problems. The more recent boundary element method (BEM) has now emerged (125) as a very powerful numerical technique for solving crack problems in fracture mechanics. It has some important advantages over the finite element method, in particular a reduction in the dimensionality of the numerical problems, and has been shown to produce more accurate stress intensity factors for both two- and three-dimensional crack problems. Although a system of integral equations rather than algebraic equations, needs to be solved, the reduction in the number of equations makes BEM calculations efficient, and further improvements are to be expected.

The integral equation to be solved for a given body with boundary Γ can be written, in the absence of body forces, as

$$c\mathbf{u} = \int_\Gamma U\mathbf{t}\,d\Gamma - \int_\Gamma T\mathbf{u}\,d\Gamma \tag{52}$$

where c is a constant; $\mathbf{u}$ and $\mathbf{t}$ are the displacement and traction fields on the boundary; U and T are the fundamental solutions which can be derived from the response of an infinite body to a localized force. The fundamental solutions have strong similarities to Green's functions discussed earlier, but do not satisfy any exterior boundary conditions. Several fundamental solutions are available (12) for bodies with internal boundaries such as cutouts and cracks.

In general, equation (52) must be solved numerically. This is done by dividing the boundary into discrete elements and describing the elastic fields and the geometry by interpolation (or shape) factors. This procedure is similar to that of finite element analysis, but only the boundary needs to be discretized. The result is a matrix equation of a form similar to that obtained in finite elements.

$$A\mathbf{x} = \mathbf{b} \tag{53}$$

where $\mathbf{x}$ is the vector of the unknown boundary displacements and tractions and $\mathbf{b}$ is the vector of known boundary conditions with known coefficients. The coefficients in $\mathbf{A}$ and $\mathbf{b}$ contain integrals over the boundary elements.

Straightforward application of the boundary element method to crack problems leads to a mathematical degeneration in the numerical formulation (singular matrix) if the two crack surfaces are considered co-planar, as in analytic studies; this was first shown by Cruse (126). In order to avoid this difficulty he suggested a number of modelling strategies, which included modelling the crack as a rounded notch with an elliptical closure. However, this model was limited to symmetric geometries and required many elements to model the tip of the rounded notch. The reported accuracy for the stress intensity factor of the centre-crack-tension-specimen was poor, with errors of around 14 percent.

Snyder and Cruse (127) introduced a special form of fundamental solution for crack problems in anisotropic media. The fundamental solution (Green's function) contained the exact form of the traction-free crack in an infinite medium, hence no modelling of the crack surfaces was required. The crack Green's function technique, although accurate, is limited to two-dimensional straight cracks. The first widely-applicable method for dealing with two co-planar crack surfaces was devised by Blandford, Ingraffea, and Liggett (128). Their approach, which is based on a multi-domain formulation was the most general devised; it can be applied to both symmetrical and non-symmetrical crack problems in both two- and three-dimensional configurations. More recently other approaches have been suggested, for instance the introduction of an alternative equation on one of the crack surfaces by Gray and Giles (129). This technique, known as the 'dual boundary element method', has been developed by Portela, Aliabadi, and Rooke (130)(131) for the calculation of stress intensity factors and crack growth direction in two-dimensional mixed-mode problems. An extension to three-dimensional problems has been developed by Mi and Aliabadi (132).

Methods of calculating stress intensity factors, based on quarter-point elements, have been used extensively in the application of the boundary element method to both two- and three-dimensional crack problems see, for example (133)(134). Another similar approach is based on the use of special crack tip elements, as in finite element analysis, for which the shape functions are modified to model the correct behaviour of the displacements and tractions near the crack tip. A more sophisticated approach based on the knowledge of the near-tip field solutions has been developed by Xanthis, Bernal, and Atkinson (135) for anti-plane strain problems, and for mode I crack problems in fracture mechanics by Aliabadi, Rooke, and Cartwright (136). This formulation was later extended to combined mode I and mode II crack problems using a multi-domain formulation by Aliabadi (137). This technique, known as the singularity subtraction technique (SST), is reviewed in reference (12).

All the standard methods for calculating stress intensity factors used in finite element analyses can be used in boundary element analysis. In particular, methods based on crack tip fields and 'J' integrals have been widely used. As in finite element analysis, the use of symmetric 'J' contours allows mixed-mode problems to be studied in both two (138) and three (139) dimensions, but without the need for symmetric internal meshes. A recent paper (140) compares some of the above methods used in boundary element and finite element analyses.

The concept of a special fundamental solution introduced by Snyder and Cruse (127) into the boundary element analysis of cracked anisotropic plane problems with traction-free cracks has been extended. Mir-Mohammad-Sadegh and Altiero (141) used the crack fundamental solution for isotropic plane problems with the indirect boundary element formulation, while Thompson (142) used the direct BEM formulation with a number of different fundamental solutions. Dowrick (143) developed a method based on a crack fundamental solution to study the effect of cracks in stiffened sheets using constant elements. Young, Cartwright and Rooke (144) followed a similar procedure as in (143) to study the effect of repair patches on cracked plane configurations using quadratic boundary elements. Mews (145) developed a crack fundamental solution for a semi-infinite straight edge crack and a central crack. In reference (145) results are presented for mode I, II, and anti-plane strain mode III stress intensity factors for a number of crack configurations including kinked-cracks. Once the boundary values of tractions $t_j(x)$ and displacements $u_j(x)$ are obtained, the stress intensity factors can be computed from the interior stresses (144) or interior displacements (145), in the limit as the internal source point approaches the crack tip.

Weight function procedures have been widely used in the boundary element evaluation of stress intensity factors, following the demonstration by Cartwright and Rooke (146) that a boundary element analysis produced more accurate stress intensity factors more efficiently than the finite element analysis of Paris *et al.* (147). This work which is based on Bueckner's fundamental fields and Betti's reciprocal theorem has been extended by Aliabadi, Rooke, and Cartwright (148) to both mode I and mode II deformation which, in this formulation, are independent. The technique has been coupled with the subtraction of the fundamental field technique as developed by Adiabadi, Cartwright, and Rooke (149). The concept of singular fields to obtain the weight function using boundary element analysis has been extended to straight-fronted cracks in three dimensions by Rooke, Cartwright, and Aliabadi (150). In this case the singular field was that due to a point force near the edge of a semi-infinite planar crack. Rooke and Aliabadi (151) have elaborated further on the use of the fundamental fields in boundary element analysis of both two- and three-dimensional problems. Bains, Aliabadi, and Rooke (152) have extended these concepts to the evaluation of stress intensity factors of cracks with curved crack fronts.

The above techniques and procedures developed for the elastostatic analysis of crack problems are currently being extended to the evaluation of dynamic stress intensity factors.

Dynamic stress intensity factors

The stress intensity factor under dynamic conditions can be defined in a similar way to that under static conditions, see equation (20); however, K will now be a function of time. It is important to distinguish between two types of crack problems, stationary cracks and moving cracks. In problems involving stationary cracks, the crack length does not change with time but the loading does. For example, the loading may be a high-speed impact (explosive) or due to high-frequency vibrations. The importance of knowing the stress intensity factor in such dynamic situations is that the peak dynamic K (K_d say) is often larger than the static K (K_s say). A typical response is as follows: initially K_d increases in value rapidly to a peak and then decays in an oscillatory fashion to the static value K_s. The height of the first peak above K_s, often known as the 'dynamic overshoot', can be significant and may even initiate premature failure, hence it is important to be able to determine its value.

The role of the stress intensity factor as a failure parameter is much more complex under dynamic conditions than under static conditions. In static problems K is used to determine the onset of crack growth or to correlate fatigue crack growth. In dynamic problems K also determines the phenomena of crack arrest, crack bending, and crack branching. Each of these phenomena have critical K values (analogous to the static toughness) associated with them. The situation is further complicated by the fact that these critical values will be functions of the crack velocity.

The analysis of the equations of elasticity is considerably more complicated when dynamic phenomena are considered. For instance, the wave nature of the disturbance, due to sudden boundary loads, must be considered as it travels through the structure. There are three types of wave that can be important, longitudinal (dilatational), transverse (shear), and Rayleigh (surface) waves. They travel through the structure at different velocities, which are determined by the elastic constants. Thus the time dependence of the response is a function of the elastic constants. The waves are reflected/refracted by boundaries and energy is redistributed between the wave types. The Rayleigh surface waves are particularly important in crack problems as they travel along the crack surfaces and are reflected at the tips giving rise to the usual singular fields, and hence a contribution to K. Travelling waves are diffracted by a crack tip; the magnitude and type of effect depends on the angle between the crack and the wave direction. If the crack is moving, then these phenomena are also functions of crack speed.

Given the complexity of the analysis it is not surprising that few analytical solutions are available, and that they are mostly limited to infinite domains (no boundaries) and semi-infinite cracks with simple loading systems. Nevertheless such solutions are important and can be useful. The solution of an impulsive point force on a semi-infinite crack in an infinite sheet, obtained by Freund (153), is a case in point. This solution has been used as a Green's function to obtain the initial response to impact loadings on finite cracks in finite structures. The results are valid up to the time when reflected waves would arrive at the crack tip. Useful estimates of the height of the first peak in the stress intensity factor can often be obtained in this way. These results and others, both theoretical and numerical, are described in detail by Freund (153) and Parton and Boriskovsky (154)(155): results for impact loading and harmonic loading on stationary cracks are described as well as moving crack problems.

Dynamic crack problems can be solved directly in the time domain, or can be transformed (Laplace, Fourier) into the frequency domain before solution. Both solution methods have been used with various numerical techniques, for example finite difference, finite element, and boundary element methods. The time domain usually requires large computational power and storage since procedures often involve the evaluation of convolution integrals (i.e., integration of a response over the loading history); this is particularly true of methods based on Betti's reciprocal theorem. The use of transforms simplifies these integrals to algebraic products and will facilitate the use of weight function techniques which are based on Betti's theorem. However, the numerical difficulties associated with the final inverse transformation should not be underestimated.

Various procedures such as displacement and traction extrapolation, and 'J' integrals have been used to obtain stress intensity factors. The introduction of a time-dependent fundamental solution into the boundary element method greatly increases the computational effort required. The usual time-independent fundamental solutions can be used, but the resulting integral equation now contains a domain integral, and much of the advantage of a boundary method is lost. However, procedures which transform the domain integrals to boundary integrals have been developed **(156)(157)** and hold promise for the efficient solution of crack problems. Many numerical and experimental techniques for obtaining stress intensity factors are reported in reference **(158)**.

Conclusions

A brief review has been given of the development of stress intensity factors, and their role in fracture mechanics analysis. The methods used to evaluate stress intensity factors range from the very simple through to large-scale numerical computation techniques. The degree of complexity of the method required tends to mirror the complexity of the structural problem. However, even for real engineering structures, the simple techniques outlined in this review are often very valuable. They can often be used to obtain bounds on the value of the stress intensity factor, and to highlight those structural (or loading) details which significantly affect the value.

The value of the techniques, described in this paper, for the analysis of cracks under fretting conditions is clear from the number in this volume, which report the use of many of them. For full details and further source references the reader is encouraged to refer to the individual papers; only a brief overview will be given here.

The paper by Faanes and Harkegard **(159)** combines several stage 2 techniques including Green's functions using approximate stress distributions and a short/long crack limit process. The Green's function concept is also used by Faanes and Fernando **(160)**. The use of dislocation distributions leading to integral equations for two-dimensional problems is described by Hills and Nowell (161) and Dai, Hills, and Nowell (162); the latter authors also study three-dimensional problems using the Eigenstrain (or body force) method which is closely related to boundary element methods.

Finite element analysis occurs several times, in particular Sheikh, Fernando, Brown, and Miller (163) calculate energy release rates and hence stress intensity factors for an advancing crack. Dubourg (164) utilizes the principles of superposition and combines finite element analysis with continuous distributions of dislocations to study crack propagation under fretting fatigue conditions. Both finite element techniques for fretting strength analysis, and boundary element techniques, for a singularity crack analysis, are demonstrated by Hattori (165). Further work on the stress singularity analysis (a generalization of the stress intensity factor to non-zero wedge angles) is described by Hattori and Nakamura (166).

Further developments of the stress intensity factor and its use are still taking place. The concept of linear elastic fracture mechanics parameters is being further extended into non-linear elasto-plastic problems. There still is much scope for development of dynamic stress intensity factors and for development in the field of contact problems with friction (non-linear) which will have applications in fretting analysis.

References

(1) GRIFFITH, A. A. (1921) The phenomena of rupture and flow in solids. *Phil. Trans.* A221, 163–198.
(2) IRWIN, G. R. (1948) Fracturing of metals, *Proceedings of ASM Symposium*, ASM, USA, pp. 147–166.
(3) OROWAN, E. (1952) Fundamentals of brittle behaviour in metals, *Fatigue and fracture of metals*, (Edited by W. M. Murray), pp. 139–167.
(4) IRWIN, G. R. (1957) Analysis of stresses and strains near the end of a crack traversing a plate, *J. appl. Mech.*, 24, 361–364.
(5) SIH, G. C. (1973) *Handbook of stress intensity factors*, Lehigh University, USA.
(6) TADA, H., PARIS, P. C., and IRWIN, G. R. (1973) *The stress analysis of cracks handbook*, (Del Research Corp., USA).
(7) ROOKE, D. P. and CARTWRIGHT, D. J. (1976) *Compendium of stress intensity factors*, HMSO, London.
(8) MURAKAMI, Y. (1993) *Stress intensity factors handbook*, Pergamon, Oxford.
(9) WESTERGAARD, H. M. (1939) Bearing pressures and cracks, *J. appl. Mech.*, 6, 49–53.
(10) WILLIAMS, M. L. (1952) Stress singularities resulting from various boundary conditions in angular corners of plates in extension, *J. appl. Mech.*, 19, 526–528.
(11) SNEDDON, I. N. (1946) The distribution of stress in the neighbourhood of a crack in an elastic solid, *Proc. R. Soc., Lon. Ser. A*, 187, 229–260.
(12) ALIABADI, M. H. and ROOKE, D. P. (1991) *Numerical fracture mechanics*, Computational Mechanics Publications, Southampton.
(13) NEUBER, H. (1946) Theory of notch stresses, (trans. J. W. Edwards), Ann Arbor, Michigan.
(14) MUSKHELISHVILI, N. I. (1953) Some basic problems of the mathematical theory of elasticity. Noordhoff, Leyden.
(15) IRWIN, G. R. (1958) Fracture, *Handbuch der physik*, Vol. VI, (Springer-Verlag, Berlin), pp. 551–590.
(16) DUGDALE, D. S. (1960) Yielding of steel sheets containing slits, *J. Mech. Phys. Solids*, 8, 100–104.

(17) BUECKNER, H. F. (1958) The propagation of cracks and the energy of elastic deformation. Tran. ASME, **80E**, 1225–1230.
(18) PARIS, P. C. (1964) The fracture mechanics approach to fatigue, *Fatigue, an interdisciplinary approach* (Edited by J. J. Burke, N. L. Reed, and V. Weiss), Syracuse University Press, pp. 107–132.
(19) EFTIS, J., SUBRAMONIAN, N., and LIEBOWITZ, H. (1977) Crack border stress and displacement equations revisited, *Engng. Fracture Mech.*, **9**, 189–210.
(20) THEIN WAH, (1984) Stress intensity factors determined by use of Westergaard's stress functions. *Engng. Fracture Mech.*, **20**, 65–73.
(21) RICE, J. R. (1968) A path independent integral and the approximate analysis of strain concentration by notches and cracks, *J. appl. Mech.*, **35**, 379–386.
(22) AAMODT, B. and BERGAN, P. G. (1976) On the principle of superposition for stress intensity factors, *Engng. Fracture Mech.*, **8**, 437–440.
(23) EMERY, A. F. and WALKER, G. E. (1968) Stress intensity factors for edge cracks in rectangular plates with arbitrary loadings, Sandia Corporation, California, Report SCL–DC–67–105.
(24) EMERY, A. F., WALKER, G. E., and WILLIAMS, J. A. (1969) A Green's function for the stress intensity factors of edge cracks and its application to thermal stresses, *J. bas. Engng.*, **91**, 618–624.
(25) LACHENBRUCH, A. H. (1961) Depth and spacing of tension cracks. *J. Geophys. Res.*, **66**, 4273–4292.
(26) EMERY, A. F. (1966) Stress intensity factors for thermal stresses in thick hollow cylinders. *J. bas. Engng*, **88**, 45–52.
(27) BALL, D. L. (1987) The development of mode I, linear-elastic stress intensity factor solutions for cracks in mechanically fastened joints, *Engng. Fracture Mech.*, **27**, 653–681.
(28) BOWIE, O. L. (1966) Analysis of edge notches in a semi-infinite region, *J. Math. Phys.*, **45**, 356–366.
(29) HARRIS, D. O. (1967) Stress intensity factors for hollow circumferentially notched round bars, *J. bas. Engng.*, **89**, 49–54.
(30) SIH, G. C. and LIEBOWITZ, H. (1968) Mathematical theories of brittle fracture, *Fracture, Vol. II, Mathematical Fundamentals*, (Academic Press, London).
(31) CHEN, Y. Z. (1988) Evaluation of K_2 values from the solution of notch problem. *Int. J. Fracture*, **38**, R61–R64.
(32) CHIANG, C. R. (1990) Evaluation of the stress intensity factors from solutions of corresponding notch problems, *Int. J. Fracture*, **42**, R61–R63.
(33) POOK, L. P. and DIXON, J. R. (1970) Fracture toughness of high strength materials: theory and practice, ISI 120, pp. 45–50.
(34) HASEBE, N. and KUTANDA, Y. (1978) Calculation of stress intensity factor from stress concentration factor. *Engng. Fracture Mech.*, **10**, 215–221.
(35) PETERSON, R. E. (1974) *Stress concentration factors*, John Wiley, New York.
(36) HARDY, S. J. and MALIK, N. H. (1992) A survey of post-Peterson stress concentration factor data, *Int. J. Fatigue*, **14**, 147–153.
(37) KOITER, W. T. (1965) Discussion on rectangular tensile sheet with symmetric edge cracks, *J. appl. Mech.*, **32**, 237.
(38) NEWMAN, J. C. (1971) An improved method of collocation for the stress analysis of cracked plates with various shaped boundaries, NASA TN–D6376.
(39) BENTHEM, J. P. and KOITER, W. T. (1973) Asymptotic approximations to crack problems, *Mechanics of fracture I – methods of analysis and solutions of crack problems*, (Edited by G. C. Sih), Noordhoff, Leyden, pp. 131–178.
(40) WILLIAMS, J. G. and ISHERWOOD, D. P. (1968) Calculation of the strain-energy release rates of cracked plates by an approximate method, *J. Strain Analysis*, **3**, 17–22.
(41) CHAN, S. K., TUBA, I. S., and WILSON, W. K. (1970) On the finite element method in linear fracture mechanics, *Engng. Fracture Mech.*, **2**, 1–17.
(42) CANNON, D. F. and ALLEN, R. J. (1974) Application of fracture mechanics to railway failures, *Railway Engng. J.*, **3**, 6–23.
(43) CARTWRIGHT, D. J. and ROOKE, D. P. (1974) Approximate stress intensity factors compounded from known solutions, *Engng. Fracture Mech.*, **6**, 563–571.
(44) ROOKE, D. P. (1986) An improved compounding method for calculating stress-intensity factors, *Engng. Fracture Mech.*, **23**, 783–792.

(45) ROOKE, D. P. (1984) Compounded stress intensity factors for cracks at fastener holes, *Engng. Fracture Mech.*, **19**, 359–374.
(46) FIGGE, I. E. and NEWMAN, Jr., J. C. (1967) Fatigue crack propagation in structures with simulated rivet forces, *Fatigue crack propagation*, ASTM STP 415, (Edited by J. C. Grosskreutz), ASTM, Philadelphia, pp. 71–93.
(47) SMITH, F. W. (1966) Stress intensity factors for a semi-elliptical surface flaw, Structural development research memorandum 17, Boeing Co.
(48) LIU, A. F. (1972) Stress intensity factor for a corner flaw, *Engng. Fracture Mech.*, **4**, 175–179.
(49) SOKOLNIKOFF, I. S. (1956) *Mathematical theory of elasticity* (Second edition), McGraw-Hill, New York, pp. 318–327.
(50) HARTRANFT, R. J. and SIH, G. C. (1973) Alternating method applied to edge and surface crack problems, *Methods of analysis and solutions of crack problems. Mechanics of fracture 1*, (Edited by G. C. Sih), Noordhoff, Leyden, pp. 179–238.
(51) ROOKE, D. P. (1986) Compounding stress intensity factors: applications to engineering structures, *Research reports in materials science*, Parthenon Press.
(52) ROOKE, D. P. and HUTCHINS, S. M. (1984) Stress intensity factors for cracks at loaded holes – effect of load distribution, *J. Strain Analysis*, **19**, 81–96.
(53) CARTWRIGHT, D. J. and ROOKE, D. P. (1979) Green's functions in fracture mechanics, *Fracture mechanics – current status, future prospects*, Pergamon, Oxford.
(54) BUECKNER, H. F. (1970) A novel principle for the computation of stress intensity factors, *Z. Angew. Math. Mech.*, **50**, 529–546.
(55) RICE, J. R. (1972) Some remarks on elastic crack-tip stress fields, *Int. J. Solids Structures*, **8**, 751–758.
(56) FETT, T., MATTHECK, C., and MUNZ, D. (1989) Approximate weight functions for 2D and 3D problems, *Engng. Analysis Bound. Elem.*, **6**, 48–63.
(57) WU, X-R. and CARLSSON, A. J. (1991) *Weight functions and stress intensity factor solutions*, Pergamon Press, Oxford.
(58) BUECKNER, H. F. (1973) Field singularities and related representations, *Methods of analysis and solutions of crack problems. Mechanics of fracture I*, (Edited by G. C. Sih), Noordhoff, Leyden, pp. 239–319.
(59) BUECKNER, H. F. (1987) Weight function and fundamental fields for the penny shaped and the half-plane crack in three-space, *Int. J. Solids Structures*, **23**, 57–93.
(60) RICE, J. R. (1985) First order variation in elastic fields due to variation in location of a planar crack front, *J. appl. Mech.*, **52**, 571–579.
(61) RICE, J. R. (1989) Weight function theory for three-dimensional elastic crack analysis, *Fracture mechanics: perspectives and directives, ASTM STP 1020*, (Edited by R. P. Wei and R. P. Gangloff), Philadelphia, pp. 29–57.
(62) GAO, H. and RICE, J. R. (1987) Somewhat circular tensile cracks, *Int. J. Fracture*, **33**, 155–174.
(63) GAO, H. and RICE, J. R. (1986) Shear stress intensity factors for a planar crack with slightly curved front, *J. appl. Mech.*, **53**, 774–778.
(64) GAO, H. and RICE, J. R. (1987) Nearly circular connections of elastic half-spaces, *J. appl. Mech.*, **54**, 627–634.
(65) BUECKNER, H. F. (1975) The weight function of the configuration of collinear cracks, *Int. J. Fracture*, **11**, 71–83.
(66) BUECKNER, H. F. (1977) The weight functions of mode I of the penny-shaped and of the elliptical crack, *Fracture mechanics and technology II*, (Edited by G. C. Sih and C. L. Chow), Sijthoff and Noordhoff International, pp. 1069–1089.
(67) BUECKNER, H. F. (1977) The edge crack in half space, *Fracture mechanics and technology II*, (Edited by G. C. Sih and C. L. Chow), Sijthoff and Noordhoff International, pp. 1091–1107.
(68) BUECKNER, H. F. (1989) Observations on weight functions, *Engng. Analysis Boundary Elements*, **6**, 3–18.
(69) GAO, H. (1988) Nearly circular shear mode cracks, *Int. J. Solids Structures*, **24**, 177–193.
(70) GAO, H. (1989) Weight functions for external circular cracks, *Int. J. Solid Structures*, **25**, 107–127.
(71) GROSS, B., SRAWLEY, J. E., and BROWN, W. F. (1964) Stress intensity factors for a

single-edge-notch tension specimen by boundary collocation of a stress function, NASA TN–D2395.

(72) ISIDA, M. (1970) On the determination of stress intensity factors for some common structural problems, *Engng. Fracture Mech.*, **2**, 61–79.

(73) CARTWRIGHT, D. J. and PARKER, A. P. (1982). Opening mode stress intensity factors for cracks in pin-loaded joints, *Int. J. Fracture*, **18**, 65–78.

(74) PEI-QING, G. (1985) Stress intensity factors for a rectangular plate with a point loaded edge crack by a boundary collocation procedure, and an investigation into the convergence of the solutions, *Engng. Fracture Mech.*, **22**, 295–305.

(75) WOO, C. W., WANG, Y. H., and CHEUNG, Y. K. (1989) The mixed-mode problem for the cracks emanating from a circular hole in a finite plate, *Engng. Fracture Mech.*, **32**, 279–288.

(76) SNEDDON, I. N. (1973) Integral transform methods, *Methods of analysis and solution of crack problems.*, *Mechanics of fracture 1*, (Edited by G. C. Sih), Noordhoff, Leyden.

(77) BILBY, B. A. and ESHELBY, J. D. (1968) Dislocations and the theory of fracture, *Fracture*, (Edited by H. Liebowitz), Vol. 1, Academic Press, New York, pp. 99–182.

(78) ROOKE, D. P. and TWEED, J. (1980) Stress intensity factors for a crack at the edge of pressurized hole, *Int. J. Engng. Sci.*, **18**, 109–121.

(79) SINGH, B. M. and DANYLUK, H. T. (1985) Stress intensity factors for two rectangular cracks in three-dimensions, *Engng. Fracture Mech.*, **22**, 475–483.

(80) NOWELL, D. and HILLS, D. A. (1987) Open cracks at or near free edges, *J. Strain Analysis*, **22**, 177–185.

(81) LE VAN, A. and ROYER, J. (1986) Integral equations for three-dimensional problems, *Int. J. Fracture*, 125–142.

(82) NISITANI, H. (1968) Two-dimensional stress problem solved using electric digital computer, *Bull. JSME*, **11**, 14–23.

(83) NISITANI, H. (1978) Solutions of notch problems by body force method, *Stress analysis of notched problems*, (Edited by G. C. Sih), pp. 1–68, Leyden, Noordhoff.

(84) NISITANI, H. (1985) Body force method for determination of stress intensity factors, *J. Aeronaut. Soc. India*, **37**, 21–41.

(85) NISITANI, H. and MURAKAMI, Y. (1974) Stress intensity factors of an elliptical crack or a semi-elliptical crack subject to tension, *Int. J. Fracture*, **10**, 353–368.

(86) MURAKAMI, Y. and NEMAT–NASSAR, S. (1983) Growth and stability of intersecting surface flaws of arbitrary shape, *Engng. Fract. Mech.*, **17**, 193–210.

(87) LISKOVETS, O. A. (1965) The method of lines, *Differential equations*, **1**, 1308–1323.

(88) GYEKENYESI, J. P. and MENDELSON, A. (1975) Three-dimensional elastic stress and displacement of finite geometry solids containing cracks, *Int. J. Fracture*, **11**, 409–429.

(89) ALAM, J. and MENDELSON, A. (1986) Elasto-plastic stress analysis of a CT specimen using method of lines, *Int. J. Fracture*, **31**, 17–28.

(90) MENDELSON, A. and ALAM, J. (1983) The use of the method of lines in 3-D fracture mechanics analyses with application to compact tension specimens, *Int. J. Fracture*, **22**, 105–116.

(91) XANTHIS, L. S. and EDWARDS, B. H. (1990) Determination of stress intensity- and higher order-factors using the method of lines and ODE solvers, *Proceedings of IMA Conference on Computational ODEs*, Oxford University Press, UK.

(92) QUINLAN, P. M. (1964) The torsion of an irregular polygon, *Proc. Roy. Soc. Lond. Ser. A*, **282**.

(93) QUINLAN, P. M. and O'CALLAGHAN, M. J. A. (1987) The edge function method (EFM) for cracks cavities and curved boundaries in elastostatics, *Topics in boundary element research*, (Edited by C. A. Brebbia), Vol. 3, Springer-Verlag, Berlin, Germany, pp. 132–167.

(94) FLEMING, J. F., GUYDISH, J. J., PENTZ, J. R., RUNNION, C. E., and ANDERSON, G. P. (1980) The finite element method versus the edge function method for linear fracture analysis, *Engng. Fracture Mech.*, **13**, 43–55.

(95) QUINLAN, P. M., GRANNELL, J. J., ATLURI, S. N., and FITZGERALD, J. E. (1982) The edge-function method for three-dimensional stress analysis, including embedded elliptical cracks and surface flaws, *Boundary element methods in engineering*, (Edited by C. A. Brebbia), Computational Mechanics Publications, Southampton, UK, pp. 457–471.

(96) GALLAGHER, R. H. (1978) A review of finite element techniques in fracture mechanics, *Proceedings of the First International Conference on Numerical methods in fracture mecha-

nics, (Edited by A. R. Luxmoore and B. D. R. J. Owen), 1–25, University College of Swansea, UK, pp. 1–25.

(97) LIEBOWITZ, H. and MOYER, Jr., E. T. (1989) Finite element methods in fracture mechanics, *Comp. Structures*, **31**, 1–9.

(98) SWEDLOW, J. L., WILLIAMS, M. L., and YANG, W. H. (1965) Elasto-plastic stresses and strains in cracked plates, Proceedings of the First International Conference on Fracture, Vol. 1, pp. 259–282.

(99) WILSON, W. K. (1971) Crack tip finite elements for plane elasticity, Westinghouse Rep. No. 71–1E7–FM–PWR–P2.

(100) OGLESBY, J. J. and LAMACKEY, O. (1972) An evaluation of finite element methods for the computation of elastic stress intensity factors, NSRDC Rep. No. 3751.

(101) TRACEY, D. M. (1971) Finite elements for determination of crack tip elastic stress intensity factors, *Engng. Fracture Mech.*, **3**, 255–266.

(102) BLACKBURN, W. S. (1973) Calculation of stress intensity factors at crack tips using special finite elements, *The mathematics of finite elements*, (Edited by J. R. Whiteman), Academic Press, London, UK. pp. 327–336.

(103) AKIN, J. E. (1976) The generation of elements with singularities, *Int. J. Numer. Methods Engng.*, **10**, 1249–1259.

(104) YAMADA, Y., EZAWA, Y., NISHIGUCHI, I., and OKABE, M. (1979) Reconsiderations on singularity of crack tip elements, *Int. J. Numer. Methods. Engng.*, **14**, 1524–1544.

(105) HENSHELL, R. D. and SHAW, K. G. (1975) Crack tip finite elements are unnecessary, *Int. J. Numer. Methods Engng.*, **9**, 495–507.

(106) BARSOUM, R. S. (1975) Further application of quadratic isoparametric finite elements to linear fracture mechanics of plate bending and general shells, *Int. J. Fracture*, **11**, 167–169.

(107) HEYLIGER, P. R. and KRIZ, R. D. (1989) Stress intensity factors by enriched mixed finite elements, *Int. J. Numer. Methods Engng.*, **28**, 1461–1473.

(108) ATLURI, S. N. (1986) *Computational methods in the mechanics of fracture*, Vol. 2. North Holland, Amsterdam.

(109) MORLEY, L. S. D. (1973) Finite element solution of boundary value problems with non-removable singularities, *Phil. Trans. Roy. Soc. London, Ser. A*, **275**, 1252.

(110) YAMAMOTO, Y. (1971) Finite element approach with the aid of analytical solutions, *Recent advances in matrix methods of structural analysis and design*, (Edited by R. N. Gallagher, Y. Yamada, and J. T. Oden), University of Alabama Press, pp. 85–103.

(111) SYMM, G. T. (1973) Treatment of singularities in the solution of Laplace's equation by an integral equation method, Report No. NAC 31, National Physical Laboratory.

(112) YAMAMOTO, Y., TOKUDO, N., and SUMI, Y. (1973) Finite element treatment of singularities of boundary value problems and its application to analysis of stress intensity factors, *Theory and practice in finite element structural analysis*, (Edited by Y. Yamada and R. H. Gallagher), University of Tokyo Press, Japan.

(113) AMARA, M., DESTUYN, der, P., and DJAOUA, M. (1980). On a finite element scheme for plane crack problems, *Numerical methods in fracture mechanics*, (Edited by D. R. J. Owen and A. Luxmoore), Pineridge Press, Swansea, UK, pp. 41–50.

(114) SINCLAIR, G. B. and MULLAN, D. (1982) A simple yet accurate finite element procedure for computing stress intensity factors, *Int. J. Numer. Methods Engng.*, **18**, 1587–1600.

(115) PARKS, D. M. (1974) A stiffness derivative finite element technique for determination of elastic crack tip stress intensity factors, *Int. J. Fracture*, **10**, 487–502.

(116) HELLEN, T. K. (1975) On the method of virtual crack extensions, *Int. J. Numer. Methods Engng.*, **9**, 187–207.

(117) PARKS, D. M. and KAMENETZKY, E. M. (1979) Weight functions from virtual crack extension, *Int. J. Numer. Methods Engng.*, **14**, 1693–1706.

(118) VANDERGLAS, M. L. (1978) A stiffness derivative finite element technique for determination of influence functions, *Int. J. Fracture*, **14**, R291–R294.

(119) VAINSHTOK, V. A. (1982) A modified virtual crack extension method of the weight functions calculation for mixed mode fracture problems, *Int. J. Fracture*, **19**, R9–R15.

(120) SHA, G. T. and YANG, C. T. (1985) Weight function calculations for mixed mode fracture problems with the virtual crack extension technique, *Engng. Fracture Mech.*, **21**, 1119–1149.

(121) ISHIKAWA, H. (1980) A finite element analysis of stress intensity factors for combined tensile and shear loading by only a virtual crack extension, *Int. J. Fracture*, **16**, R243–R246.

(122) SHAM, T. L. (1987) A unified finite element method for determining weight functions in two and three dimensions, *Int. J. Solids Structures*, **23**, 1357–1372.

(123) SHAM, T. L. and ZHOU, Y. (1989) Computation of three-dimensional weight functions for circular and elliptical cracks, *Int. J. Fracture*, **41**, 51–57.

(124) SHAM, T. L. and ZHOU, Y. (1989) Weight functions and fundamental fields in layered media, *Engng. Analysis Bound. Elements*, **6**, 38–47.

(125) *Boundary elements – abstract and newsletter*, Vol. 1, Computational Mechanics Publications, Southampton, UK.

(126) CRUSE, T. A. (1972) Numerical evaluation of elastic stress intensity factors by the boundary-integral equation method. The surface crack: physical problems and computational solutions, (Edited by J. L. Swedlow), ASME, New York, USA, pp. 153–170.

(127) SNYDER, M. D. and CRUSE, T. A. (1975) Boundary-integral equation analysis of cracked anisotropic plates, *Int. J. Fracture*, **11**, 315–328.

(128) BLANDFORD, G. E., INGRAFFEA, A. R., and LIGGETT, J. A. (1981) Two-dimensional stress intensity factor computations using the boundary element method, *Int. J. Numer. Methods Engng.*, **17**, 387–404.

(129) GRAY, L. J. and GILES, G. E. (1988) Application of the thin cavity method to shield calculations in electroplating. *Proceedings of the Tenth International Conference on BEM*, (Edited by C. A. Brebbia), Vol. 2, Computational Mechanics Publications, Southampton, UK, pp. 441–452.

(130) PORTELA, A., ALIABADI, M. H., and ROOKE, D. P. (1993) Dual boundary element analysis of linear elastic crack problems, Advanced formulations in boundary element methods, (Edited by M. H. Aliabadi and C. A. Brebbia), Computational Mechanics Publications, Southampton, UK, pp. 1–30.

(131) PORTELA, A., ALIABADI, M. H., and ROOKE, D. P. (1993) Dual boundary element analysis of fatigue crack growth, *Advances in boundary element methods for fracture mechanics*, (Edited by M. H. Aliabadi and C. A. Brebbia), Computational Mechanics Publications, Southampton, UK, pp. 1–46.

(132) MI, T. and ALIABADI, M. H. (1992) Dual boundary element method for analysis of 3-D crack problems. *Boundary elements XIV, Stress analysis and computational aspects*, (Edited by C. A. Brebbia, J. Dominguez, and F. Paris), Vol. 2, Computational Mechanics Publications, Southampton, UK, pp. 315–329.

(133) MARTINEZ, J. and DOMINGUEZ, J. (1984) On the use of quarter-point boundary elements for stress intensity factor computations, *Int. J. Numer. Methods Engng.*, **20**, 1941–1950.

(134) GANGMING, L. and YONGYUAN, Z. (1988) Improvement of stress singular element for crack problems in three-dimensional boundary element method, *Engng Fracture Mech.*, **31**, 993–999.

(135) XANTHIS, L. S., BERNAL, M. J. M., and ATKINSON, C. (1981) The treatment of singularities in the calculation of stress intensity factors using the integral equation method, *Comp. Methods Appl. Mech. Engng.*, **26**, 285–304.

(136) ALIABADI, M. H., ROOKE, D. P., and CARTWRIGHT, D. J. (1987) An improved boundary element formulation for calculating stress intensity factors: application to aerospace structures, *J. Strain Analysis*, **22**, 203–207.

(137) ALIABADI, M. H. (1987) An enhanced boundary element method for determining fracture parameters, *Proceedings of Fourth International Conference on numerical methods in fracture mechanics*, Pineridge Press, UK, pp. 27–39.

(138) ALIABADI, M. H. (1990) Evaluation of mixed-mode stress intensity factors using the path-independent J-integral. *Boundary elements XII, Applications in stress analysis, potential and diffusion*, (Edited by M. Tanaka *et al.*), Vol. 1, Computational Mechanics Publications, Southampton, UK, pp. 281–291.

(139) RIGBY, R. H. and ALIABADI, M. H. (1992) Boundary element analysis of three-dimensional crack problems, Localized damage II, *Computational methods in fracture mechanics*, (Edited by M. H. Aliabadi, H. Nisitani, and D. J. Cartwright), Vol. 2, Computational Mechanics Publications, Southampton, UK, pp. 91–105.

(140) PORTELA, A. and ALIABADI, M. H. (1989) On the accuracy of boundary and finite element techniques for crack problems in fracture mechanics, Advances in boundary elements, (Edited by C. Brebbia and J. J. Connor), Vol. 1, Computational Mechanics Publications, Southampton, UK, pp. 123–137.

(141) MIR-MOHAMMAD-SADEGH, A. and ALTIERO, N. J. (1979) Solution of the problem of a crack in a finite plane region, using an indirect boundary-integral method, *Engng. Fracture Mech.*, **11**, 831–837.

(142) THOMPSON, R. M. (1985) *The BIE method applied to the derivation of stress concentration and stress intensity factors*, PhD Thesis, Royal Military College of Science, Shrivenham, UK.

(143) DOWRICK, G. (1990) Boundary effects for a reinforced cracked sheet using the boundary element method, *Theoret. appl. Fracture Mech.*, **12**, 251–260.

(144) YOUNG, A., CARTWRIGHT, D. J., and ROOKE, D. P. (1988) The boundary element for analysing repair patches on cracked finite sheets, *Aero J.*, **92**, 416–421.

(145) MEWS, H. (1987) Calculation of stress intensity factors for various crack problems with the boundary element method, *Proceedings of the Ninth Integral Conference on BEM*, (Edited by C. A. Brebbia, W. L. Wendland, and G. Kuhn), Vol. 2, Computational Mechanics Publications, Southampton, UK. pp. 259–278.

(146) CARTWRIGHT, D. J. and ROOKE, D. P. (1985) An efficient boundary element model for calculating Green's function in fracture mechanics, *Int. J. Fracture*, **27**, R43–R50.

(147) PARIS, P. C., McMEEKING, R. M., and TADA, H. (1976) The weight function method for determining stress intensity factors, *Cracks and fracture, ASTM STP 601*, ASTM, Philadelphia, pp. 471–489.

(148) ALIABADI, M. H., ROOKE, D. P., and CARTWRIGHT, D. J. (1987) Mixed-mode Bueckner weight functions using boundary element analysis, *Int. J. Fracture*, **34**, 131–147.

(149) ALIABADI, M. H., CARTWRIGHT, D. J., and ROOKE, D. P. (1989) Fracture-mechanics weight-functions by the removal of singular fields using boundary element analysis, *Int. J. Fracture*, **40**, 271–284.

(150) ROOKE, D. P., CARTWRIGHT, D. J., and ALIABADI, M. H. (1987) Boundary elements combined with singular fields for three-dimensional cracked solids, *Proceedings of the Fourth International Conference on numerical methods in fracture mechanics*, (Edited by A. R. Luxmoore *et al.*), Pineridge Press, UK, pp. 15–26.

(151) ROOKE, D. P. and ALIABADI, M. H. (1989) Weight functions for crack problems using boundary element analysis, *Engng. Analysis with boundary elements*, **6**, 19–29.

(152) BAINS, R., ALIABADI, M. H., and ROOKE, D. P. (1993) Stress intensity factor weight functions for cracks in 3-D finite geometries, *Advances in boundary element methods for fracture mechanics*, (edited by M. H. Aliabadi and C. A. Brebbia), Computational Mechanics Publications, Southampton, UK, pp. 201–268.

(153) FREUND, L. B. (1990) *Dynamic fracture mechanics*, Cambridge University Press, Cambridge, UK.

(154) PARTON, V. Z. and BORISKOVSKY, V. G. (1989) Dynamic fracture mechanics, *Stationary Cracks*, Vol. 1, Hemisphere, New York, USA.

(155) PARTON, V. Z. and BORISKOVSKY, V. G. (1990) Dynamic fracture mechanics, *Propagating Cracks*, Vol. 2, Hemisphere, New York, USA.

(156) NARDINI, D. and BREBBIA, C. A. (1985) Boundary integral formulation of mass matrices for dynamic analysis, *Topics in boundary element research, Time-dependent and vibration problems*, (Edited by C. A. Brebbia), Vol. 2, Springer-Verlag, Berlin, Germany, pp. 191–208.

(157) PARTRIDGE, P. W. and BREBBIA, C. A. (1993) The dual reciprocity method, *Advanced formulations in boundary element methods*, (Edited by M. H. Aliabadi and C. A. Brebbia), Computational Mechanics Publications, Southampton, UK, pp. 31–75.

(158) The Albert S. Kobayashi anniversary volume, *Engng Fracture Mech.*, **23**, 320.

(159) FAANES, S. and HARKEGARD, G. (1994) Simplified stress intensity factors near the contact surface in fretting fatigue, *Fretting Fatigue*, Mechanical Engineering Publications, London, pp. 73–81. *This volume.*

(160) FAANES, S. and FERNANDO, U. S. (1994) Life prediction in fretting fatigue using fracture mechanics, *Fretting Fatigue*, Mechanical Engineering Publications, London, pp. 149-159. *This volume.*

(161) HILLS, D. A. and NOWELL, D. (1994) A critical analysis of fretting fatigue experiments, *Fretting Fatigue*, Mechanical Engineering Publications, London, pp. 171–182. *This volume.*

(162) DAI, D. N., HILLS, D. A., and NOWELL, D. (1994) Stress intensity factors for three-dimensional fretting fatigue cracks, *Fretting Fatigue*, Mechanical Engineering Publications, London, pp. 171–182. *This volume.*

(163) SHEIKH, M. A., FERNANDO, U. S., BROWN, M. W., and MILLER, K. J. (1994) Elastic stress intensity factors for fretting cracks using the finite element method, *Fretting Fatigue*,

Mechanical Engineering Publications, London, pp. 83–101. *This Volume.*

(164) DUBOURG, M. C. (1994) A theoretical model for the prediction of crack field evolution, *Fretting Fatigue*, Mechanical Engineering Publications, London, pp. 135–147. *This volume.*

(165) HATTORI, T. (1994) Fretting fatigue problems in structural design field, *Fretting Fatigue*, Mechanical Engineering Publications, London, pp. 437–451. *This volume.*

(166) HATTORI, T. and NAKAMURA, M. (1994) Fretting fatigue evaluation using stress singularity parameters at contact edge, *Fretting Fatigue*, Mechanical Engineering Publications, London, pp. 453–460. *This volume.*

D. N. Dai*, D. A. Hills*, and D. Nowell*

Stress Intensity Factors for Three-Dimensional Fretting Fatigue Cracks

REFERENCE Dai, D. N., Hills, D. A., and Nowell, D., **Stress intensity factors for three-dimensional fretting fatigue cracks,** *Fretting Fatigue,* ESIS 18 (Edited by R. B. Waterhouse and T. C. Lindley) 1994, Mechanical Engineering Publications, London, pp. 59–71.

ABSTRACT The phenomenon of fretting fatigue must be interpreted by reference to a three-dimensional crack model if the early stages of crack growth are to be properly understood. This paper employs the Eigenstrain technique to establish a model for planar cracks of arbitrary shape growing in a contact stress field which may in addition contain residual stresses and lead to partial crack closure. Several sample cases are addressed, including semi-elliptical cracks propagating under two- and three-dimensional Hertzian stress fields, and approximate crack shapes which give rise to constant stress intensity factors are determined. The Eigenstrain method is shown to have considerable advantages over finite element techniques for this type of analysis.

Introduction

In an accompanying paper **(1)**, the point is made that fretting fatigue differs from plain fatigue principally at the stage in a component's life when the crack is no bigger than the characteristic contact dimension. Thus, in order to be able to understand and quantify the influence of the contact on the crack's propagation a careful study of the crack is needed, as it exists at this time in a very steep stress gradient. The question of the conditions for crack initiation is not addressed here. It is assumed that this phase has already occurred, and that the crack is fully formed, and at a scale where it can be described as existing in a continuum (that is, it must span several grains), and the crack tip behaviour must be fully described by the stress intensity factor. This imposes certain conditions on the overall nature of the contact; the grain size must be much smaller than the characteristic contact dimension, and the contact pressure must be such that the extent of any plasticity is confined both to asperities within the contact and to a small crack tip zone compared with the zone described by the singular solution. These prescriptions are not necessarily restrictive, but should be borne in mind when applying the results. As the contact pressure implied by these requirements is relatively low, it is probable that a significant proportion of the cracks' lives will be taken up in the initiation and early growth stages. It follows that some care needs to be exercised in evaluating the stress intensity factors, as, using Paris' fatigue law, the growth rate is roughly proportional to their fourth power.

A number of features make the evaluation of stress intensity factors more demanding than those for many configurations and these have largely

* Department of Engineering Science, University of Oxford, Parks Road, Oxford, OX1 3PJ, UK.

restricted previous solutions to two dimensions. There are four main difficulties

(1) Fretting fatigue cracks inevitably exist in a very steep stress gradient, caused by the contact loading. This needs to be incorporated into the solution with great care if accurate results are to be obtained.
(2) The contact stress field is largely compressive in nature so that crack face closure may well occur at some stages of the loading cycle. Such closure may well take place remote from the crack tip so that any solution obtained must be checked for the possibility of closure anywhere along the crack faces rather than simply ensuring that $K_1 > 0$.
(3) Fretting fatigue cracks often turn from a 45 degree initial angle to being normal to the free surface as they move from stage 1 to stage 2 growth. This produces a distinct kink which should feature in the geometry modelled.
(4) Fretting fatigue is clearly a surface phenomenon and it is anticipated that residual stresses arising from the manufacturing process of the specimen or component will occur precisely in this region. These stresses may well be quite significant compared to the contact stresses and often vary even more rapidly.

All of these features are challenging to model and the chosen method of analysis must be capable of incorporating them with the minimum of difficulty. One possible method of attack which has been used at Oxford with some success in the analysis of two-dimensional configurations is to use the dislocation density method. This technique can readily cope with the steep stress gradient (2), any kink (3), crack closure (4), and residual stresses (5) without difficulty. The basis of the solution is the state of stress induced by a single dislocation in a half-plane (6). The state of stress in the absence of the crack is first calculated and the unsatisfied tractions $\tilde{\sigma}(x)$ which arise on the crack face are determined. These are calculated by distributing an array of displacement discontinuities along the line of the crack; the density $b(\xi)$ of this array may be determined from an integral equation of the form

$$\tilde{\sigma}(x) + \int_0^a K(x, \xi)b(\xi) \, d\xi = 0 \qquad (1)$$

The kernel of this equation, $K(x, \xi)$ is the state of stress induced at a point x by a single dislocation located at ξ and which has a singularity of the form l/r, where r is the distance from x to ξ. The equation may readily be inverted by means of powerful numerical quadratures (7). The principal limitation of the technique is only that it is incapable of extension to three dimensions, and this means that is difficult to follow early progress of the crack since even in a nominally two-dimensional contact configuration (such as a cylinder on a flat) individual cracks will initiate at a point and grow as thumbnails before they coalesce to form an essentially two-dimensional flaw.

The three-dimensional analogue of the dislocation density method is the Eigenstrain technique. Development of this method has been somewhat slower than its two-dimensional counterpart, principally because the integral equation which results has a hypersingular kernel $(1/r^3)$, which is difficult to handle both analytically and numerically. Much of the early development was carried out by Murakami and co-workers (8)(9) who initially looked at open cracks, but also developed some solutions for partially closed cracks (10). In this paper the basic procedure is reviewed and recent developments are indicated, including an improved integration method, and techniques for handling crack closure and sliding.

The Eigenstrain method

The Eigen, or transformation, strain method has its origins in Eshelby's work in modelling the influence of voids, inclusions, and cracks in the 1950s (11). At that time the computational power to realize the potential of the method was not available, but recent developments have made the technique not only feasible but computationally very attractive compared with the finite element method for the solution of crack and similar problems, in the neighbourhood of a free surface or interface (12). As with the two-dimensional dislocation density method, the technique relies on a description of the influence of a distributed displacement discontinuity $b_m(\xi_1, \xi_2)$ over the region of the crack, Fig. 1. Here, m can take the values 1, 2, 3, which physically correspond to the opening and two tangential components of displacement discontinuity at that point. The integral equation, requiring that the crack faces remain free of tractions, may be written in the form

$$\tilde{\sigma}_{3n}(x_1, x_2) + \int_S K_{nm}(x_1, x_2, \xi_1, \xi_2) b_m(\xi_1, \xi_2)\, \mathrm{d}S = 0 \qquad (2)$$

where $\tilde{\sigma}_{3n}(x_1, x_2)$ is the traction distribution arising over the plane of the crack in its absence. The kernel for the problem $K_{nm}(x_1, x_2, \xi_1, \xi_2)$ is available in closed form for a strain nucleus near a free surface (13) by manipulation of Mindlin's solution for a point force near the surface of a half-space (14). However, although the integral equation (2) looks deceptively like the simpler 'distributed dislocation' form (1) it has two features which make it much less tractable, viz., it is a surface rather than a line integral, and the kernel $K_{nm}(x_1, x_2, \xi_1, \xi_2)$ has a $1/r^3$ rather than a $1/r$ type singularity. The first problem is remedied by splitting the domain of the crack up into a finite number of elements within each of which the form of $b_m(\xi_1, \xi_2)$ is prescribed, and hence for which the influence function may be explicitly evaluated. The second problem has proved particularly challenging as the integral exists only in the so-called Hadamard finite part sense, but some progress has recently been made in its interpretation (13). A typical mesh used to discretize the crack faces is shown in Fig. 2.

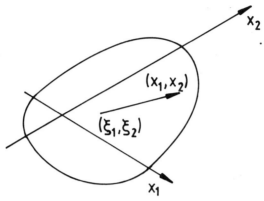

Fig 1 A crack lying on the $x_3 = 0$ plane; stress induced at point (x_1, x_2) by a displacement discontinuity at (ξ_1, ξ_2)

In the plane (dislocation density) model of cracks, a solution to the integral equation is chosen in the form of the product of an unknown function and a so-called fundamental solution which incorporates the required end-point behaviour of the integral: in particular, that the stress state be square root singular at the crack tip. It should be recalled that this corresponds to the displacements near the crack tip falling to zero like $\sqrt{r}$, and it is precisely this form of the weight function which is used in the representation of $b_m(\xi_1, \xi_2)$, equation (2). The distance r is interpreted as the shortest distance from the point in question to the nearest point on the crack front. This permits ready abstraction of the crack tip stress intensity factor.

Convergence of the solution is usually quite rapid, and computational effort is a fraction of that expended formulating the same problem by finite elements; this is at least partly because three aspects of the problem are implicit in the form of the solution, i.e, the semi-infinite extent of the solid, the presence

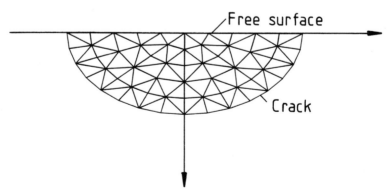

Fig 2 Typical mesh employed in the analysis of a semi-elliptical crack in a half space using the Eigenstrain technique. Note that in contrast to the finite element method, only the crack faces need to be meshed

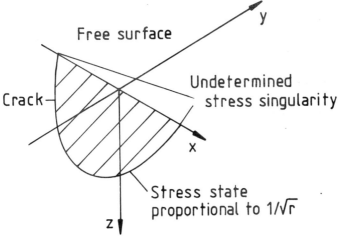

Fig 3   A surface-breaking three-dimensional crack; the problem of determining the stress singularity at the free surface

of the nearby free boundary, and the displacement field in the neighbourhood of the crack front. The results presented in this paper were obtained with meshes employing in the region of 100 elements. Meshes, of this degree of refinement give results within 1 percent of accepted solutions for standard cases such as surface-breaking semi-elliptical cracks under uniform far field tension. However, there is one aspect of the solution (and indeed any solution where the crack tip intersects a free surface) which calls for comment, and that is the behaviour of the crack near the free surface in the case of surface-breaking problems, Fig. 3. As the free surface is approached a boundary condition must arise where the traction components of stress vanish. This, in turn, gives rise to a complex variation in the state of stress where the usual crack tip asymptotics associated with the Westergaard solution (15) do not hold, and hence neither does the classic square-root singular behaviour. The singularity is weaker at the surface (16), but it is not possible, at the moment, to predict its extent. The assumed form of the solution, therefore, forces an incorrect behaviour on the solution which in turn hinders convergence of the results in the neighbourhood of the free surface and consequently makes interpretation of the results in this region difficult. As mentioned above, this phenomenon is not, of course, peculiar to the Eigenstrain method but is a difficulty which must be borne in mind when interpreting any stress intensity factor solutions for surface-breaking cracks.

Application to the fretting problem

Sphere on flat configuration

An attractive geometry for carrying out fretting fatigue tests is to use a sphere on flat configuration since there are no end effects or alignment difficulties and in principle highly repeatable results should be obtained. However, the results from such tests can be difficult to interpret as the contact stress field is truly three-dimensional and complex crack shapes develop (17). The problem we wish to solve is idealized in Fig. 4. A sphere is pressed normally into a block of material subject to a tensile stress σ_0, which is uniform over (x, y), remote from the contact. A residual stress field, σ_R is also present, which, like the bulk tension, is constant, and so the stick-slip pattern predicted by the Mindlin–Cattaneo solution (18), is unaffected. The normal traction distribution is given by

$$p(r) = -p_0\sqrt{\{1 - (r/c)^2\}} \tag{3}$$

$$p_0 = 3P/(2\pi c^2) \tag{4}$$

where c is the radius of the contact circle and the radius of the stick zone, c', is given by

$$\left(\frac{c'}{c}\right)^3 = 1 - \frac{Q}{fP} \tag{5}$$

Here Q is the peak shear force, P the normal load and f the coefficient of friction. Given this boundary value condition the interior stress field is readily found (18). A crack is now installed in the half-space $z > 0$ at some point 'behind' the contact where the region of maximum tension occurs. In an earlier solution to this problem (17) it was assumed that the crack was two-dimensional, but the solution was evaluated at different values of y, Fig. 4. This is tantamount to assuming that the crack front shape has a large radius of curvature compared with the characteristic contact size, c. A more refined analysis where the crack is represented by a semi-ellipse of depth a and transverse semi-extent (semi-axis), b will now be carried out.

The first set of results presented (Fig. 5) show the effect of varying the position of a semi-circular crack of constant size. Stress intensity factors are plotted for points at the bottom of the crack and at the surface. (In practice,

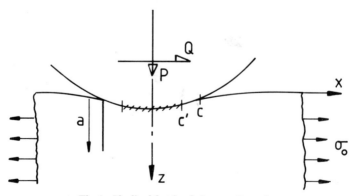

Fig 4 Idealized fretting fatigue configuration

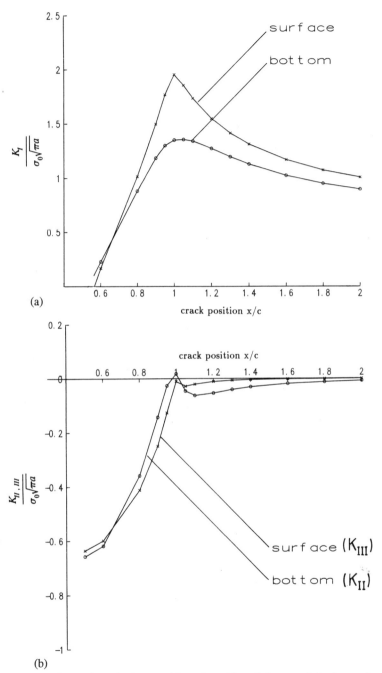

(a)

(b)

Fig 5 Variation of stress intensity factors with crack position x/c for a semi-circular crack normal to the free surface, $a/c = 0.1$, $Q/fP = 0.89$, $p_0/\sigma_0 = 3.0$, $f = 0.7$, no residual stress. (a) Opening mode (K_I); (b) shear mode K_{II} (bottom) and K_{III} (surface)

stress intensity factors are evaluated at the centre of the element closest to the surface. This avoids the difficulty mentioned earlier concerning the different order of stress singularity at the surface.) As with the case of two-dimensional cracks, it can be seen that the greatest opening mode stress intensity occurs when the crack is located at the edge of the contact disc ($x = -c$) and is larger at the surface than at the bottom. However, the shear mode stress intensity almost vanishes at this point, whereas it takes a relatively large value for cracks located towards the centre of the contact. However, cracks located here have small values of K_1 and approach closure at the crack front. Closure effects are not taken into account in the current analysis but may be incorporated into the Eigenstrain method (19).

The second set of results is shown in Fig. 6. Here, the crack is located at the edge of the contact ($x/c = -1.0$), where the opening stress intensity is at its maximum. For two sample cases the crack depth, a, is held constant and the width of the crack, b, is varied in order to investigate the variation of the stress intensity factors with crack aspect ratio. Both crack depths show similar results; for cracks with high ellipticity (b/a large) the stress intensity factor at the bottom of the crack exceeds that at the surface points. If crack growth takes place according to a Paris Law-type criterion this suggests that the crack will grow preferentially from the bottom and hence the ellipticity will decrease. Conversely for low ellipticities the surface stress intensity factors are higher and an increase in ellipticity with growth is suggested. In the steady state we might expect that cracks adopt a shape which gives uniform stress intensity factor along the crack front (the so-called iso-K shape). From the results presented here it might be expected that an ellipticity of between 1.8 and 2.0 might be expected under the fretting conditions analyzed although it should be noted that stress intensity factor variation along the entire crack front has not yet been probed and a more complex shape than a semi-ellipse may well be needed in order to achieve exact iso-K conditions. Ellipticities observed in Kuno's experiments (17) are about 2.0 although the cracks obtained were not planar.

Cylinder on flat configuration

It has been usual in previous analyses (e.g. (20)) to consider cracks resulting from the two-dimensional cylinder on flat configuration as two-dimensional. However, it is clear that individual cracks will initiate from a point and thus in the early stages of growth they will be fully three-dimensional, albeit propagating in a two-dimensional stress field. Later in the life of a specimen several such cracks may well amalgamate to form an essentially two-dimensional through thickness crack, but in order to investigate the early stages of crack propagation further, it is essential to develop a three-dimensional crack model, even for cases of two-dimensional stress fields. Clearly the initial shape as initiation proceeds will be critically dependent on the exact shape of the

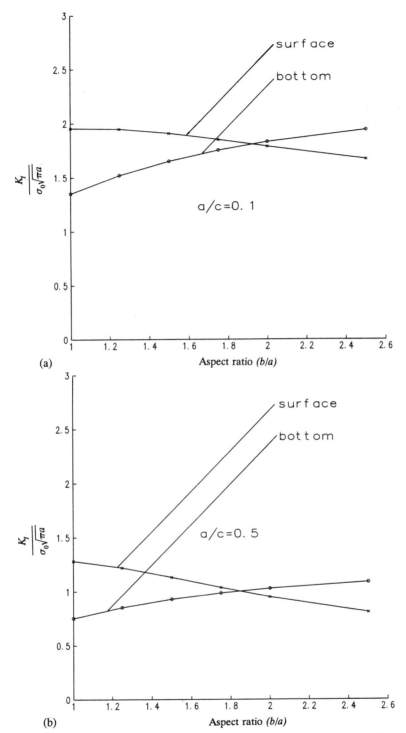

Fig 6 Variation of opening mode stress intensity factor with crack aspect ratio b/a for a semi-elliptical crack normal to the free surface, $Q/fP = 0.89$, $p_0/\sigma_0 = 3.0$, $f = 0.7$, no residual stress. (a) $a/c = 0.1$; (b) $a/c = 0.5$

embryo defect. Very soon after initiation, however, crack growth becomes regulated by the stress intensity factor, and as discussed above it is reasonable to assume that the crack front develops into one giving a constant stress intensity factor along its length. There is much experimental evidence to suggest that real cracks do indeed behave in this fashion (21)–(23). One way to predict the shape adopted by the crack is to use a time marching solution, as was done in the work just cited. However, this is very time consuming, and one simple way of progressing would be to assume that the crack grows in the approximate shape of a thumbnail, which might be idealized as a semi-elliptical shape. It is then possible to determine the ellipticity of the crack so as to achieve the best approximation to the actual crack shape.

Figure 4 may be used to illustrate the problem we wish to consider in more detail, but this time the contact is interpreted as two-dimensional, extending over the region $-c < x < c$. The stick–slip solution is again due to Mindlin–Cattaneo, the pressure distribution is again given by equation (3) but the peak pressure and stick zone semi-width are now given by

$$p_0 = 2P/(\pi c) \tag{6}$$

$$\left(\frac{c'}{c}\right)^2 = 1 - \frac{Q}{fP} \tag{7}$$

The stress state induced by this geometry is also known in closed form (18). A crack is again installed at the trailing edge of the contact where the zone of maximum tension is found. To determine the optimum ellipticity various trial shapes are employed and the crack is discretized using a mesh of the type shown in Fig. 2. For each crack depth the variation of stress intensity factor along the crack front is evaluated for a range of different ellipticities. Figure 7 depicts some sample results of the variation of stress intensity factor with angular position along the crack front θ ($\theta = 90$ degrees corresponds to the bottom of the crack). As with the case of spherical contact, choosing too high an eccentricity gives a higher stress intensity factor at the bottom of the crack, whereas too low an eccentricity leads to higher surface values. The best approximation to constant K is chosen by determining the value of b/a at which the stress intensity factors at the elements closest to the surface and at the bottom of the crack are equal. As can be seen from Fig. 7 this is sufficient to ensure that the variation of K with position is less than 5 percent. Closer matching could clearly be achieved by departing from the assumption of a semi-elliptical crack and using more parameters to describe the shape.

For small cracks an aspect ratio of approximately 1.6 leads to an iso-K condition but as the crack grows larger, a more semi-circular shape is adopted. With the two-dimensional stress field caused by cylindrical contact the parts of the crack front close to the surface always remain affected by the contact stress field and thus the ellipticity giving rise to an iso-K crack front differs from that required in a uniform stress field ($b/a \approx 1.24$). Crack shapes are self-similar for

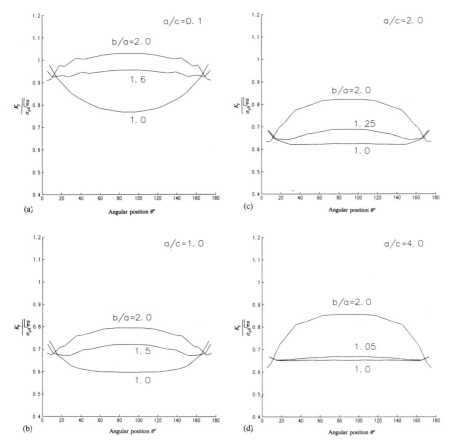

Fig 7 Variation of opening mode stress intensity with angular position along the crack front for a
semi-elliptical crack normal to the free surface under the stress field caused by a cylindrical
fretting pad. $Q/fP = 0.75$, $p_0/\sigma_0 = 1.0$, $f = 0.75$, no residual stress. (a) $a/c = 0.1$;
(b) $a/c = 1.0$; (c) $a/c = 2.0$; (d) $a/c = 4.0$

a/c larger than 4.0, and the normalized stress intensity factor produced is
approximately 0.67. This compares with a value of 0.73 for an iso-K crack in a
uniform stress field. Hence the effect of the typical contact stress field studied is
to generate a more semi-circular crack shape and to reduce the stress intensity
factor for larger cracks. In real fretting fatigue problems several different
cracks of this type would be generated at different points along the trailing
edge of contact but these would eventually merge to form a through thickness
crack.

Figure 8 summarizes the development of a single crack under the typical
two-dimensional Hertzian stress field studied here. Small cracks of aspect ratio
$b/a \approx 1.6$ grow to form almost semi-circular defects as a/c increases.

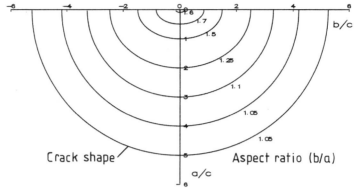

Fig 8 Evolution of crack shape giving rise to constant opening mode stress intensity factor with crack size. $Q/fP = 0.75$, $p_0/\sigma_0 = 1.0$, $f = 0.75$, no residual stress

Conclusions

It has been demonstrated that a full three-dimensional crack model is necessary in order to analyse fretting fatigue crack growth under the influence of three-dimensional stress fields, which were desirable from an experimental point of view, and in the early stages of crack growth for two-dimensional contact configurations. The Eigenstrain method has proved to be a useful tool for carrying out such analysis. For a similar degree of accuracy, computation time is much less than the finite element method and this enables parametric studies to be undertaken readily. Although not considered here in detail, the Eigenstrain method permits ready incorporation of crack face closure and near surface residual stresses into the analysis. In this paper we have assumed a semi-elliptical crack shape in order to estimate the aspect ratio adopted by fretting fatigue cracks under constant stress intensity factor conditions. A more sophisticated analysis could be developed based on the present technique but using more parameters to describe the crack shape and employing a perturbation method. The authors, therefore, believe that the Eigenstrain method has distinct advantages over finite element techniques for three-dimensional crack analysis aimed at gaining a fundamental understanding of the fretting fatigue process.

References

(1) HILLS, D. A. and NOWELL, D. (1993) A critical analysis of fretting experiments, *Fretting Fatigue*, Mechanical Engineering Publications, London, pp. 171–182.
(2) NOWELL, D. and HILLS, D. A. (1987) Open cracks at or near free edges, *J. Strain Analysis*, **22**, 177–185.
(3) LI YINGZHI and HILLS, D. A. (1990) Stress intensity factor solutions for kinked surface cracks, *J. Strain Analysis*, **25**, 21–27.
(4) HILLS, D. A. and NOWELL, D. (1989) Stress intensity calibrations for closed cracks, *J. Strain Analysis*, **24**, 37–43.

(5) WILKS, M. D. B., NOWELL, D., and HILLS, D. A. (1993) The evaluation of stress intensity factors for plane cracks in residual stress fields, *J. Strain Analysis*, **28**, 145–152.
(6) KELLY, P., HILLS, D. A., and NOWELL, D. (1992) The stress field due to a straight dislocation near an interface, Oxford University Engineering Laboratory Internal Report, No. OUEL 1909/92.
(7) ERDOGAN, F., GUPTA, G. D., and COOK, T. S. (1973) Numerical solution of singular integral equations, *Methods of analysis and solutions of crack problems*, (Edited by G. C. Sih), Noordhott, Leyden, pp. 368–425.
(8) MURAKAMI, Y., KANETA, M., and YATSUZUKA, H. (1985) Analysis of surface crack propagation in lubricated rolling contact, *Trans. ASLE*, **28**, 60–68.
(9) KANETA, M., MURAKAMI, Y., and YATSUZUKA, H. (1985) Mechanism of crack growth in lubricated rolling/sliding contact, *Trans ASLE*, **28**, 407–414.
(10) KANETA, M., SUETSUGU, M., and MURAKAMI, Y. (1986) Mechanism of surface crack growth in lubricated rolling/sliding spherical contact, *J. appl. Mech.*, **53**, 354–360.
(11) ESHELBY, J. D. (1956) The continuum theory of lattice defects, *Solid state physics* (Edited in F. Seitz and D. Turnbull), Vol. 3, Academic Press, New York, pp. 79–144.
(12) MURAKAMI, Y. and NEMAT-NASSER, S. (1983) Growth and stability of interacting surface flaws of arbitrary shape, *Engng Fracture Mech.*, **17**, 193–210.
(13) DAI, D. N., NOWELL, D., and HILLS, D. A. (1993) Eigenstrain methods in three-dimensional problems: an alternative integration procedure, *J. mech. Phys. Solids*, **41**, 1003–1017.
(14) MINDLIN, R. D. (1936) Force at a point in the interior of a semi-infinite solid, *Physics*, **7**, 195–202.
(15) WESTERGAARD, H. M. (1937) Bearing pressures and cracks, *J. appl. Mech.*, **6**, A49–A53.
(16) BENTHEM, J. P. (1977) State of stress at the vertex of a quarter infinite crack in a half space, *Int. J. Solids Structures*, **13**, 479–492.
(17) KUNO, M., WATERHOUSE, R. B., NOWELL, D., and HILLS, D. A. (1989) Initiation and growth of fretting fatigue cracks in the partial slip regime, *Fatigue Fracture of Engng Mater. Structures*, **12**, 30–37.
(18) HILLS, D. A., NOWELL, D., and SACKFIELD, A. (1993) *Mechanics of elastic contacts*, Butterworth-Heinemann, UK.
(19) DAI, D. N., NOWELL, D., and HILLS, D. A. (1993) Partial closure and slip of three dimensional cracks, *Int. J. Fracture*, **63**, 89–99.
(20) NOWELL, D., HILLS, D. A., and O'CONNOR, J. J. (1987) An analysis of fretting fatigue, Proceedings of IMechE. Conference on Tribology.
(21) SOBOYEJO, W. O, KISHIMOTO, K., SMITH, R. A., and KNOTT, J. F. (1989) A study of the interaction and coalescence of two coplanar fatigue cracks in bending, *Fatigue Fracture Engng. Mater. Structures*, **12**, 167–174.
(22) SMITH, R. A. and COOPER, J. F. (1989) A finite element model for the shape development of irregular planar cracks, *Int. J. Pressure Vessel Piping*, **36**, 315–326.
(23) GILCHRIST, M. D. and SMITH, R. A. (1991) Finite element modelling of fatigue crack shapes, *Fatigue Fracture Engng Mater. Structures*, **14**, 617–626.

S. Faanes* and G. Härkegård†

Simplified Stress Intensity Factors in Fretting Fatigue

REFERENCE Faanes, S. and Härkegård, G., **Simplified stress intensity factors in fretting fatigue**, *Fretting Fatigue*, ESIS 18 (Edited by R. B. Waterhouse and T. C. Lindley) 1994, Mechanical Engineering Publications, London, pp. 73–81.

ABSTRACT Stress fields and stress intensity factors of an edge crack are studied in a semi-infinite body exposed to distributed contact loads at the boundary. General and asymptotic solutions to the stress intensity factors are determined based on the stress distribution of the uncracked body. Closed form asymptotic solutions are used to derive explicit conditions for the growth of fretting fatigue cracks.

Introduction

Fretting fatigue occurs in mechanical joints showing small relative movements between contact surfaces. Since the fatigue strength of a component may be severely weakened by fretting, there is an obvious need for tools capable of predicting the conditions under which a fretting crack will propagate.

In (1), Rooke and Jones developed a numerical procedure to determine the stress intensity factor of an edge crack in a semi-infinite body under general surface tractions. In the present paper, a slightly different approach to find the stress intensity factors is presented. In particular, the solutions for constant surface pressure and surface shear stress are made easily available for engineering applications through closed form asymptotic solutions. Based on these solutions, and a simple criterion for fatigue crack growth, approximate expressions for the fretting fatigue limit are derived.

Contact stress field

Consider an elastic half-space bounded by the yz plane and with the x axis directed into the solid. The half-space is loaded by normal and tangential tractions, $p(y)$ and $q(y)$, acting on the surface between the straight lines $y = 0$ and $y = b$. The surface tractions are assumed to be independent of the z coordinate but may vary arbitrarily in the y direction. A cross-section of the solid is shown in Fig. 1.

In the following section, the opening mode stress intensity factor for a straight surface crack in the xz plane with its front parallel to the z axis is determined. For this purpose, the distribution of σ_y in the xz plane of the

* Department of Applied Mechanics, The Norwegian Institute of Technology, Trondheim, Norway.
† ABB Power Generation Limited, Baden, Switzerland.

FRETTING FATIGUE

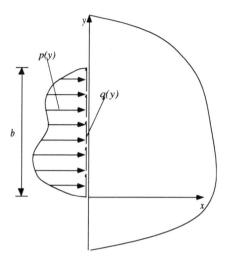

Fig 1   Elastic half-space arbitrarily loaded over a narrow strip

uncracked solid is required. According to Johnson (2)

$$\sigma_y(x) = -\frac{2x}{\pi} \int_0^b \frac{p(y)y^2 \, dy}{(x^2 + y^2)^2} + \frac{2}{\pi} \int_0^b \frac{q(y)y^3 \, dy}{(x^2 + y^2)^2} \qquad (1)$$

Since the width b of the loaded strip is the only characteristic length of the current problem, it is natural to introduce dimensionless coordinates $\xi = x/b$ and $\eta = y/b$ into equation (1)

$$\sigma_y(\xi) = -\frac{2\xi}{\pi} \int_0^1 \frac{p(\eta)\eta^2 \, d\eta}{(\xi^2 + \eta^2)^2} + \frac{2}{\pi} \int_0^1 \frac{q(\eta)\eta^3 \, d\eta}{(\xi^2 + \eta^2)^2} \qquad (2)$$

Now assume that surface tractions can be sufficiently well approximated by expanding them in truncated power series

$$p(\eta) = \sum_{i=0}^n p_i \eta^i, \quad q(\eta) = \sum_{i=0}^n q_i \eta^i \qquad (3)$$

Substituting (3) into (2) yields

$$\sigma_y(\xi) = -\frac{2\xi}{\pi} \sum_{i=0}^n p_i I_{i+2}(\xi) + \frac{2}{\pi} \sum_{i=0}^n q_i I_{i+3}(\xi) \qquad (4)$$

where

$$I_i(\xi) = \int_0^1 \frac{\eta^i \, d\eta}{(\xi^2 + \eta^2)^2} \qquad (5)$$

The integral can be expressed by the recurrence formula

$$I_i(\xi) = J_i(\xi) - \frac{i-1}{i-3} \xi^2 I_{i-2}(\xi) \quad i \geqslant 4 \tag{6a}$$

$$J_i(\xi) = \frac{1}{(i-3)(1+\xi^2)} \tag{6b}$$

$$I_2(\xi) = \frac{1}{2}\left[\frac{1}{\xi}\arctan\frac{1}{\xi} - \frac{1}{1+\xi^2}\right] \tag{6c}$$

$$I_3(\xi) = \frac{1}{2}\left[\log\left(\frac{1}{\xi^2}+1\right) - \frac{1}{1+\xi^2}\right] \tag{6d}$$

From equations (4) and (6) it can be concluded that as $\xi \to 0$, the asymptotic stress field is characterized by a logarithmic singularity, viz

$$\sigma_y(\xi) = \frac{2q_0}{\pi}(-\log \xi) \tag{7a}$$

which is induced by the constant shear load q_0 only.

A constant surface pressure p_0 causes a compressive stress which approaches the constant value

$$\sigma_y(\xi) = -\tfrac{1}{2}p_0 \tag{7b}$$

as $\xi \to 0$.

Determination of the stress intensity factor

Following Rooke and Jones (1), the opening mode stress intensity factor for a surface crack extending in the xz plane of Fig. 1 to a depth $x = a\ (=\alpha b)$ can be calculated as

$$K_I = \frac{1}{\sqrt{(\pi a)}} \int_0^a \sigma_y(x/b)g(x/a)\,dx \tag{8}$$

Introducing the new dimensionless coordinate $u = x/a$, equation (8) is modified to read

$$K_I = \sqrt{\left(\frac{\alpha b}{\pi}\right)} \int_0^1 \sigma_y(\alpha u)g(u)\,du \tag{9}$$

where $g(u)$ is the Green's function due to Hartranft and Sih (3) given by

$$g(u) = \frac{2\{1 + f(u)\}}{\sqrt{(1-u^2)}} \tag{10}$$

$$f(u) = (1-u^2)(0.2945 - 0.3912u^2 + 0.7685u^4 - 0.9942u^6 + 0.5094u^8) \tag{11}$$

Stress intensity factors for constant surface stresses: general solutions

There are several reasons to give particular attention to the case of constant surface shear stress. As was concluded above, this load case gives rise to a

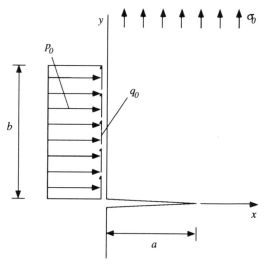

Fig 2 Elastic half-space with surface crack under constant contact shear stress q_0 and bulk stress σ_0

stress singularity, see equation (7a), which is likely to be of great importance to the nucleation and early growth of fretting fatigue cracks under cyclic load conditions. In many situations encountered in practice, there is insufficient information about the contact stress field to allow a more detailed determination of the stress intensity factor than that based on a constant shear stress equal to the average shear stress over the contact zone. As the crack grows deeper, K_1 becomes increasingly insensitive to the precise surface stress distribution and is basically determined by the resultant contact force.

Thus, for the configuration shown in Fig. 2 with $\sigma_0 = 0$, combining equations (4), (6c), (6d) and (9) yields the dimensionless stress intensity factors for $q_0 \neq 0$ and $p_0 \neq 0$ respectively

$$\frac{K_{I_q}}{q_0\sqrt{b}} = \frac{2}{\pi} \sqrt{\left(\frac{\alpha}{\pi}\right)} \int_0^1 \left[\log\left\{\frac{1}{(\alpha u)^2} + 1\right\} - \frac{1}{1 + (\alpha u)^2} \right] \frac{1 + f(u)}{\sqrt{(1 - u^2)}} \, du \qquad (12a)$$

$$\frac{K_{I_p}}{p_0\sqrt{b}} = -\frac{2}{\pi} \sqrt{\left(\frac{\alpha}{\pi}\right)} \int_0^1 \left[\arctan\frac{1}{\alpha u} - \frac{\alpha u}{1 + (\alpha u)^2} \right] \frac{1 + f(u)}{\sqrt{(1 - u^2)}} \, du \qquad (12b)$$

By numerical integration, one obtains K_1 as a function of $\alpha = a/b$ as shown in Fig. 3.

Stress intensity factors for a constant shear stress: asymptotic solutions

As demonstrated in the following, the asymptotic solutions of equation (12) for very deep and very shallow cracks can be obtained in closed form. This is quite useful in checking the accuracy of the numerical solution and in making quick, engineering predictions of the behaviour of fretting fatigue cracks.

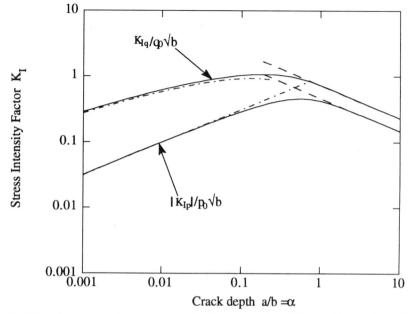

Fig 3 Dimensionless stress intensity factors for the configuration of Fig. 2 with $\sigma_0 = 0$ as a function of crack depth

Consider first the case of a *deep crack*, $a \gg b$. The stress field from a concentrated shear force $F_y = q_0 b$ is such that the resultant force normal to the xz plane is $\frac{1}{2}q_0 b$. Since the stress field only contributes for $u \ll 1$, the mode I stress intensity factor can be obtained directly from equation (8) by substituting the singular normal stress, $\frac{1}{2}q_0 b \cdot \delta(x)$, for σ_y i.e.

$$K_{1_q} = \frac{q_0 b}{2\sqrt{(\pi a)}} g(0) = 1.2945 \frac{q_0\sqrt{b}}{\sqrt{(\pi \alpha)}} \tag{13a}$$

Similarly, a concentrated normal force $F_x = p_0 b$ gives rise to a stress field, the resultant force of which normal to the xz plane is $(1/\pi)p_0 b$. The following can be derived from this.

$$K_{1_p} = -\frac{p_0 b}{\pi\sqrt{(\pi a)}} g(0) = -1.2945 \frac{2}{\pi} \frac{p_0\sqrt{b}}{\sqrt{(\pi \alpha)}} \tag{13b}$$

The close agreement between these asymptotic expressions and the general solution is confirmed by Fig. 3. Thus for $\alpha > 1$ equations (13a) and (13b) deviate from equations (12a) and (12b) by less than 0.1 percent and 10.1 percent, respectively.

For a *shallow crack*, $a \ll b$, the general expression according to equation (9) is used as a starting point. Since the factor $1 + f(u)$ varies moderately across the integration interval in comparison with the other factors of the integrand, it is suggested to substitute $1 + f(u)$ for a constant value Γ. The optimum choice of Γ will obviously depend on the stress distribution $\sigma_y(x)$ being considered. For a constant stress, $\sigma_y(x) = \sigma_0$, the stress intensity factor of an edge crack is given by

$$K_I = 1.1215\sigma_0\sqrt{(\pi a)} \tag{14}$$

which is satisfied identically by choosing $\Gamma = 1.1215$.

Approximating $1 + f(u)$ by the constant Γ, the asymptotic behaviour of equation (12), as $\alpha = a/b \ll 1$, can be expressed as

$$\frac{K_{I_q}}{q_0\sqrt{b}} = \frac{4\Gamma}{\pi}\sqrt{\left(\frac{\alpha}{\pi}\right)} \int_0^1 \frac{-\log \alpha - \log u - \frac{1}{2}}{\sqrt{(1 - u^2)}}\, du$$

$$= \Gamma\sqrt{\left(\frac{\alpha}{\pi}\right)}\left(\log \frac{1}{\alpha^2} + \log 4 - 1\right) \tag{15a}$$

$$\frac{K_{I_p}}{p_0\sqrt{b}} = -\frac{4\Gamma}{\pi}\sqrt{\left(\frac{\alpha}{\pi}\right)} \int_0^1 \frac{(\pi/2) - \alpha u}{\sqrt{(1 - u^2)}}\, du = -\Gamma\sqrt{\left(\frac{\alpha}{\pi}\right)}\left(\frac{\pi}{2} - \frac{2}{\pi}\alpha\right) \tag{15b}$$

The dominating term of equation (15a), $\log(1/\alpha^2)$, originates from the constant shear stress term $\log \alpha$ and the purely numerical terms nearly cancel. For the constant surface pressure, the dominating term is the constant $\pi/2$. Hence, the stress intensity factors should be well approximated by

$$\frac{K_{I_q}}{q_0\sqrt{b}} = \Gamma\sqrt{\left(\frac{\alpha}{\pi}\right)} \log \frac{1}{\alpha^2} \tag{16a}$$

$$\frac{K_{I_p}}{p_0\sqrt{b}} = -\Gamma\sqrt{\left(\frac{\alpha}{\pi}\right)}\frac{\pi}{2} = -\tfrac{1}{2}\Gamma\sqrt{(\pi a)} \tag{16b}$$

for $\alpha \ll 1$ with $\Gamma = 1.1215$. This is demonstrated graphically in Fig. 3. For $\alpha < 0.1$, the deviation between the asymptotic solution, equations (16a) and (16b), and the general solution, equations (12a) and (12b), is less than 10.5 percent and 8.5 percent, respectively.

A direct comparison between equations (7) and (16), the asymptotic distribution of σ_y in the uncracked solid and the asymptotic expression for K_I, respectively, shows that K_I can be written as

$$K_I = \Gamma\sigma_y(a)\sqrt{(\pi a)} \tag{17}$$

This is *formally* equal to the stress intensity factor of an edge crack in a semi-infinite solid subject to a nominal stress at infinity equal to the stress σ_y (a) at the location of the crack tip in the *uncracked* solid; cf. the 'tip stress' method as described by Rooke et al. (4).

Prediction of the fretting fatigue limit

According to linear elastic fracture mechanics, a crack in a cyclically loaded component will grow, if the associated stress intensity range, ΔK, exceeds the threshold range, ΔK_{th}. Thus the critical condition can be written as

$$\Delta K = K_{max} - K_{min} = \Delta K_{th}(R) \tag{18}$$

where ΔK_{th} depends on the load ratio

$$R = \frac{K_{min}}{K_{max}} \tag{19}$$

Equation (18) may (formally) be extended to include short cracks by introducing an 'intrinsic' crack length a_0 (6). As pointed out by Härkegård (7), a_0 cannot be regarded as a material constant, since it is influenced by the actual crack configuration and loading system. Thus, combining equations (17) and (18) yields an 'apparent' stress intensity range corrected for short crack effects, viz.

$$\Delta K^* = \Gamma \Delta \sigma_y(a) \sqrt{\{\pi(a + a_0)\}} = \Delta K_{th}(R) \tag{20}$$

The intrinsic crack length is determined by the requirement that the critical stress range for $a = 0$ must equal $\Delta \sigma_f$, the plain fatigue limit, which implies

$$a_0(R) = \frac{1}{\pi} \left\{ \frac{\Delta K_{th}(R)}{\Gamma \Delta \sigma_f(R)} \right\}^2 \tag{21}$$

Alternatively, equation (20) can be written as

$$\Delta K = \Delta K_{th}^*(a, R) = \frac{\Delta K_{th}(R)}{\sqrt{(1 + a_0/a)}} \tag{22}$$

where the right-hand side may be interpreted as the apparent, crack-size corrected stress intensity threshold.

Consider the case of a purely alternating bulk stress $\pm \frac{1}{2}\Delta\sigma_0$ superimposed on the cyclic surface shear stress $\pm \frac{1}{2}\Delta q_0$ and the steady surface pressure p_0. Depending on the relative order of magnitude of these stresses, the resulting stress intensity range, ΔK, and R values as a function of crack depth, a, may vary considerably. Therefore, the following analysis is limited to the extreme case of $\Delta q_0 \gg \Delta\sigma_0$, which means that the surface shear stress will dominate ΔK for small crack depths ($a \ll b$), the bulk stress for large crack depths ($a \gg b$). In order to keep a fretting crack growing, the resulting stress intensity range must always exceed the (apparent) threshold, i.e. $\Delta K > \Delta K_{th}^*$. A limiting state is shown in Fig. 4, where the ΔK curve for the combined surface shear and bulk stresses just touches the threshold curve from above at the two points $\alpha = \alpha_1$ and $\alpha = \alpha_2$.

Fatigue limit of short cracks under surface shear loading

By assuming $a \ll b$, it is sufficient to study the asymptotic behaviour of ΔK according to equation (16a). The critical condition for fatigue crack growth, equation (22), then reads

$$\Gamma \sqrt{\left(\frac{\alpha}{\pi}\right)} \log \frac{1}{\alpha^2} = \frac{\Delta K_{th}}{\Delta q_0 \sqrt{b}} \frac{1}{\sqrt{(1 + \alpha_0/\alpha)}} \tag{23}$$

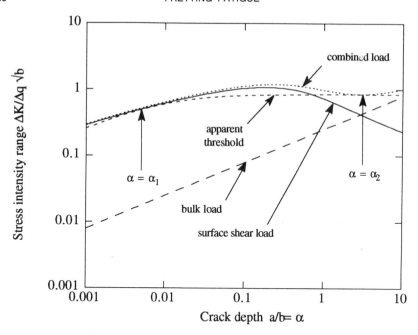

Fig 4   **Graphic determination of the fretting fatigue limit**

The limiting condition, where the load (left-hand side) and the threshold (right-hand side) curves touch, is expressed by the additional requirement

$$\frac{d\Delta K}{da} = \frac{d\Delta K_{th}^{*}}{da} \tag{24}$$

Combining equations (23) and (24) yields the algebraic equation

$$\alpha = \frac{\alpha_0}{\log\left(1/\sqrt{\alpha}\right) - 1} \tag{25}$$

which can be solved (iteratively) for the point of tangency, $\alpha = \alpha_1$, as long as $\alpha_0 < \frac{1}{2}e^3 = 0.0249$. The corresponding surface shear fatigue limit is obtained by combining equations (23) and (25)

$$\Delta q_{ff} = \frac{\Delta K_{th}(R = -1)}{4\Gamma\sqrt{(\alpha_1 b/\pi)}(1 + \alpha_0/\alpha_1)^{3/2}} \tag{26}$$

Within its area of applicability, i.e. $\alpha_0 < \frac{1}{2}e^3$ (and $\alpha_1 < 1/e^3$), equation (26), which is based on the asymptotic solution for ΔK, overestimates Δq_{ff} by less than 7.3 percent in comparison with the result obtained by using the exact solution for ΔK.

Fatigue limit of long crack under combined surface shear and bulk loading

In the absence of an alternating bulk load, a fretting fatigue crack, initiated and propagated by a surface shear stress range $\Delta q_0 > \Delta q_{\text{ff}}$, will eventually become non-propagating. The depth, at which the crack is arrested, is given by equations (13a) and (22). Under combined surface shear and bulk loading, there is a minimum stress intensity range at a crack depth determined by setting the ranges due to equations (13a) and (14) equal

$$\alpha_2 = \frac{1.2945}{\pi\Gamma} \frac{\Delta q_0}{\Delta \sigma_0} \tag{27}$$

By further assuming the combined load stress intensity range for $a = \alpha_2 b$ to be equal to the threshold range, one obtains the stress range just required to propagate the fatigue crack beyond $a = \alpha_2 b$

$$\Delta\sigma_{\text{ff}} = \frac{1}{4 \cdot 1.2945 \cdot \Gamma} \frac{\{\Delta K_{\text{th}}(R)\}^2}{\Delta q_0 b} \tag{28}$$

Conclusions

Stress intensity factors for edge cracks subjected to fretting forces at the surface have been determined and the results confirm presently available calculation techniques.

The strength of the procedure presented is its ability to handle small ratios between the crack length and the fretting load length, and the asymptotic solutions may serve as benchmark tests for numerical calculations of stress intensity factors in fretting.

A simple crack growth criterion based on the stress intensity factors and threshold behaviour of short cracks has been developed, and two mechanisms of crack arrest in fretting are outlined.

References

(1) ROOKE, D. P. and JONES, D. A. (1979) Stress intensity factors in fretting fatigue, *J. Strain Analysis*, **14**, 1–6.
(2) JOHNSON, K. L. (1985) *Contact mechanics*, Cambridge University Press, Cambridge.
(3) HARTRANFT, R. J. and SIH, G. C. Alternating method applied to edge and surface crack problems, *Methods of analysis and solutions of crack problems*, (Edited by G. C. Sih), Chapter 4, Noordhoff, Leyden.
(4) ROOKE, D. P., BARATTA, F. I., and CARTWRIGHT, D. J. (1981) Simple methods of determining stress intensity factors, *Engng Fracture Mech.*, **14**, 397–426.
(5) ELBER, W. (1971) *The significance of fatigue crack closure*, ASTM STP 486, ASTM, Philadelphia, pp. 230–242.
(6) EL HADDAD, M. H., SMITH, K. N., and TOPPER, T. H. (1979) *A strain based intensity factor solution for short fatigue cracks initiating from notches*, ASTM STP 677, ASTM, Philadelphia, pp. 274–289.
(7) HÄRKEGÅRD, G. (1982) An effective stress intensity factor and the determination of the notched fatigue limit, *Fatigue thresholds* (Edited by J. Bäcklund, A. F. Blom, and C. J. Beevers), EMAS.

M. A. Sheikh*, U. S. Fernando*, M. W. Brown* and K. J. M. Miller*

Elastic Stress Intensity Factors for Fretting Cracks using the Finite Element Method

REFERENCE   Sheikh, M. A., Fernando, U. S., Brown, M. W., and Miller, K. J. M., **Elastic stress intensity factors for fretting cracks using the finite element method,** *Fretting Fatigue,* ESIS 18 (Edited by R. B. Waterhouse and T. C. Lindley) 1994, Mechanical Engineering Publications, London, pp. 83–101.

ABSTRACT   In fretting fatigue, intensification of normal stress and friction at the leading edge of the contact patch tend to initiate cracks at a very early stage of life. The fracture process is seen to be dominated by the gradual growth of cracks. Hence the fracture mechanics methodology is considered to be appropriate for the characterization of the fretting fracture process. In situations where only microscopic yielding occurs at the contact surface, linear elastic fracture mechanics is adequate and the elastic stress intensity factors may be employed for the prediction of fretting fatigue life. A finite element procedure for the determination of these stress intensity factors for fretting cracks has been established. A fretting fatigue specimen with contact pads has been modelled to simulate the individual loading components, i.e., axial load, normal load, and friction and analysed for various pad geometries and crack configurations. Three distributions of normal load, namely uniform, triangular, and anti-parabolic are assumed. Results are obtained for unit stresses corresponding to all the loading components so that superposition can be applied to analyse any given loading combination.

Notation

a	Crack length
Δa	Crack extension
S	Fretting pad span
b	Width of the fretting pad
E	Modulus of elasticity
v	Poisson's ratio
σ_A	Axial stress
σ_N	Normal pad pressure
σ_F	Frictional stress
μ	Coefficient of friction
K_I	Mode I stress intensity factor
K_{II}	Mode II stress intensity factor
ΔK	Stress intensity factor range
LC	Assumed load distribution case
da/dn	Fatigue crack growth rate
n	Number of cycles
θ	Crack angle

* SIRIUS, University of Sheffield, Mappin Street, Sheffield S1 3JD, UK.

A Axial load
N Normal pad load
F Friction force
G Griffith's energy release rate
G_I Mode I component of G for coplanar crack extension
G_{II} Mode II component of G for coplanar crack extension
x, y Cartesian coordinates
u_x, u_y Displacements in x and y directions, respectively

Introduction

Cracks are frequently initiated as a result of contact between metallic surfaces where cyclic stresses exist – a process known as 'fretting' (1). Unlike ordinary fatigue failure at low stress levels, the presence of intense cyclic surface friction forces in a fretting environment tends to initiate cracks at a very early stage. Thus the major part of lifetime is spent on the propagation of the cracks, so that the fracture mechanics approach provides the basic tool for the prediction of fatigue life. In practical situations where only microscopic yielding occurs at the contact surface, linear elastic fracture mechanics may be adequate to characterize the stress field, so that elastic stress intensity factors can be employed for the prediction of fretting fatigue life (2).

In order to support experimental fretting research, a wide ranging finite element study of fretting–crack driving forces for various combinations of opening and shear modes has been undertaken. Some results of this study are summarized in this paper.

The characteristic crack path in fretting indicates that cracks grow in mixed-mode under combined mode I (tensile) and mode II (shear), to a size of the order of 0.5 to 1.0 mm before they turn to a mode I plane and propagate across the specimen to failure. Stress intensity factors (K_I and K_{II}) are, therefore, required for normal and inclined edge cracks. A model has been developed for a specimen with fretting cracks (3), which is based on the finite element package 'TOMECH' developed at the University of Sheffield. TOMECH provides a facility for evaluating the stress intensity factors for a mixed-mode situation using a number of different techniques. Here, a crack advance technique has been employed, whereby the crack is advanced by an incremental amount permitting the calculation of the strain energy release rates G_I and G_{II}, for each mode (4). The main advantage of this method is that the stress intensity factors can be obtained from the values of G_I and G_{II} for a range of crack lengths by successively advancing the crack through a number of small increments.

A quarter-domain of the fretting specimen was modelled using symmetry (5). The geometrical, material, and loading data were selected in order to simulate the experimental tests. The crack was located at the leading edge of the fretting pad where normal pressure and frictional stresses were intense.

From elastic analysis, values of K_I and K_{II} were obtained for different combinations of pad span, axial and normal loading, crack length, and crack angle. Three types of distribution, namely uniform, triangular, and anti-parabolic were considered for the normal load. The coefficient of friction between the contacting surfaces was assumed to be unity for the prescription of the frictional load.

Since, under elastic conditions, superposition of stresses can be used to obtain the stress intensity factors for any given loading situation where the cracks are fully open, the analysis was performed for individual unit loading components (i.e., axial, normal, and frictional) for various crack configurations. These results are presented in this paper, and the effects of pad and specimen geometry are discussed.

The above results can be correlated to the crack growth data obtained from the experimental tests in order to assess the validity of LEFM parameters for the interpretation of fatigue crack growth behaviour (6). Similarly they may be compared to other published results, particularly where Green's function methods are adopted, which also depend on linear superposition (2)(7).

Two-dimensional plane strain fretting model

Geometry

The geometry of the fretting specimen (Fig. 1) is represented as a full domain model of the working section, as shown in Fig. 2 (all dimensions in mm). Only a quarter of the domain (OBCD) needs to be modelled due to symmetry, as shown in Fig. 3. The width of the fretting pad (b), through which normal load is applied, is taken as 1.27 mm, whereas three values of the fretting span (S) are considered as follows:

$S1 = 16.5$ mm
$S2 = 34.35$ mm
$S3 = 6.35$ mm

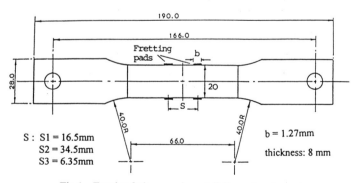

Fig 1   Fretting fatigue specimen (all dimensions in mm)

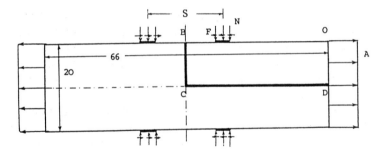

Fig 2   Fretting specimen – full domain model

The specimen is 8 mm thick, for which a plane strain analysis was deemed appropriate.

Material properties

The specimen is made of a 4 percent copper aluminium alloy (BS–L65) for which the following material data is specified:

E (Modulus of elasticity) = 74.0 GPa
v (Poisson's ratio)   = 0.33

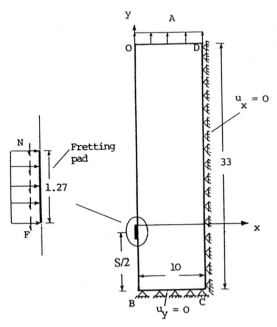

Fig 3   Fretting specimen – quarter domain model

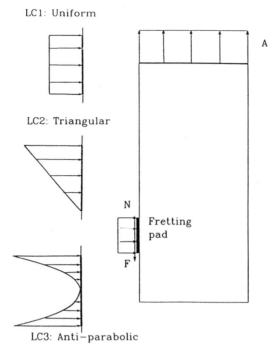

LC1: Uniform

LC2: Triangular

N

Fretting
pad

F

LC3: Anti−parabolic

Fig 4   Assumed distributions of normal load and friction on the fretting pad

Loading conditions

Referring to Fig. 3, the three modes of loading are defined. The axial stress, σ_A, corresponds to the cyclic load amplitude (A) to which the specimen is subjected. For the present model the axial stress is always taken as tensile with a value of 1 GPa. Secondly the normal load on each pad ($N = 10.16$ kN distributed over 8 mm specimen thickness) produced a constant average pressure of $\sigma_N = 1.0$ GPa. Thirdly a corresponding friction stress is induced by the axial displacements, where $\sigma_F = \mu\sigma_N$. The coefficient of friction for the present analysis was taken to be unity; $\mu = 1.0$. Three distributions of σ_N and σ_F, namely, uniform (LC1), triangular (LC2), and anti-parabolic (LC3), were considered, as shown in Fig. 4.

Referring to Fig. 3, two boundary conditions are specified for the quarter domain model. Along the edge CD, $u_x = 0$ from symmetry, and on BC, $u_y = 0$ where (u_x, u_y) refer to the displacement field in x–y coordinate system.

Crack representation

The crack location in the quarter domain model is shown in Fig. 5. In this analysis, the length of the crack (a) was varied over a range from 0.1 to 2.0 mm.

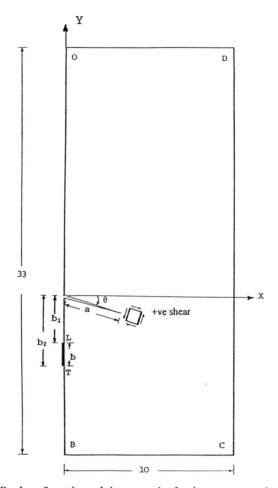

Fig 5   Crack configuration and sign convention for shear stresses at the crack site

The position of the crack is defined with reference to point 'L' which is the leading edge of the fretting pad contact patch, by the two parameters:

b_1 = distance between the crack and the leading edge of the fretting pad (L);
b_2 = distance between the crack and the trailing edge of the fretting pad (T).

In the present analysis, the crack was always assumed to initiate at the leading edge of the fretting pad, so that $b_1 = 0$, and $b_2 = b$, as defined in Fig. 5.

The orientation of the crack is defined by the angle θ which the crack makes with the positive x axis, being taken as positive is the clockwise direction towards the fretting contact patch. A range of θ values were analysed, with three values of θ, zero, 22.5, and 45 degrees being reported in the present paper.

Finite element model

A typical finite element model of the fretting specimen ($S = 16.5$ mm) is shown in Fig. 6. It consists of 1690 eight-noded isoparametric quadrilateral elements with 5401 nodes. This model was generated by using the finite element package TOMECH (8) developed in the Department of Mechanical and Process Engineering, University of Sheffield. Due to a high stress gradient near the contact boundary and the requirement that the crack extension (Δa) be small, a fine mesh was employed in the fretting pad region.

Analysis

LEFM mixed-mode analysis

For a coplanar crack extension in an elastic material, the strain energy release rate (G) can be expressed as the sum of the mode I and mode II components (9), so that

$$G = G_{\mathrm{I}} + G_{\mathrm{II}} \tag{1}$$

and

$$G_{\mathrm{I}} = K_{\mathrm{I}}^2/E', \quad G_{\mathrm{II}} = K_{\mathrm{II}}^2/E', \tag{2}$$

where $E' = E/(1 - v^2)$ for plane strain.

The finite element package TOMECH provides the facility for evaluating the stress intensity factors in a mixed-mode loading situation, by first advancing the crack through an incremental amount (Δa) and then calculating the strain energy release rates, G_{I} and G_{II}, for each mode. The stress intensity factors (K_{I} and K_{II}) are directly calculated from equation (2). The main advantage of this method is that the stress intensity factors can be evaluated for a number of crack lengths by successively advancing the crack through various increments. This procedure is found to be very economical in the use of computer CPU time.

Results

It is apparent that under linear elastic loading superposition of stresses could be employed to obtain K_{I} and K_{II} values for any given loading combination. The fretting specimen was, therefore, analysed for individual unit load cases; axial load (σ_{A}), normal load (σ_{N}) and friction (σ_{F}).

Three finite element models corresponding to the three fretting pad spans (S1, S2, and S3) were generated for a uniform distribution of the axial load (σ_{A}). For the normal load (σ_{N}) and friction (σ_{F}), nine such models were developed for each span and the three load distributions (LC1, LC2, and LC3). For each model a crack was initiated at the leading edge of the fretting pad (Fig. 5) and analysed for various crack configurations given by the crack angle (θ). The

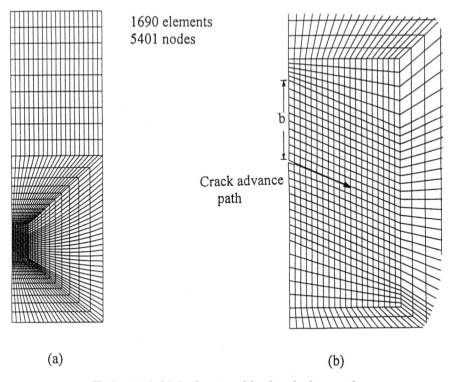

1690 elements
5401 nodes

Crack advance
path

(a)   (b)

Fig 6   A typical finite element model and crack advance path

opening and sliding mode stress intensity factors (K_I and K_{II}) are plotted against the crack length (a) and are presented in Figs 7–11. It must be noted that whilst the positive values of K_I indicate crack opening, the negative values of K_I are meaningless in isolation; they can only be used in a superposition with positive values of K_I so that the total K_I is positive and the crack is guaranteed open along its whole length. Positive or negative values of K_{II}, on the other hand, give the sense of the shear stresses and crack tip shear displacements at the crack singularity, as defined in Fig. 5.

Discussion

In Fig. 7, K_I and K_{II} values are plotted against the crack length (*a*) for a uniform unit axial stress (σ_A). These are given for all the three spans and three crack configurations given by the crack angle $\theta = 0$, 22.5, and 45 degrees. These results are, in general, as expected with K_I decreasing and K_{II} increasing as the crack angle increases. The results for spans S1 (16.5 mm) and S2 (34.35 mm) are in very close agreement. The results for span S3 (6.35 mm) agree well with the other two spans for a crack length of up to about 1.0 mm but beyond that the difference progressively increases with the crack length and K_I values

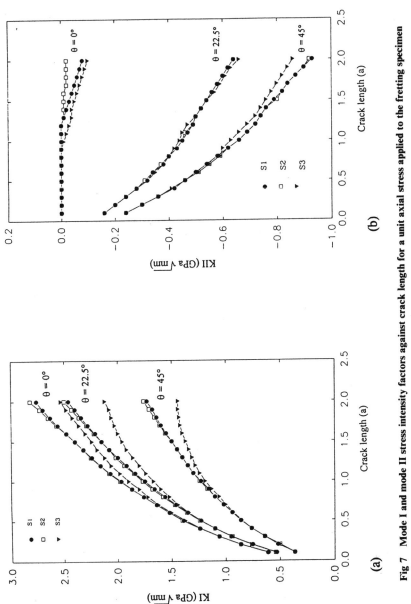

Fig 7 Mode I and mode II stress intensity factors against crack length for a unit axial stress applied to the fretting specimen

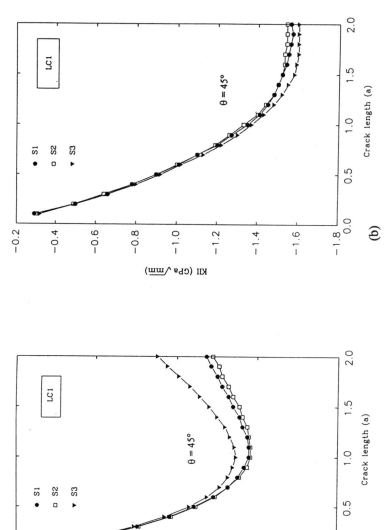

Fig 8 Mode I and mode II stress intensity factors against crack length for a unit normal stress applied through the fretting pad to the fretting specimen

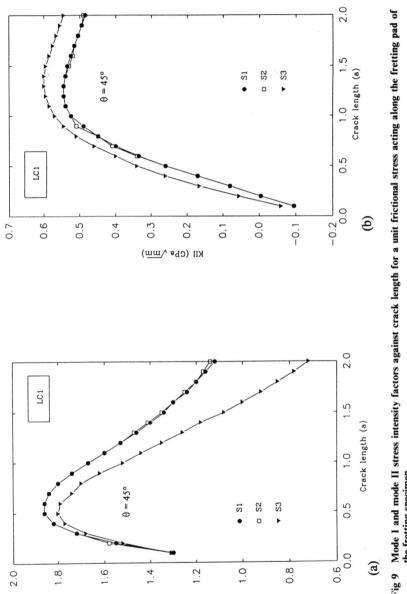

Fig 9 Mode I and mode II stress intensity factors against crack length for a unit frictional stress acting along the fretting pad of the fretting specimen

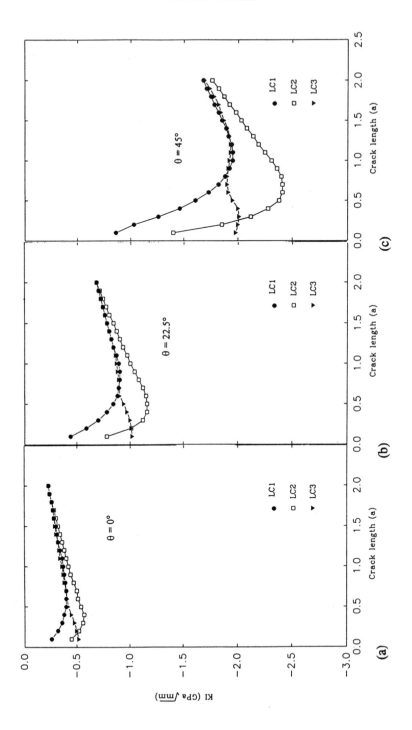

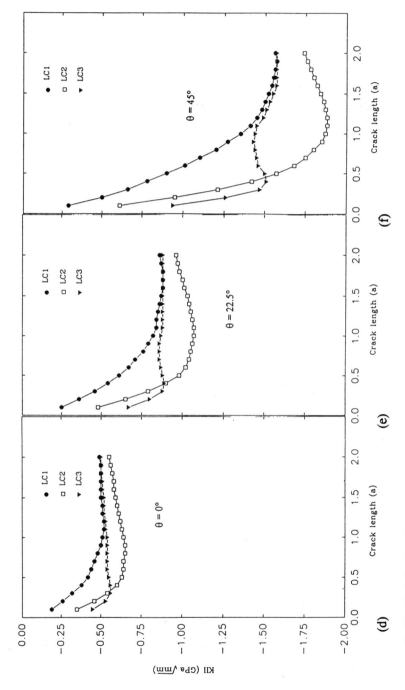

Fig 10 Mode I and mode II stress intensity factors against crack length for various distributions of normal load and crack configurations

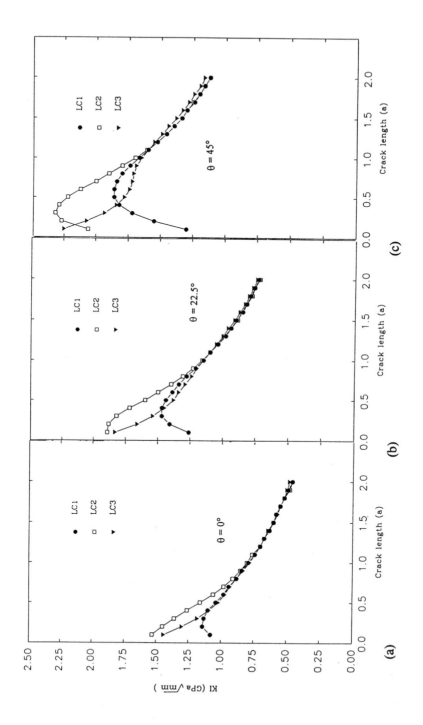

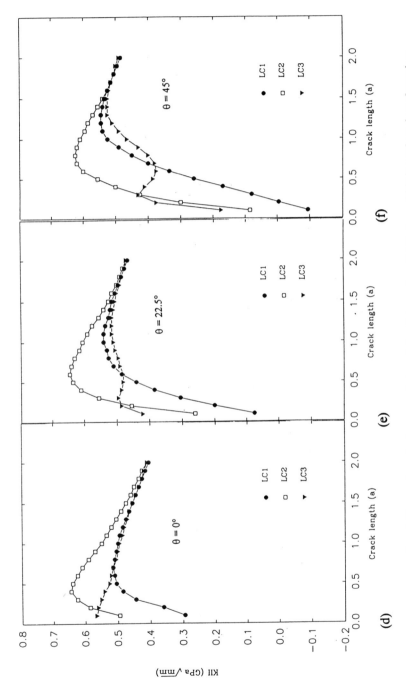

Fig 11   Mode I and mode II stress intensity factors against crack length for various distributions of friction and crack configurations

for span $S3$ are found to be consistently lower. For crack angle $\theta = 45$ degrees where mode II becomes dominant, the magnitudes of K_{II} values are also lower for span $S3$.

In Figs 8 and 9, K_I and K_{II} values are plotted for load case $LC1$ and a crack angle of 45 degrees. The results are compared for all the three spans employed in the analysis. Figure 8 shows the variation of K_I and K_{II} with the crack length for the normal load (σ_N). For a small crack length ($a < 0.8$ mm) the results for K_I are in close agreement for all the spans, but beyond that span $S3$ gives progressively lower magnitudes of K_I values. For K_{II}, slightly higher values (in magnitude) of K_{II} are obtained for span $S3$ whereas the values for the other two spans ($S1$ and $S2$) are, again, in close agreement. Figure 9 makes the comparison of K_I and K_{II} values for the three spans for a unit applied shear stress (σ_F). These results show lower values of K_I and higher values of K_{II} for span $S3$. For K_I the difference progressively increases with the crack length above $a = 0.5$ mm. For K_{II}, the difference in K_{II} values between $S3$ and the other two spans remains constant for the entire range of the crack length. Similar patterns of behaviour have been observed for other load cases and crack angles. This could be due to the crack tip stress field being influenced by the adjacent pad located at a distance of only about 6 mm in case of span $S3$.

Figure 10 shows six plots of K_I and K_{II} against the crack length (a) for a unit normal pressure (σ_N) on the fretting pad for a given span ($S1 = 16.5$ mm). These plots are drawn separately for each crack angle ($\theta = 0, 22.5, 45$ degrees) and show the variation of K_I and K_{II} for the three load distribution cases, ($LC1$, $LC2$, and $LC3$). It is observed that both K_I and K_{II} increase with the crack angle, as expected. As regards the load cases, load case $LC3$ (anti-parabolic) gives the largest value of K_I, almost twice as large as given by the uniform distribution ($LC1$) for a short crack ($a = 0.1$ mm). As the crack length increases beyond 0.2 mm, the value of K_I given by load case $LC2$ (triangular) is consistently higher than the other two load cases before attaining its peak value at a crack length of around 0.5 mm. For higher crack lengths ($a > 0.6$ mm) the value of K_I given by load cases $LC1$ and $LC3$ are in good agreement. Beyond its peak, K_I values for load case $LC2$ also start to converge, although slowly for higher crack angles. It is observed that the crack length for which a peak value of K_I is obtained increases with the crack angle for all the load cases. For a given crack angle, however, this peak value is attained first by load case $LC3$, then by $LC2$, and finally by $LC1$.

Similar observations can be made for K_{II} where load case $LC3$ gives the largest value for crack lengths of up to about 0.3 mm – beyond that it is consistently higher for load case $LC2$. For higher angles (say $\theta = 45$ degrees) load case $LC3$ has two peaks – one near each end of the fretting pad. This is expected as the crack tip moves towards the trailing edge of the fretting pad where the load intensity is high due to the nature of this distribution.

Figure 11 shows plots for K_I and K_{II} for a unit applied shear stress (σ_F) on the fretting pad. Again, each of the three crack angles are considered and the results are presented for span S1. It is observed that the values of K_I increase with the crack angle (θ). For load case LC3 K_I appears to peak at a very short crack length ($a < 0.1$ mm). Since the minimum crack length value considered for the present analysis is 0.1 mm, this behaviour is not clearly visible from the results presented. Analysis with crack lengths less than 0.1 mm would be needed to confirm this observation. This, however, suggests a considerable increase in the value of K_I due to friction, once a small crack is nucleated from fretting. Such increases in K_I can be a reason for the critical damage influence of friction on the early part of fretting crack growth. The crack length for which a peak value of K_I is obtained increases slightly with the crack angle for each load case. For crack lengths beyond about 0.8 mm, all the load cases produce similar K_I values for each of the crack angles considered, which gradually tend towards zero for both σ_N and σ_F contributions.

Regarding K_{II} values, they decrease with the crack angle (θ) for each load case but only up to a certain crack length. Beyond this crack length the numerical values of K_{II} for all crack angles are found to be within a narrow range. For each crack angle the value of K_{II} is largest for load case LC3 for short cracks. As the crack grows, LC2 gives the largest values but the differences diminish as the crack grows further. A very interesting observation can be made here for load case LC1 which produced a negative K_{II} value for the highest crack angle ($\theta = 45$ degrees) and a small crack length of less than 0.2 mm. This suggests that the shear deformation of the crack site is opposite to its normal positive sense. This could also be true for load cases LC2 and LC3, but to confirm this would require analysis with crack lengths below 0.1 mm. It should be noted that K_{II} values for crack angle $\theta = 45$ degrees are negative for load components σ_A and σ_N and the cumulative effect of the three loads on K_{II} could considerably increase mode II dominance.

Conclusions

A numerical procedure using the finite element method has been developed to evaluate stress intensity factors for fretting cracks. Plane strain models of fretting fatigue specimens have been developed to simulate the individual specimen/pad loading components and mode I and mode II elastic stress intensity factors have been evaluated for axial load, normal load, and friction. Three pad geometries as defined by the fretting pad span, and various crack configurations have been considered for the analysis.

The results obtained could be used for analysing stable crack growth under fretting fatigue conditions for any given loading combination, by applying superposition. In the current fretting fatigue test programme (6), the numerically-obtained values of K_I and K_{II} for each individual loading component are curve fitted using polynomial equations. For each case, functions of K_I and K_{II} are derived consisting of two independent variables, crack length (a) and crack angle (θ). These equations are, in turn, used together with experimentally-recorded axial, normal, and friction load data to evaluate stress intensity factors at each point in the hysteresis loop to produce ΔK against a and da/dn against ΔK (cyclic range of K) data for a given test in order to perform life prediction analysis.

The results, as presented and discussed in the previous section, compare the effect of pad geometry (span) and the normal load distribution on K_I and K_{II}. For all loading components the two larger spans ($S1$ and $S2$) have produced results which are in close agreement. Results for the smaller span ($S3$) show a deviation for longer crack lengths, more so for K_I than for K_{II}. This is due to the interference of the crack tip stress field of the adjacent pad which is in close proximity. As regards the load distributions for normal pressure and friction, load case $LC3$ gives higher values of K_I and K_{II} for small cracks ($a < 0.2$ mm) with load case $LC2$ taking over for the intermediate crack length range (up to about 1.5 mm). Finally all load distribution produce similar or converging values. For load case $LC3$ peak values of K_I and K_{II} appear to arrive at a very short crack length and further analysis for crack lengths below 0.1 mm would be required to determine them.

The results presented in this paper correspond to the tensile part of the loading cycle. For this case, both axial load and friction result in the opening of the crack (positive K_I) whereas the normal load tends to close the crack. For the compressive part of the cycle all three loading components will enhance crack closure.

As for mode II, which is dominant at a higher crack angle ($\theta = 45$ degrees); axial and normal loads produce a negative value of K_{II} for the tensile part of the loading cycle. Friction, on the other hand, gives a positive value for this crack configuration but for a very short crack the peak value may be negative. Again, further analysis for short crack lengths ($a < 0.1$ mm) would be required to confirm this. But if true, the superposition of all the loading components for K_{II} would considerably enhance the possibility of early crack growth in mode II, before entering the mixed-mode regime and finally to mode I.

Acknowledgements

The work presented in this paper was supported by the Science and Engineering Research Council, UK. Discussions with RAE, Farnborough are gratefully acknowledged.

References

(1) WATERHOUSE, R. B. (1972) Fretting corrosion, Pergamon Press, Oxford.
(2) ROOKE, D. P. and JONES, D. A. (1979) Stress intensity factors in fretting fatigue, J. Strain Analysis, 14, 1–6.
(3) EDWARDS, P. R. and COOK, R. (1978) Frictional force measurements on fretted specimens under constant amplitude loading, TR 78019, Royal Aircraft Establishment, Farnborough, UK.
(4) OWEN, D. R. J. and FAWKES, A. J. (1983) Engineering fracture Mechanics, Pineridge Press, Swansea, UK.

(5) SHEIKH, M. A, FERNANDO, U. S., BROWN, M. A., and MILLER, K. J. (1993) Finite element modelling and evaluation of stress intensity factors in fretting specimens, Proceedings of the Conference on Modern Practice in Stress and Vibration Analysis, Institute of Physics Stress Analysis Group.

(6) FERNANDO, U. S., FARRAHI, G. H., and BROWN, M. W. (1994) Fretting fatigue crack growth behaviour of BS L65 aluminium alloy under constant normal load, *Fretting Fatigue*, Mechanical Engineering Publications, London, pp. 183–195. *This volume.*

(7) ROOKE, D. P., RAYOPROLU, D. B., and ALIABADI, M. H. (1992) Crack-Line and edge Green's functions for stress intensity factors of inclined edge cracks, *Int. J. Fatigue Fracture of Engng Mater and Structures*, **15**, 441–461.

(8) DIX, A. J. (1991), TOMECH: user manuals (First Edition), SIRIUS, University of Sheffield, UK.

(9) KFOURI, A. P. and MILLER, K. J. (1985) Crack separation energy rates for inclined cracks in a biaxial field of an elastic–plastic material, *Multiaxial fatigue, ASTM STP 853*, ASTM, Philadelphia.

M. W. Moesser*, S. Adibnazari*†, and D. W. Hoeppner*

Finite Element Model of Fretting Fatigue with Variable Coefficient of Friction over Time and Space

REFERENCE Moesser, M. W., Adibnazari, S., and Hoeppner, D. W., **Finite element model of fretting fatigue with variable coefficient of friction over time and space**, *Fretting Fatigue*, ESIS 18 (Edited by R. B. Waterhouse and T. C. Lindley) 1994, Mechanical Engineering Publications, London, pp. 103–109.

ABSTRACT In order to accurately determine the life of a component subjected to fretting, the subsurface stress state must be known. Many solutions for the subsurface stress assume a constant coefficient of friction at the contact. However, fretting fatigue wear processes significantly alter the coefficient of friction over both the number of cycles and the position under the contact area. The purpose of this paper is to present one method attempted to determine if the effect the variable coefficient of friction has on the subsurface stress is worth investigating. An iterative finite element method is developed to accomplish this.

Observations from the methods and models used thus far have shown that if the fretting condition is pressure-dependent, the area of maximum tensile stress moves from the extreme edge of the contact towards the centre of contact. This could result in an initially pressure-dependent condition becoming deflection-dependent. This behaviour is not commonly observed in practice and may be eliminated by including the effect of early adhesion. For the deflection-dependent case it is shown that a variation in the coefficient of friction can significantly alter the stress state of a fretting contact. The model predicts that the maximum tensile stress at the extreme edge of contact is reduced over time. However, it is hypothesized that even after a high number of cycles, the stress intensity of a crack near the immediate edge may still be higher than the stress intensity of cracks near the maximum tensile stress.

Introduction

Most solutions to the state of stress under a fretting contact assume a constant coefficient of friction at the fretting interface. This is despite the fact that it is widely known that the coefficient of friction varies with both position and time in common fretting conditions. This paper presents an exploration of one possible method to account for this variation in the coefficient of friction by using an iterative finite element procedure. Another purpose of this work is to determine if the effects of variable coefficient of friction are significant. This question is critical to the continuing work of the authors on modelling fretting fatigue of riveted aircraft joints.

The mechanisms by which the coefficient of friction varies during fretting change over the life of the specimen. During the first few cycles of fretting, the coefficient of friction is relatively low as surfaces slide on surface contaminant films. Upon further cycles, asperities of opposing surfaces adhesively weld and

* Department of Mechanical Engineering, University of Utah, Salt Lake City, Utah 84112, USA.
† Now at Sharif University of Technology, Tehran, Iran.

notably increase the coefficient of friction. Eventually incomplete metal transfer, surface smearing, and oxidation result in a build-up of oxidized particles between the surfaces. These particles can behave similar to ball bearings and reduce the coefficient of friction. The surfaces roll on a bed of particles instead of sliding across one another.

The purpose of many investigations on the change of coefficient of friction often centres around an attempt to determine either how particles are formed, or which nucleation processes are dominant during a particular stage of the process. This paper focuses on the influence these changes in coefficient of friction have on the stress state near the surfaces.

Method

The finite element program ANSYS was used throughout this investigation. Temporal variation was accounted for by having a variable coefficient of friction for each contact element. The process began by constructing a finite element model of a fretting condition with a coefficient of friction associated with virgin surfaces. This model was solved. New coefficients of friction were calculated, guided by general trends observed in the literature. This information was transferred to a new command file. The model was then rebuilt with the new coefficients of friction.

When expanding this technique to several iterations it is necessary to update files containing the load history of each contact element. For example, the entire model may be at a simulated 50 000 cycles, but a particular contact element may not have started slipping until after 30 000 cycles. This contact element would, therefore, have seen a life of 20 000 cycles.

Also, convergence of fine mesh models requires a method of altering the coefficients of friction in a manner which does not make the model become unstable. A diagram of the geometry and boundary conditions used for the displacement controlled results are presented in Fig. 1.

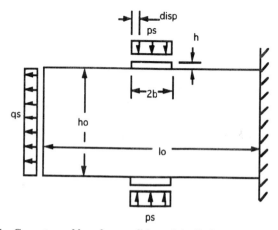

Fig 1   Geometry and boundary conditions of the displacement controlled case

Verification of the model

Several load cases were used to verify the finite element model. The most detailed case was accomplished with the results of a paper by Chung, Oran, and Hoeppner (1); they performed an elasticity theory-based study of the effect fretting contact stresses have upon the subsurface stress state. Good correlation with the predictions of the elasticity model were found.

Calibration of the model

In order to use this model to predict the fretting stress state for a particular material, the major variables which influence the coefficient of friction under fretting conditions have to be determined experimentally. Relative slip displacement, normal pressure, and the number of cycles are variables currently under consideration. A review of the literature shows that there is very little information in this area. As Hills *et al.* describe (2), when most investigators refer to testing they have performed to determine the coefficient of friction, they have actually performed tests to determine the ratio of tangential force to normal force applied to a fretting pad. The data necessary for the model described herein is the static coefficient of friction, defined as the ratio of tangential stress to normal stress to initiate relative motion.

Some difficulty exists in obtaining this type of information experimentally. The tests to determine the material response will themselves be subject to some variation in pressure and relative slip. Two possible methods may minimize the uncertainty associated with this effect. The first tailors the test apparatus to minimize variation in contact conditions. Uncertainty associated with this approach may be acceptable if the contact geometry to be studied with the finite element model has a much greater variation in contact conditions than the material test. A second approach models the material test geometry, assumes a material behaviour, then alters those assumptions until the model's predictions agree with the material tests. Although both methods have significant uncertainty associated with them, it is doubtful whether this uncertainty is greater than that introduced by ignoring the variation in coefficient of friction altogether.

The assumed material response for these preliminary studies was based upon general information in the literature. The data presented in this paper were based upon the assumption that the number of cycles was the predominant variable affecting the coefficient of friction, with relative slip displacement assuming a lesser role.

Predictions of displacement controlled model

Figures 2 and 3 show the results assuming a constant coefficient of friction (first iteration) and a variable coefficient of friction for the same displacement controlled load case ($ps = 100$ MPa, $qs = 200$ MPa, $h = 2$ cm, $b = 1$ cm,

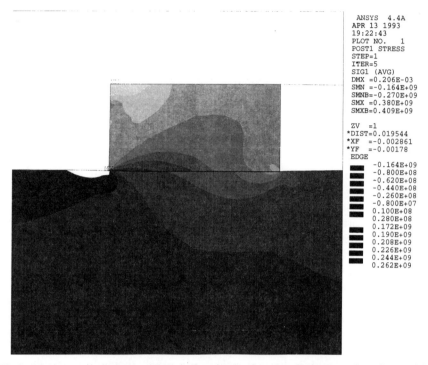

Fig 2   Maximum principal stress (SIG1, in Pascals), first iteration, displacement-dependent model

$lo = 12$ cm, $ho = 9$ cm). From the figures it can be observed that there is a significant reduction of the subsurface maximum tensile stress level at the edge of contact, and that the tensile stress level further from the edge increases. However, it is likely that the stress levels further from the edge of contact are not an important consideration. Cracks usually nucleate and propagate near the edge of contact much earlier than further towards the centre of contact. By the time the stress state has reduced at the edge of contact, the length of the crack is sufficient so that the stress intensity at a crack near the edge is greater than an assumed crack in the region of maximum tensile stress.

Work is currently focused on expanding this method to cover the entire life of a specimen under fretting conditions. The task is to develop a method which will avoid an oscillatory behaviour observed when several iterations are made with fine meshes. On the first iteration there is a high coefficient of friction and thus relatively small displacements. Low coefficients of friction are computed for the next iteration. On the second iteration the low coefficients of friction result in large displacements. From these large displacements a large coefficient of friction is computed for the next iteration. This cycle continues indefinitely. At present it is thought that using a differential method may prevent this oscillation.

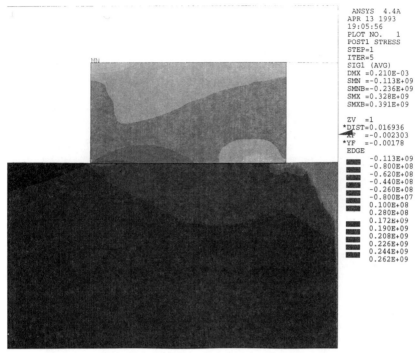

```
ANSYS   4.4A
APR 13 1993
19:05:56
PLOT NO.   1
POST1 STRESS
STEP=1
ITER=5
SIG1 (AVG)
DMX =0.210E-03
SMN =-0.113E+09
SMNB=-0.236E+09
SMX =0.328E+09
SMXB=0.391E+09

ZV  =1
*DIST=0.016936
*XF  =-0.002303
*YF  =-0.00178
EDGE
        -0.113E+09
        -0.800E+08
        -0.620E+08
        -0.440E+08
        -0.260E+08
        -0.800E+07
        0.100E+08
        0.280E+08
        0.172E+09
        0.190E+09
        0.208E+09
        0.226E+09
        0.244E+09
        0.262E+09
```

Fig 3   Maximum principal stress (SIG1, in Pascals), second iteration, displacement-dependent model

Predictions of force controlled model

When the movement of the upper block was pressure dependent, it was found that if the moduli of the materials were low relative to the applied stresses, a gradual movement of the maximum tensile stress was observed, see Fig. 4 (gross mesh models). The contacting surfaces at the edge of contact had relative displacement early on. This resulted in a reduction of the coefficient of friction near the edge of contact, thus redistributing the load away from that area. This redistribution increased the stress in regions where no slip had occurred previously, and on the next iteration slip was more likely there. These regions then, in turn, had reduced coefficients of friction and further redistributed the stresses. If the ratio of tangential stress to normal stress on the upper block were slightly less than the initial coefficient of friction, gross movement of the upper block would eventually occur. This implies that under some service conditions, an initially pressure-dependent fretting state could eventually become a displacement-dependent state. In practice it is often observed that if slip does not occur across the entire contact area on the first load cycle, the slip/no slip boundary does not move far beyond the edge of contact under subsequent cycles.

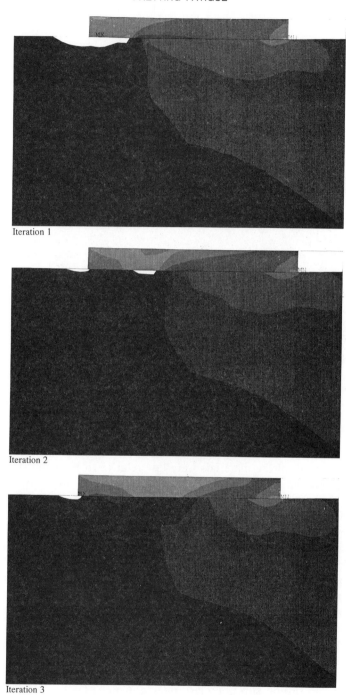

Iteration 1

Iteration 2

Iteration 3

Fig 4   Three iterations of the pressure-dependent case, maximum principal stress shown (SIG1, in Pascals)

This discrepancy might be due to the increase in coefficient of friction from adhesion observed during the first few cycles of fretting by many investigators. During the first few cycles, approximately up to fifty cycles, Berthier *et al.* and others have reported that for some materials the coefficient of friction during fretting can increase rapidly (3). However, it was decided not to include the behaviour of the coefficient of friction at very low cycles due to the assumptions the authors have made about the purpose of this model. It was assumed that fretting fatigue is a four-stage process; crack nucleation by asperity contact; accelerated crack growth by fretting contact stresses; crack propagation through the bulk of the specimen largely independent of the contact stress; and final fast fracture. Since crack nucleation occurs on an asperity scale, the stress levels are assumed to be at plastic conditions in most service conditions. As crack nucleation occurs early in life, it was thought that the coefficient of friction during these cycles need not be modelled. However, the results obtained with the current model indicated that including the effects of the early increase in coefficient of friction from adhesion may be needed to predict movement of the slip/no slip boundary.

Summary and conclusions

The method presented has proved to be adequate for these preliminary studies. The finite element method model used to test the variable coefficient of friction method has been verified with the results of an elasticity solution. It was also found that for a displacement controlled fretting fatigue load case, the model predicted that the areas of maximum stress slowly moved from the extreme edge of contact inward. However, this process may occur slowly enough that the stress intensity of cracks at the edge of contact may still be higher than the stress intensity of cracks near the area of maximum tensile stress.

It was found that under pressure-dependent fretting condition, the model predicted the maximum tensile stress would slowly move from the edge of contact. It may be possible for an unstable condition to occur in the model during which the maximum interface shear stresses slowly move inward from the edge of the contact. This behaviour may be eliminated by considering the effects of early adhesion.

Acknowledgements

The authors wish to give their special thanks to the Federal Aviation Administration for financial support of this project, and to Dr David Macferran for his technical help with the finite element analyses.

References

(1) CHUNG, J. S., ORAN, C., and HOEPPNER, D. W. (1981) Elastic stresses in fretting fatigue, *J. Engng Mech. Div.*, 387–403.

(2) HILLS, D. A., NOWELL, D., and O'CONNOR, J. J. (1988) On the mechanics of fretting fatigue, *Wear*, **125**, 129–146.

(3) BERTHIER, Y., VINCENT, L., and GODET, M. (1989) Fretting fatigue and fretting wear, *Tribology Int.*, **22**, 235–242.

C. Gerdes*, H. Bartsch*, and G. Härkegård*

Two-Dimensional Modelling of Fretting Fatigue

REFERENCE Gerdes, C., Bartsch, H., Härkegård, G., **Two-dimensional modelling of fretting fatigue**, *Fretting Fatigue*, ESIS 18 (Edited by R. B. Waterhouse and T. C. Lindley) 1994, Mechanical Engineering Publications, London, pp. 111–124.

ABSTRACT The present paper deals with the influence of various design parameters on the fretting fatigue strength of turbine blade attachments. A testing machine was designed and experiments conducted, which provides data on fretting-induced cracking. The first phase of the design effort involved simulating the turbine environment in a laboratory setting. The second phase considered the geometry of the test specimens and the contact forces. Finally, the testing machine with its instrumentation necessary to control, monitor, and record the testing conditions was designed.

The actual turbine blade attachment was simulated by a simple 'T' shaped root. Tests were carried out on austenitic and ferritic test specimens with different slip ranges (LCF), surface pressures, and superimposed alternating bending (HCF). The initiation and growth of fretting fatigue cracks was observed. The coefficient of friction of the blade root/rotor contact surface at elevated temperature was determined.

Introduction

The lifetime of a mechanical component may be reduced to a fraction of its intended design life as a result of fretting fatigue, i.e, the initiation and growth of fatigue cracks caused by small alternating displacements between two bodies in mechanical contact. Although the actual wear damage is usually quite small, the fatigue strength of a specimen subject to simultaneous fretting and cyclic stress may decrease to less than 50 percent of the fatigue strength of the non-fretted specimen. Fretting fatigue failures can occur in different components, namely in fitted bolts of flange connections (1) as well as in blade and wedge attachments of turbine and generator rotors (2). Fretting fatigue in the dovetail joints of gas turbine blades has been the subject of several investigations (3)(4). At elevated temperatures, fretting fatigue has been studied for a titanium alloy used for jet engine compressor components (5)(6) and for Ni-base alloys for the blades and rotor discs of gas turbines (7). This is an important area of study, since the observed trend to higher inlet temperatures in gas and steam turbines may aggravate fretting fatigue conditions.

The most common theory of fretting fatigue damage is based on the following assumptions of mechanical and chemical effects, according to Waterhouse and Nishioka (8)(9).

* ABB Power Generation Limited, Baden, Switzerland.

- Mating surfaces come into contact at high asperities.
- Oscillatory slip causes cyclic shear stresses. Together with high Hertzian stresses, they induce local plastic deformation at asperities.
- Microwelding (adhesion) and fracture of asperities result in fretting debris and material transfer. Under corrosion conditions, the debris is harder than the base metal and may cause abrasion.
- Simultaneous action of the phenomena mentioned above initiates fatigue micro-cracks in the fretting region.
- Crack propagation is possible if supported by an external stress field (possibly accelerated by a wedging action of the fretting debris).

Test objectives and design considerations

The goal of the investigation was to conduct experiments identifying critical parameters and their influence on fretting-induced cracking.

The turbine operating parameters to be simulated were:

- temperature (510°C);
- normal contact stress (surface pressure) due to centrifugal loading;
- bending stresses due to dynamic steam loading (HCF cycles);
- relative movement of contact surfaces due to thermal cycling of the rotor (LCF cycles).

The design of the blade root and rotor test specimens was dictated by independent requirements, the first being that the test specimens bear physical resemblance to a real blade fixation. The second requirement was that it should be possible to induce and record the parameters influencing the fretting fatigue strength.

The blade root shape chosen was a simple 'T' profile with equal and symmetric contact areas on two hooks. Figure 1 shows the blade root/rotor specimen arrangement and the terminology used. The total nominal contact area is 330 mm^2. The rotor specimen was 'U' shaped with hooks to match the blade root.

In Fig. 2, the principle design of the fretting test machine is demonstrated. Attached to the bottom of the frame is a hydraulic cylinder, which, by way of a load cell and an extension arm, applies a load to the blade root/rotor contact areas, thereby simulating the centrifugal force.

At the juncture of the blade extension arm, at 90 degrees to the vertical axis, dynamic bending stresses up to 41.4 MPa were applied by way of a force shaker at 75 Hz at a predetermined displacement, such that the required bending stresses were induced at the blade root (HCF-load). Accelerometer displacement (see Fig. 2) was correlated to surface strain, which was measured by strain gauges applied on the extension arm of the blade root, as a means of calibration. Good agreement was observed between calculated and measured values.

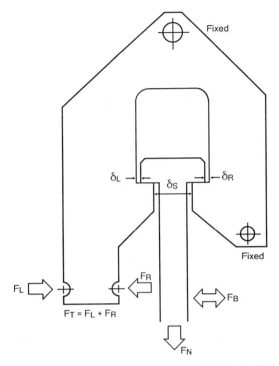

Fig 1   Blade root/rotor specimen arrangement used for fretting fatigue tests

δ_L Left displacement	F_L Left hydraulic cylinder force
δ_R Right displacement	F_R Right hydraulic cylinder force
δ_S Servo displacement	F_N Normal axial force
F_T Total tangential force	F_B Bending force

With one side of the rotor specimen fixed to a common mechanical ground, a pair of 100 mm hydraulic cylinders alternately pushed (in opposite directions) on the free leg of the rotor specimen to produce the transverse motion at the contact areas simulating the slippage at the contact areas (LCF load). The rate of cycling was designed to be between 0.1 and 1 Hz and was controlled by a pressure compensated flow control valve with micrometer control. Precise control and recording of the transverse motion at the contact area was perhaps the most critical and demanding aspect of the test. The range of motion was from 50 to 125 μm with a resolution better than 1 μm. The transducers had to survive extended exposure at 510°C as well as frequent mechanical cycling. The solution to this problem was a capacitive probe.

Electric cartridge resistance heaters were used to heat the rotor and blade specimen to the test temperature. The different heat conduction characteristics of the two materials ultimately required one heater in the blade and six in the rotor specimen to achieve a nearly uniform temperature across the contacting surfaces.

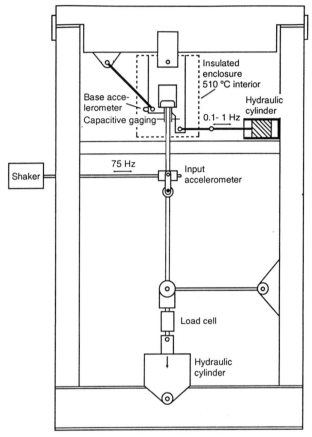

Fig 2   Fretting test machine

Once the test was running, the task of the operator was to periodically examine the blade specimen for cracking. Crack growth analysis was made possible by periodically stopping the test for approximately 15 mins, thus creating a beachmark on the surface of the crack.

Fretting test plan

Fretting fatigue tests

Two different types of blade material were used: austenitic (16Cr13Ni steel) and ferritic (12Cr steel). The rotor material was always 1CrMoV steel. The specified minimum mechanical properties for these materials are given in Table 1 for room temperature (RT) and 500°C.

The basic parameters identified for the experiments were: temperature, axial load, fretting slip range, and dynamic stress range. The temperature was kept constant at 510°C. The parameters varied were surface pressure, servo range

Table 1 Specified minimum mechanical properties for test materials at room temperature (RT) and 500°C

Component	Material	Yield strength $R_{p0.2}$ (MPa)		Tensile strength R_m (MPa)		Elongation A_5 (percent)
		RT	500°C	RT	500°C	RT
Blade	16Cr13Ni (equivalent to DIN 1.4962/ X12 CrNiWTi 16 13)	440	355	590	500	20
Blade	12 Cr (equivalent to DIN 1.4926/ X21 CrMoV 12 1)	700	375	850	490	13
Rotor	1CrMoV (equivalent to DIN 1.6979/ 20 CrMoNiV 4 7)	590	375	680	470	16

(LCF), dynamic bending stress (HCF), and blade material. The test matrices for the austenitic and ferritic material are given in Tables 2(a) and 2(b), respectively.

Table 2(a) Fretting test matrix for austenitic steel at 510°C

LCF servo range $\Delta\delta_s$ (μm)	Surface pressure σ_p (MPa)	Number of test specimens for HCF stress range (MPa)		
		0	20.7	41.4
75	160	–	–	3
100	200	3	3	3
125	100	–	–	3
	160	–	–	3
	200	3	2	2
150	200	3	3	3

Table 2(b) Fretting test matrix for ferritic steel at 510°C

LCF servo range $\Delta\delta_s$ (μm)	Surface pressure σ_p (MPa)	Number of test specimens for HCF stress range (MPa)		
		0	20.7	41.4
75	200	–	–	6
125	200	–	–	3

Friction tests under fretting fatigue conditions

One of the important parameters influencing initiation of fretting fatigue cracks is the interface coefficient of friction. The objectives of the friction tests were:

(1) measurement of the frictional force for different material combinations, representative of blade root/rotor surface pressures and small slip amplitudes at high temperatures (510°C);

(2) measurement of changes in the frictional force as fretting slip cycles are applied at the contact surface.

The test equipment to be used for the friction tests is the same as that used for the fretting tests described above. The material combinations, servo ranges, and surface pressures used in testing up to 1000 LCF fretting cycles are shown in Table 3. Two additional tests were performed with superimposed vibration.

Table 3   Results of friction tests at 510°C

| | | Coefficient of friction F_T/F_N | | | |
| | | Austenitic blade* Surface pressure σ_p | | Ferritic blade* Surface pressure σ_p | |
LCF servo range $\Delta\delta_s$ (μm)	HCF stress range (MPa)	100 MPa	200 MPa	100 MPa	200 MPa
75	0	(0.60)	0.38	0.42	0.38
125	0	0.53	0.52	0.64	0.55
125	20.7	–	0.16	–	–
125	41.4	–	0.14	–	–

* Rotor material 1 Cr Mo V.

Fretting fatigue results and observations

Fretting fatigue crack initiation

Figure 3 shows the fretting fatigue results for the austenitic steel at constant surface pressure $\sigma_p = 200$ MPa (compare Table 2(a)). The number of LCF fretting cycles to crack initiation (N_i) is plotted against slip range (Δs). The Δs value was taken as the slip range of that side ($\Delta\delta_L$ for left or $\Delta\delta_R$ for right-hand side, respectively) where the fretting fatigue crack was initiated. A characteristic 'U' shape of the N_i–Δs curve was observed. An influence of dynamic stress on crack initiation could not be distinguished.

In Fig. 4, the fretting fatigue results are compared for the austenitic and the ferritic steel tested under similar conditions (compare Tables 2(a) and 2(b)). The austenitic and ferritic specimens show a comparable number of cycles to crack initiation.

Figure 5 shows the influence of surface pressure on the fretting fatigue behaviour of the austenitic steel. The data points for 200 MPa are always below those for 160 MPa. It may also be noted that tests with a surface pressure of 100 MPa did not produce cracks within 30 000 cycles.

Fatigue crack propagation

The distance between successive beachmarks, which had been produced by heat tinting, was determined by fractographic examination. The propagation of the fretting fatigue induced cracks could only be studied for crack depths larger than 1 mm. Crack growth rates between 10^{-7} and 10^{-6} m/cycle were observed.

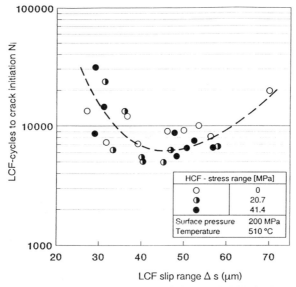

Fig 3 Influence of HCF stress range on fretting fatigue crack initiation in austenitic steel

Coefficient of friction

Typical time histories for the total slip $\delta_L + \delta_R$, elastic deformation δ_E and coefficient of friction F_T/F_N are shown in Figs 6 and 7. A typical hysteresis loop (F_T/F_N versus $\delta_L + \delta_R$) is shown in Fig. 8.

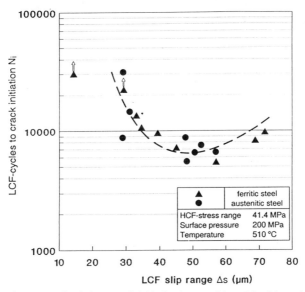

Fig 4 Fretting fatigue crack initiation in austenitic and ferritic steel

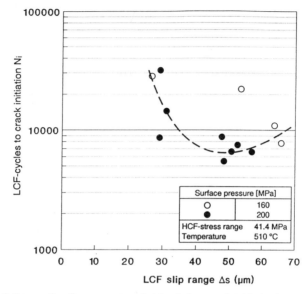

Fig 5   Influence of surface pressure on fretting fatigue crack initiation in austenitic steel

Two basic phenomena with respect to frictional behaviour are described in the available literature (e.g., Edwards **(10)**).

(1) Micro-slip is characterized by frictional forces approximately proportional to the slip due to elastic deformation of the contacting asperities during an overall sliding motion. This gives rise to a steep slope in the displacement and force time histories.

(2) During macro-slip, sliding occurs between contacting asperities.

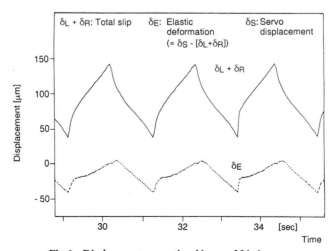

Fig 6   Displacement versus time history of friction test

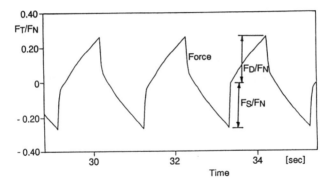

Fig 7   Tangential force versus time history of friction test

Hence, the total tangential force (F_T) can be considered to be composed of micro-slip (static) tangential force (F_S) and macro-slip (dynamic) tangential force (F_D). This was confirmed by our test results (Figs 7 and 8).

The force ratio (F_T/F_N) for each loop is plotted as a function of the number of fretting cycles (example in Fig. 9). The test results of all friction tests are summarized in Table 3.

Discussion

Fretting fatigue crack initiation

The major factors affecting the fretting phenomenon are:

- surface pressure magnitude and distribution;
- relative slip amplitude;
- interface coefficient of friction;
- temperature and environment.

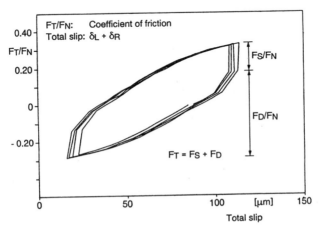

Fig 8   Hysteresis loop of friction test

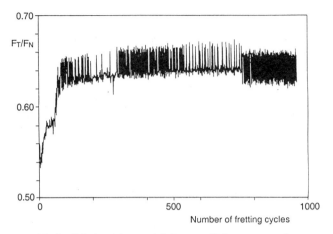

Fig 9   Frictional (tangential) force coefficient versus cycles

Surface pressure magnitude and distribution
The surface pressure is an important parameter and has a direct influence on fretting fatigue life. King (2), Nishioka (9), and Gaul (11) have reported that increased surface pressure reduces the fretting fatigue life, other conditions remaining the same. The effect was confirmed by our test series.

The surface pressure distribution may render certain locations in the contact area preferred sites for fretting damage – usually the edges of the contact area (slip/no-slip boundary).

Relative slip amplitude
The interface between two elastic contacting bodies may be divided into no-slip and slip regions. When the normal and tangential loads are such that there is a non-vanishing no-slip area, the maximum slip occurs at the edge of this contact area and is referred to as 'micro-slip'. If the tangential force is further increased, 'macro-slip' or sliding will occur. Yielding may occur at the edge of the contact area.

Fretting failures are usually observed under macro-slip conditions, although fretting damage may also occur under micro-slip conditions. Under micro-slip conditions, the external stress field and the unnotched fatigue strength determine the susceptibility for failure.

There is a critical value of the slip range, at which fretting fatigue life shows a minimum. In the tests performed in this programme, the critical slip range was between 40 and 50 μm (see Figs 3 to 5). This is in agreement with experiments by Nishioka (9) and Gaul (11), where critical slip ranges of 40 and 50 μm were observed at surface pressures about 500 and 30 MPa, respectively.

The existence of a minimum fretting fatigue life as a function of slip range can be explained as follows. The increasing slip causes an increased fretting wear and increased frictional shear stresses and hence decreased fatigue life. At higher amplitudes, the micro-cracks are worn away before they can propagate. Wear debris may contribute to a lowered coefficient of friction, thereby causing an increase in life.

Temperature and environment
According to Waterhouse (12), temperature and environment are vital to the fretting phenomenon. The chemical interaction between the three constituents, the two contacting surfaces and the environment, is not very well understood. For example, a higher temperature may assist in the formation of a glaze (oxide) at the interface under an oxidizing environment, thereby reducing the coefficient of friction. However, the fatigue strength itself may be reduced at this higher temperature. Therefore, the overall fretting fatigue strength cannot be predicted in such a case.

Coefficient of friction
At small slip amplitudes, two friction phenomena have been reported to exist by Edwards (10) and Endo (13). The first is the 'micro-slip friction' (elastic-slip friction), which has been attributed to the elastic deformation of contacting asperities. The micro-slip gives rise to steep slopes in the displacement–time plots at near-zero sliding speeds. The second kind is the 'macro-slip friction', which arises during actual sliding motion and can be related to the static coefficient of friction. Both coefficients tend to increase as fretting occurs.

The two components of the frictional force (macro-slip and micro-slip) can be clearly seen from the F_T/F_N versus slip plot in Fig. 8. The micro-slip coefficient (F_S/F_N) varies between 0.2 and 0.3 and is nearly independent on the surface pressure (see Fig. 10(a)). For the same slip range, the surface pressure has no influence on the macro-slip coefficient (F_D/F_N) either (see Fig. 10(b)). However, this coefficient is proportional to the slip range with a proportionality factor of 0.2 to 0.3 per 100 μm.

The two material combinations show similar coefficients of friction. It may be argued, however, that an insufficient number of specimens have been tested to justify a general conclusion.

For most cases, the frictional force increased in the first few hundred cycles and then became constant (see Fig. 9). This behaviour is similar to that reported by Endo (13) and Nishioka (9).

Application of vibratory loads to the blade specimen leads to a marked reduction in the frictional forces, see Table 3. A similar result has been reported by Lenkiewicz (14). However, only two tests were conducted and no general conclusion can be drawn as to the quantitative effect of applied dynamic loads.

Conclusions

(1) Cracks invariably initiate at the edge of the contact zone.
(2) Reduction in surface pressure is very effective in increasing lifetime.

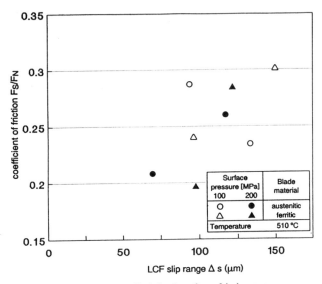

Fig 10a   Micro slip behaviour from friction tests

(3) For very small LCF slip ranges (below 25 μm), no fretting fatigue cracking was observed. The life versus slip range curve exhibits a characteristic
'U' shape. The falling part of the curve suggests an increase in fretting
wear with increasing slip, causing increased friction (shear stress) and
hence a decrease in life. The rising part suggests that fretting wear causes
fatigue cracks to wear out rapidly and hence cause an increase of life.

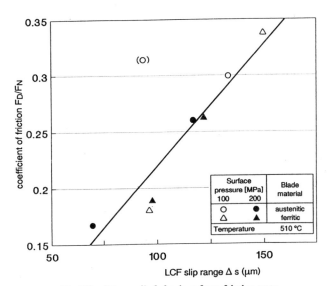

Fig 10b   Macro slip behaviour from friction tests

(4) Crack growth rate has been estimated to be between 10^{-7} and 10^{-6} m/cycle for the conditions tested.

(5) The influence of high-cycle dynamic stress could be neglected as far as crack initiation is concerned. It may effect crack growth rates.

(6) Austenitic and ferritic specimens show similar results for similar conditions of surface pressure and slip.

(7) The effective coefficient of friction at 510°C for austenitic steel blade and 1CrMoV rotor for 75 to 150 μm slip range is in the range 0.4–0.6.

(8) The micro-slip coefficient of friction at zero sliding speeds is in the range 0.2–0.3. The dependence of the surface pressure is negligible.

(9) The macro-slip coefficient of friction is proportional to the slip range up to 150 μm.

(10) The frictional force increases in the first few hundred fretting cycles and attains a constant value after that.

(11) Application of vibratory loads reduces the frictional force by a factor of about three for the cases tested. Not enough tests were conducted to quantify the reduction as a function of magnitude of applied vibratory loads.

Acknowledgement

Thanks are due to Stress Technology Incorporated (Rochester, USA) for carrying out and documenting the fretting fatigue tests to the authors' full satisfaction.

References

(1) VAKHTEL, V. Yu. (1962) Reduction of frictional-corrosion effect on fatigue, *Russian Eng. J.*, **49**, **2**, 16–17.

(2) KING, R. N. and LINDLEY, T. C. (1982) Fretting fatigue in a 3.5 NiCrMoV rotor steel, *Advances in fracture research*, Vol. 2, pp. 631–641.

(3) MORTON, P. G., GOODMANN, P. J., and KAWECKI, Z. M. (1969–70) *Proc. Instn. Mech. Engrs.*, **184**, 66–74.

(4) JERGEUS, H. A. (1978) Contact problems and load transfer in mechanical assemblages, Euromech, Linköping Institute of Technology. pp. 109–113.

(5) HAMDY, M. M. and WATERHOUSE, R. B. (1979) The fretting-fatigue behaviour of Ti-6Al-4V at temperatures up to 600°C, *Wear*, **56**, 16.

(6) SCHÄFER, R. and SCHÜTZ, W. (1990) Fretting fatigue strength of Ti-6Al-4V at room and elevated temperatures and ways of improving it, *High temperature surface interactions*, AGARD, Conference Proceedings No. 461, pp. 11/1–11/15.

(7) HAMDY, M. M. and WATERHOUSE, R. B. (1979) The fretting fatigue behaviour of a nickel-basis alloy (inconel 718) at elevated temperatures, *Wear of materials*, ASME, New York, p. 351.

(8) WATERHOUSE, R. B. (1972) *Fretting corrosion*, Pergamon, Oxford.

(9) NISHIOKA, K. and HIRAKAWA, K. (1969) Fundamental investigations of fretting fatigue, *Bull. JSME*, **12**, 692.

(10) EDWARDS, P. R. (1981) The application of fracture mechanics to predicting fretting fatigue, *Fretting fatigue*, (Edited by R. B. Waterhouse), Applied Science Publishers.

(11) GAUL, D. J. and DUQUETTE, D. J. (1980) The effect of fretting and environment on fatigue crack initiation and early propagation in a quenched and tempered 4130 steel, *Met. Trans.*.

(12) WATERHOUSE, R. B. (1981) Fretting at high temperatures, *Tribology International*, **14**, 203–207.

(13) ENDO, K., GOTO, H., and FUKUNAGA, T. (1974) Behaviours of frictional force in fretting fatigue, *JSME*, **17**, 647–654.

(14) LENKIEWICZ, W. (1969) The sliding friction process – effect of external vibrations, *Wear*, **13**, 99–108.

S. Adibnazari*† and D. W. Hoeppner*

The Role of Normal Pressure in Modelling Fretting Fatigue

REFERENCE Adibnazari, S. and Hoeppner, D. W., **The role of normal pressure in modelling fretting fatigue,** *Fretting Fatigue,* ESIS 18 (Edited by R. B. Waterhouse and T. C. Lindley) 1994, Mechanical Engineering Publications, London, pp. 125–133.

ABSTRACT The damage threshold and the pressure threshold are two fretting fatigue characteristics that could be utilized in conjunction with different design approaches to model fretting fatigue failure. These characteristics not only help in understanding the role of normal pressure on different stages of fretting fatigue, but also help in modelling fretting fatigue failure for complex situations such as in joints.

One of the main challenges in modelling fretting fatigue failure is establishing a method to precisely determine the end of the crack nucleation stage. This paper discusses the damage threshold and its relationship to crack nucleation. Factors that relate to the normal pressure threshold are also discussed. These characteristics are used to explain the role of normal pressure in fretting fatigue as well as to establish procedures that could act as guidelines for modelling this important failure mode.

Introduction

Fretting fatigue failure is a time-dependent process that occurs in different stages, i.e., damage production, damage growth, crack nucleation, and crack propagation. Prediction of fretting fatigue failure requires an understanding of the mechanism(s) involved and how well they could be modelled. Many studies **(1)(9)** have been conducted on the effects of different variables on fretting fatigue failure and its four stages. These studies have led to different models **(1)(8)–(13)** which have been used to predict the occurrence of fretting fatigue failure.

Unfortunately, the technical community is far from being able to predict the fretting fatigue life of components. This is because of the lack of understanding as well as the complexity of the subject. Thus, the approach most designers have been taking is to 'prevent' fretting from occurring by using different techniques rather than considering fretting fatigue failure in the analysis part of a design. However, fretting fatigue, like other types of failure, has some characteristics such as the damage threshold **(14)–(16)** (the minimum number of cycles required to cause a reduction in fatigue life due to fretting) and pressure theshold **(17)(18)** (a certain pressure between the contacting surfaces greater than which the fretting fatigue life shows little change) that could aid in understanding the mechanisms involved. These characteristics could also be used to study the role of different variables such as normal pressure, fatigue stress, environment, and material in modelling fretting fatigue as well as establishing procedures that could act as guidelines for designing against this failure.

* Department of Mechanical Engineering, University of Utah, Salt Lake City, Utah 84112, USA.
† Now at Shariff Institute of Technology, Tehran, Iran.

It is obvious that fretting fatigue life reduction would not occur if it were not for normal pressure and the contacting surfaces. Thus, it is imperative for the design community to understand the role of normal pressure in fretting fatigue. It is the objective of this paper to discuss the role of normal pressure in modelling fretting fatigue using the pressure threshold, the damage threshold, and other characteristics of fretting fatigue.

Characteristics of fretting fatigue

Fretting usually reduces the fatigue life of components by introducing damage which is believed to lead to early crack nucleation. An important step in modelling fretting fatigue is to determine when fretting damage becomes sufficient

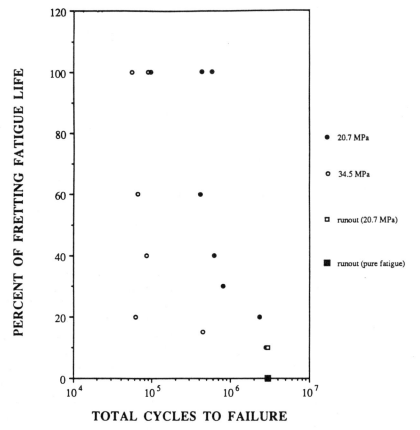

Fig 1   Interrupted fretting fatigue test results for the maximum fatigue stress of 137.9 MPa and normal pressures of 20.7 and 34.5 MPa at $R = +0.1$, 7075–T6 aluminium alloy. Each test was interrupted (i.e., the normal load was removed) at the number of cycles that corresponded to the desired percentage of the mean uninterrupted fretting fatigue life for the same normal load. Then the test was continued with the same fatigue loading, but no normal load, to determine the total cycles to failure

to have an effect on the fatigue life of components. The corresponding number of cycles is the damage threshold. Tests were conducted at the Quality and Integrity Design Engineering Center of the University of Utah on 7075–T6 aluminium alloy. The fretting fatigue test apparatus and experimental details are described extensively in references (11) and (19). Each test was conducted using a flat, dog-bone-shaped specimen which was subjected to tensile fatigue loading ($R = 0.1$). Fretting pads, one of the same material as the specimen and one of a phenolic (relatively non-fretting) material, were applied on opposite sides of the test section of the specimen. The fretting motion resulted from the strain caused by the fatigue loading. The results of these tests (Fig. 1) and the results obtained at other labs (14)(15) show the existence of a damage threshold. The damage threshold concept makes it possible to investigate the

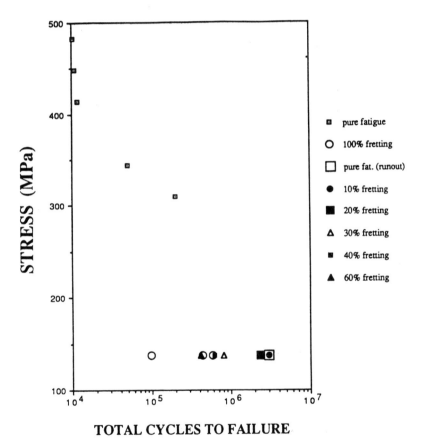

TOTAL CYCLES TO FAILURE

Fig 2   Maximum fatigue stress versus life plot for a maximum fatigue stress during fretting fatigue tests of 137.9 MPa and normal pressure of 20.7 MPa at $R = +0.1$, 7075–T6 aluminium alloy. The damage threshold was at about 20 percent of the fretting fatigue life. Data for corresponding pure fatigue (no fretting) tests with maximum fatigue stresses as high as 480 MPa are shown for comparison

processes that occur during the damage production and growth stages of fretting fatigue. The damage threshold can also be used to identify the start of the final portion of the crack nucleation process, when fretting damage is sufficient to allow cracking to occur that eventually would lead to fracture even if the fretting normal load were removed. The authors have found (11) that for aluminium alloy 7075–T6 tested under certain conditions, cracks appeared after the damage threshold was reached (Figs 2 and 3).

A normal pressure threshold exists that is similar to the fretting fatigue damage threshold (17)(18). The pressure threshold concept, shown in Fig. 4, basically states that above a certain normal pressure fretting fatigue endurance shows little change due to a further increase of normal pressure. It is believed that for a normal pressure higher than the pressure threshold there exists a

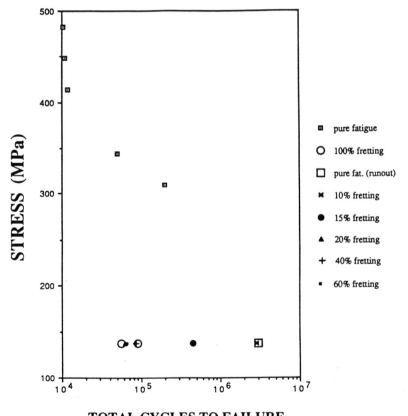

TOTAL CYCLES TO FAILURE

Fig 3   Maximum fatigue stress versus life plot for a maximum fatigue stress during fretting fatigue tests of 137.9 MPa and normal pressure of 34.5 MPa at $R = +0.1$, 7075–T6 aluminium alloy. The damage threshold was at less than 15 percent of the fretting fatigue life. Data for corresponding pure fatigue (no fretting) tests with maximum fatigue stresses as high as 480 MPa are shown for comparison

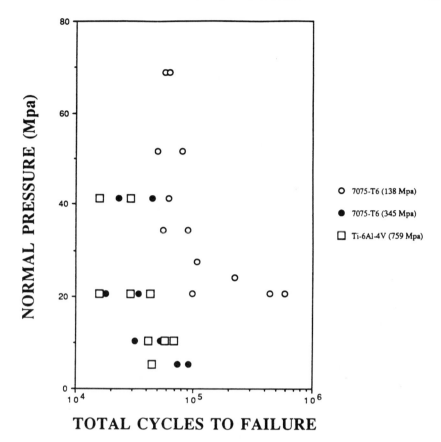

Fig 4   Fretting fatigue pressure threshold for 7075–T6 aluminium alloy and Ti–6Al–4V titanium
alloy at various maximum fatigue stresses, a stress ratio of + 0.1 and a frequency of 30 Hz,
laboratory environment

critical frictional stress that causes the fretting damage to grow faster and
reach the damage threshold more quickly than it would with a lower normal
pressure. After the damage threshold has been reached, the stress states caused
by the normal pressure and fatigue stress might cause a compressive stress at
some level below the surface which retards crack propagation. There is also a
possibility that if the contact pressure is increased above the pressure threshold then the contact condition changes because the slip amplitude reaches a
limiting value and the slip contact condition changes to mixed stick-slip which
causes the fretting fatigue life of the components to show little change.

   Another characteristic of fretting fatigue is the behaviour of cracks propagating under fretting conditions. Fretting fatigue crack propagation takes
place in two stages, i.e., stages I and II, Fig. 5. Stage I crack propagation is
oblique to the fretted surface and the experimental results (19) on 7075–T6

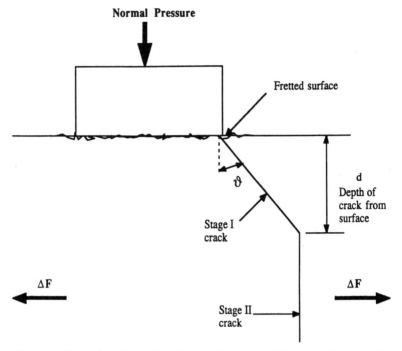

Fig 5   Schematic illustration of a fretting fatigue crack as observed during testing reported in references (11) and (19). Values of ϑ and d, determined during this testing, are shown in Table 1 for a maximum fatigue stress of 137.9 MPa and various normal pressures

aluminium alloy specimens show that the angle of inclination and its depth depends on the magnitude of normal pressure, Table 1. As it appears, the angle of inclination of oblique stage I cracks decreases as the normal pressure increases, and the depth of stage I cracks increases as the normal pressure increases. Stage II crack propagation is perpendicular to the applied fatigue stress. Due to different behaviour of the stage I and stage II crack propagation, different models might be needed at different stages of crack propagation to predict fretting fatigue failures.

Table 1   Stage I crack depths (d) and angles from the perpendicular to the specimen fretted surface (ϑ) that were found during testing reported in (11) and (19)

Max. fatigue stress (MPa)	Normal pressure (MPa)	ϑ^* (°)	d^* (mm × 10⁻²)
137.9	10.3	43	3.3
137.9	20.7	40	14.2
137.9	34.5	32	19.8
137.9	51.7	20	19.1

* These terms are defined in Fig. 5

The characteristics mentioned above certainly are not the only character-istics of fretting fatigue. There are other characteristics like damage pro-duction, fracture surface characteristics, etc. that are as important. However, the damage thesehold, the pressure threshold, and crack propagation behaviour are three fretting fatigue characteristics that could be utilized in conjunction with different design approaches, such as damage-tolerant design, in order to model fretting fatigue failure and also to carry out sensitivity studies to deter-mine which factors are more important to fretting fatigue failure.

The role of normal pressure in modelling fretting fatigue

Fretting fatigue life of components can be modelled in two stages

Total fretting fatigue life = crack nucleation + crack propagation   (1)

For 7075–T6 aluminium alloy the total fretting fatigue life can be written as

Total fretting fatigue life = cycles to the damage threshold + fretting
influenced fatigue crack propagation + fatigue crack propagation   (2)

The role of normal pressure in modelling fretting fatigue life is different above and below the pressure threshold. Above the pressure threshold, it can be considered that fretting fatigue life of 7075–T6 aluminium alloy and Ti–6Al–4V titanium alloy is not a function of ΔP, i.e., variation of normal pressure does not significantly affect the total life. Therefore, keeping all other variables constant, one relationship might describe the behaviour and determine the fretting fatigue life.

Below the pressure threshold, fretting fatigue life of 7075–T6 aluminium alloy and Ti–6Al–4V titanium alloy is a function of normal pressure, its varia-tion, and other variables. The introduced model should address the effect of normal pressure on the damage threshold and on the crack propagation. It is believed that the normal pressure and the damage threshold have an inverse relationship, as shown above. Thus, the higher the normal pressure is, the shorter the life to crack nucleation. For example, the increase of normal pres-sure from 20.7 MPa to 34.5 MPa for 7075–T6 aluminium alloy, Figs 2 and 3, reduces the damage threshold from about 20 percent of fretting fatigue life for 20.7 MPa to less than 15 percent for 34.5 MPa. Thus life to crack nucleation has been reduced based on total specimen lives. This not only illustrates the role of normal pressure, but it shows that fretting fatigue may not be solely a nucleation controlled process as claimed by some investigators. The results also show that after a certain amount of fretting damage, contact pressure had no further effect on life. Thus, it is possible to conclude that the stress state created by the normal pressure has more effect on the nucleation and 'early' crack propagation.

The life used to propagate a nucleated crack to failure consists of life to propagate a stage I crack into a stage II crack and then to failure. It appears that the stress state caused by normal pressure influences stage I and not stage II crack propagation. The different behaviour of stage I and stage II crack propagation illustrates that two models might be needed to predict crack propagation in fretting fatigue. The model for stage I crack propagation is a function of the normal pressure, variation of normal pressure, geometry of the crack and the component, etc. This model should include both 'short' and long crack behaviour. The model for stage II crack propagation does not depend on normal pressure but rather it depends on fatigue stress, frequency, material, environment, etc. The outcomes of these models could be combined to give a more accurate prediction of fretting fatigue crack propagation. Other factors may be important to the crack propagation phase. Also, cracks may be nucleated such that stage I does not occur. Undoubtedly, certain stress, microstructural, and environmental conditions would favour this. More effort is needed to ascertain those factors that clearly control the crack nucleation and crack propagation stages of fretting fatigue.

Conclusions

In order to model fretting fatigue failure effectively, the effect of normal pressure on fretting fatigue failure should be understood. For example, it is often believed that an increase in contact pressure will decrease the life of the components. This is a general statement which would lead to models that wrongly predict failure at normal pressures above the pressure threshold for 7075–T6 aluminium alloy and Ti–6Al–4V titanium alloy.

Different characteristics of fretting fatigue such as the damage threshold, the pressure threshold, and crack propagation behaviour could be used to explain the mechanism(s) of fretting fatigue failure as well as to model it. In this paper, the damage threshold was related to the final portion of the crack nucleation stage. This finding illustrates the fact that most of the fretting fatigue life is used to propagate a crack rather than nucleate it. This verifies the value of the damage-tolerant approach and limited applicability of stress life and/or strain life approach to fretting fatigue life prediction/modelling. These characteristics could also be used to conduct sensitivity studies to determine which factors are important to fretting fatigue failure and its different stages. It is clear that much additional research is needed to clarify those factors that govern the normal pressure and damage threshold.

References

(1) HOEPPNER, D. W. (1992) Mechanisms of fretting fatigue and their impact on test methods development, *Standardization of fretting fatigue test methods and equipment, ASTM STP 1159*, (Edited by M. Helmi Attia and R. B. Waterhouse), ASTM, Philadelphia, pp. 23–31.
(2) BILL, R. C. (1982) Review of factors that influence fretting wear, *Material evaluation under fretting conditions, ASTM STP 780*, ASTM, Philadelphia, pp. 165–182.
(3) DOBROMIRSKI, J. M. (1992) Variables of fretting process: are there 50 of them?

Standardization of fretting fatigue test methods and equipment, ASTM STP 1159, (Edited by M. Helmi Attia and R. B. Waterhouse), ASTM, Philadelphia, pp. 60–66.

(4) WATERHOUSE, R. B. (1972) The effect of fretting corrosion in fatigue crack initiation, Corrosion fatigue: chemistry, mechanics, and microstructure, NACE–2 Conference, National Association of Corrosion Engineers, p. 607.

(5) BERTHIER, Y., VINCENT, L., and GODET, M. (1988) Fretting wear mechanisms and their effects on fretting fatigue, *J. Tribology*, **110**, 517–524.

(6) SODERBERG, S., BRYGGMAN, U., and McCULLOUGH, T. (1986) Frequency effects in fretting wear, *Wear*, **110**, 19–34.

(7) ATTIA, M. H. and D'SILVA, N. S. (1985) Effect of mode of motion and process parameters on the prediction of temperature rise in fretting wear, *Wear*, **106**, 203–224.

(8) SATO, K., FUJII, H., and KODAMA, S. (1986) Crack propagation behavior in fretting fatigue, *Wear*, **107**, 245–262.

(9) WATERHOUSE, R. B. (1981) Theories of fretting fatigue, *Fretting fatigue*, (Edited by R. B. Waterhouse) Applied Science Publishers, London, 203–219.

(10) CHUNG, J. S., ORAN, C., ASCE, M., and HOEPPNER, D. W. (1981) Elastic stresses in fretting fatigue, *J. Engng Mech. Div.*, 387–403.

(11) ADIBNAZARI, S. and HOEPPNER, D. W. (1992) Study of fretting fatigue crack nucleation in 7075–T6 aluminium alloy, *Wear*, **159**, 257–264.

(12) SWITEK, W. (1984) Early crack propagation in fretting fatigue, *Mech. Maters*, **3**, 257–267.

(13) EDWARD, P. R., RYMAN, R. J., and COOK, R. (1977) Fracture mechanics prediction of fretting fatigue, Proceedings of 15th ICAF,

(14) WHARTON, M. H., TAYLOR, D. E., and WATERHOUSE, R. B. (1973) Metallurgical factors in the fretting-fatigue behavior of 70/30 brass and 0.7 percent carbon steel, *Wear*, **23**, 251–260.

(15) HOEPPNER, D. W. and GOSS, G. L. (1974) A fretting fatigue damage threshold, *Wear*, **27**, 61–70.

(16) ADIBNAZARI, S. and HOEPPNER, D. W. (1992) Characteristics of the fretting fatigue damage threshold, *Wear*, **159**, 43–46.

(17) ADIBNAZARI, S. and HOEPPNER, D. W. (1992) A fretting fatigue normal pressure threshold concept, *Wear*, **160**, 33–35.

(18) NAKAZAWA, K., SUMITA, M., and MARUYAMA, N. (1992) Effect of contact pressure on fretting fatigue of high strength steel and titanium alloy, *Standardization of fretting fatigue test methods and equipment, ASTM STP 1159*, (Edited by M. Helmi Attia and R. B. Waterhouse), ASTM, Philadelphia, pp. 115–125.

(19) ADIBNAZARI, S. (1991) Investigation of fretting fatigue mechanisms on 7075–T6 aluminum alloy and Ti–6Al–4V titanium alloy, PhD Thesis, University of Utah.

M. C. Dubourg* and V. Lamacq*

A Theoretical Model for the Prediction of Crack Field Evolution

REFERENCE Dubourg, M. C. and Lamacq, V., **A theoretical model for the prediction of crack field evolution**, *Fretting Fatigue*, ESIS 18 (Edited by R. B. Waterhouse and T. C. Lindley), 1994, Mechanical Engineering Publications, London, pp. 135–147.

ABSTRACT An elastic half-space or a finite body with multiple cracks, is subjected to constant amplitude tensile/compressive cyclic loading. Cracks are perpendicular to the surface. Interactions between cracks are considered. Contact conditions at crack interfaces, i.e. the division of open, slip, and stick zones, are taken into account. It is shown that crack propagation is considerably slowed down. Further load transfer favours the systematic propagation of external cracks to the detriment of central ones, which are rapidly trapped.

Notation

a^i	Crack i size
a_0^i	Starting size of crack i
b_{1x}, b_{1y}	Burger's vectors for crack 1
d	Distance between two consecutive cracks
f	Interfacial crack coefficient of friction
k	$3-4v$: plane strain $(3-v)/(1+v)$: plane stress
K_{ij}	Stress kernels expressed in reference axis (i, j)
K_I, K_{II}	Stress intensity factor in mode I and II
m	Number of cracks
$\delta u_t, \delta u_n$	Slip and opening between crack faces
$\sigma^T, \sigma^{MC}, \sigma^F$	Resultant, continuous, and crack stress fields
$\sigma_{nn}^{T'}, \sigma_{nt}^{T'}$	Normal and shear resultant stresses
μ	Shear modulus
v	Poisson's ratio
Γ_1	Crack 1 profile

Introduction

In many real cases, both single and multiple cracks are initiated from the contacting surfaces of two bodies (1) (Fig. 1). A wide range of crack profiles is observed: surface breaking or embedded cracks, straight or branched, inclined or perpendicular to the contacting surfaces. Unlike the uniaxial tension test, these cracks experience non-proportional loadings inducing complex contact conditions at the crack interface, including frictional locking. Some of these cracks may either be arrested or continue to propagate leading to spall

* Laboratoire de Mécanique des Contacts – CNRS URA 856, INSA Bâtiment 113, 20 avenue Albert Einstein, 69621 Villeurbanne cedex, France.

$\underset{\longmapsto}{216\,\mu m}$

Fig 1   **Multiple cracks initiated from contacting surfaces (1). Normal force: 50 daN; displacement amplitude $\pm 35\,\mu m$; static bulk traction $\sigma_s = \sigma d/2$; frequency 15 Hz, $N = 10^6$ cycles**

detachment for instance, or change direction and branch deeply into the surface, thereby causing catastrophic failure.

Conditions that lead to the self-arrest or propagation of some cracks in this field and the prediction of crack path direction are open problems. In the field of linear elasticity the fatigue crack behaviour, measured in terms of critical crack lengths, prediction of crack path, rates of crack growth, and calculation of the residual strength of cracked structural components, is determined solely by the stress intensity factor values which depend on the stress field at crack tips, which itself is governed by the contact conditions at crack interface. It is, therefore, of great importance to know if the crack faces are in contact, and if they slide with respect to each other. Considerable research effort has been devoted to this field. Keer and his co-workers **(2)(3)** have devised a method that uses a distribution of dislocations to represent a two-dimensional crack in rolling/sliding contacts. Stress intensity factors (SIFs) in mode I and II are calculated, and subsurface and surface breaking cracks, inclined to the surface or parallel to it, are analysed under various loading conditions, taking into account frictional locking **(4)–(12)**. The presence of a lubricant between crack faces is studied too **(13)**. A general model was proposed by Dubourg and Villechaise **(14)–(18)** to analyse fatigue crack behaviour in an isotropic elastic medium, whatever the loading conditions and the crack geometries. The model is based on a modified dislocation theory and on the contact problem solution between crack faces as a unilateral contact problem with friction. This systematic approach avoids assumptions about the contact zone division that limit the application field of the previous models **(4)–(12)**. Interactions between multiple cracks are analysed with this model and the key parameters identified **(15)–(16)**. It was shown that interactions result in significant SIF variations, both decreasing and increasing.

Crack network propagation, taking into account interactions between cracks and frictional locking is studied here. Cracks are perpendicular to the surface and subjected to a constant amplitude tensile/compressive loading cycle. It is a preliminary study to the complex problem of propagation under non-proportional loadings. The model is based on the dislocation theory and on the unilateral contact analysis with friction. It is shown that crack propagation is considerably slowed down due to interactions between cracks.

Fatigue crack modelling including crack closure

A theoretical two-dimensional linear elastic model of multiple frictional contact fatigue cracks was developed **(14)–(18)** to determine the stress and displacements fields in cracked solids and the stress intensity factors K_I and K_{II}. This model was based on the dislocation theory and on the unilateral contact analysis with friction. Friction between the crack lips is introduced using, for lack of anything better, Coulomb's law. Stress and displacement fields (σ^T, δu, δv) are given by superposing the uncracked solid (σ^{MC}) and the crack (σ^F, δu_n, δu_t) responses to the load in a way which σ^T satisfies the boundary conditions along the faces of the presumed cracks. These boundary conditions are expressed as (Fig. 2):

– in a contact zone: (no penetration of the contacting crack faces is possible)

$\delta u_n = 0, \quad \sigma^T_{nn} < 0$

– in an open zone: (the distance between the crack faces is positive)

$\sigma^T_{nn} = 0, \quad \delta u_n > 0$

– in a backward slip zone: (slip directed from crack tip to mouth; right crack face taken as reference)

$\sigma^T_{nt} = f * \sigma^T_{nn}, \quad \delta u_t * \sigma^T_{nt} > 0$

– in a forward slip zone: (slip directed from crack mouth to tip; right crack face taken as reference)

$\sigma^T_{nt} = -f * \sigma^T_{nn}, \quad \delta u_t * \sigma^T_{nt} > 0$

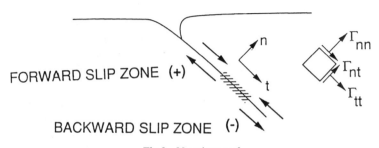

FORWARD SLIP ZONE (+)

BACKWARD SLIP ZONE (-)

Fig 2   Notations used

– in a stick zone: (no displacement difference between crack faces)

$$\delta u_t = 0, \quad |\sigma_{nt}^T| < f * |\sigma_{nn}^T|$$

The continuum stress σ^{MC} in the uncracked solid may be obtained numerically (e.g. finite element analysis) or analytically in the case of a half-plane. The finite element method is suitable for complex geometries and boundary conditions. The finite element mesh is refined along the virtual crack profiles, but is considerably rougher away from the vicinity of cracks. Stresses along virtual cracks are then redistributed along the crack discretization points through a linear interpolation.

The crack response corresponds to displacement discontinuities along its faces, opening and slip, that generate stresses. These displacement zones are modelled with continuous distributions of dislocations bx and by, a method pioneered by Keer (2)(3). Single distributions of dislocations bx and by are considered along each crack. It is assumed that by and bx are square root singular at crack tips and at crack mouths for embedded cracks. The correct behaviour of the stress field along cracks is thus guaranteed. The strength of these singularities is then driven numerically to zero in the case of a contact zone or a stick zone at the crack tip (14). Consistent equations come from corresponding boundary conditions ($\delta u_n = 0$, $\delta u_t = 0$). This method gives single stress and displacement expressions for the whole crack, independent of the final contact division

$$\sigma_{ij}^F = \frac{2\mu}{\pi(k+1)} \sum_{l=1}^{m} \left(\int_{\Gamma_l} b_{lx}(\xi) K_{ij}^x(x, y_\eta, \xi) \, d\xi \right.$$
$$\left. + \int_{\Gamma_l} b_{ly}(\xi) K_{ij}^y(x, y_\eta, \xi) \, d\xi \right) \qquad i, j = x, y$$

$$\delta u_t = \int_{\Gamma_l} b_{lx}(\xi) \, d\xi$$

$$\delta u_n = \int_{\Gamma_l} b_{ly}(\xi) \, d\xi$$

The stress kernels, K_{ij}^x, K_{ij}^y, expressed in (15), define the dislocation influence at point ξ on point (x, y); they depend on the distance between these two points. Stress expressions are singular integral equations, solved following Erdogan et al. (19). Discretized stress and displacement expressions are obtained. The $2NI$ unknowns are the bx and by values at the discretization points, where NI is defined by $NI = \sum_{i=1}^{m} p_i$, where p_i is the number of discretization points for crack i.

Difficulties in determining the contact division between crack faces stem from the fact that the boundary conditions are formulated in terms of both equations and inequalities. The algorithm developed was suggested by the work of Kalker (20) for the rolling contact between two bodies. Initially the

Panagiotopoulos's process **(14)** was used: the normal problem (determination of the contact area and the resultant normal stress) and the tangential problem (determination of the slip and stick zones over the contact area and the resultant shear stress) were alternatively calculated until convergence (state of all points remains unchanged from one iteration to the next) occurred. The Kombi process is used now **(21)(22)**. At the beginning of each load step, the cracks are assumed to be closed and adherent. The traction bound g is estimated to be the value of the continuum stress field at the load step: $g = \sigma_{nn}^{MC}$. Corresponding equations are solved on the basis of this estimated traction boundary. The equations then become $\sigma_{nt}^T = \pm f^*g$. The algorithm determines the resultant normal and shear stresses and also the opening and slip at crack interfaces. The solutions are then tested to see if they satisfy the boundary conditions in terms of the inequalities:

– for Qi belonging to a contact zone,

   if $\sigma_{nn}^T(Qi) > 0$, Qi is set to the open zone

– for Qi belonging to a stick zone,

   if $|\sigma_{nt}^T(Qi)| > f \times |\sigma_{nn}^T(Qi)|$, Qi is set in a slip zone

Equations are solved again if any modification has taken place. If not,

– for Qi belonging to a slip zone,

   if $\delta u_t(Qi) \times \sigma_{nt}^T(Qi) < 0$, Qi is set in a stick zone

– for Qi belonging to an open zone,

   if $\delta u_n(Qi) < 0$, Qi is set in a contact zone.

Equations are solved again if any modification has taken place. When convergence is reached, then, on the basis of the tangential traction, the traction bound g is re-estimated ($g = \sigma_{nn}^T$), and the whole contact problem solved again. 1 to 3–4 iterations on the traction bound are needed to attain the convergence. This process is slower than Panagiotopoulos's method, but is more reliable. The contact problem solution between crack faces as a unilateral contact problem with friction automatically gives the contact area division, slip, stick and open zones, and the suitable distributions of dislocations. Load cycles are described with an incremental description which takes into account the load history as hysteresis is generated by friction at the crack interface.

This model was used to determine the stress intensity factors experienced at crack tips under various loading conditions, sliding or rolling contact conditions, bulk tractions, etc. Results obtained **(15)(16)** show that it is particularly important to account for interactions between cracks, as they strongly modify the stress field near the crack tips and consequently the SIFs. Stress intensity factors of a crack situated in a crack field cannot be extrapolated from those obtained from a single identical crack subjected to identical loading conditions as interaction phenomena are too complex and are likely to result in significant drops or increases in SIFs, depending on the distance between cracks, but also on the relative crack position with respect to the loading zone, the interfacial crack coefficient of friction, and the loading mode conditions (mode I, II, or mixed).

Fatigue crack growth modelling

The model is extended to include crack propagation. This requires input of relevant crack propagation data and stress history for the formulation of criteria for crack path direction prediction and laws for crack growth rate. Depending on the loading conditions (proportional or non-proportional loadings, with or without overloads), crack growth behaviour is different **(23)(24)**.

The work presented here is the first step in that direction. Loading conditions and crack geometries are thus rather simple, the aim being to demonstrate special crack interaction effects on crack propagation. Crack network propagation under classical tensile and compressive constant amplitude loadings is considered. Mode II effects are not taken into account for crack propagation as K_{II} values are negligible compared to K_I values for the loading and geometrical conditions considered. Cracks are perpendicular to the half-plane surface. Their direction is therefore compatible with the loading and no branching will occur. A fatigue crack propagates when the stress intensity factor variation ΔKI during the load cycle exceeds the threshold value ΔK_{th}. The crack extension da, corresponding to the stress intensity factor variation ΔK_I, is obtained according to a Paris law obtained for a single crack situated in a similar material subjected to identical loading conditions.

Crack growth simulation is performed this way:

- Number of cycles: $N = 0$; starting crack sizes $a^i = a_0^i$, $i = 1, m$ m: number of cracks
- Calculation of stress intensity factor range during the load cycle: $\Delta KI^i = f(a^i, \Delta\sigma)$
- Calculation of the crack extension da according to Paris law: $da^i/dN = g(\Delta KI^i, \Delta\sigma)$
- Crack extension Δa of 1 percent of the current crack size is performed: $\Delta a^i = 1$ percent$*a^i$; this corresponds to a step of ΔN^i cycles with

$$\Delta N^i = \Delta a^i/(da^i/dN)$$

where

$$\Delta N = \text{Min }(\Delta N^i)$$

$$a^i = a^i + \Delta N^* (da^i/dN)$$

$$N = N + \Delta N$$

It is possible to evaluate (a) the number of cycles to failure and (b) the number of cycles of a starting size to a permissible size.

Results

The purpose of this study is to show special crack interaction effects on crack propagation. Simple loading conditions and crack geometries are retained. Cracks are perpendicular to the surface, regularly spaced, whatever their length and their number. Propagation is described following a Paris law obtained for a titanium alloy and a ratio $R = 0.1$: $da/dN = 9.11 * 10^{-11} \Delta KI^{3.46}$ (24), ΔKI expressed in $hbar\sqrt{(mm)}$ and da/dN in mm/cycle. ΔK_{th} is equal to 26 $hbar\sqrt{(mm)}$ or 8.2 MPa$\sqrt{m}$. Four cases are analysed; these are presented in Table 1.

Network of three cracks

Case 1 is first presented in detail. Three identical cracks are considered, the starting size of which is 5 mm. The friction coefficient f is chosen equal to 0.2. The distance d between two consecutive cracks varies from 2.5 mm to 7 mm. The evolutions of the size of cracks 1 and 2 (3 behaves like 1) and the SIF range versus the number of cycles are presented in Fig. 3 for a distance d equal to 7 mm. As the load is shared between the three cracks, the SIF experienced at the crack tips is considerably reduced with respect to that experienced for a single crack under similar loading conditions.

Crack 2 is particularly unloaded – its SIF range is smaller than that experienced by cracks 1 and 3, so consequently, its extension will be slower. Four periods are noted from the beginning of the propagation up to the catastrophic failure (Fig. 3).

Period I   This period is related to the first thousand cycles. During this time, the three cracks are open from mouth to tip at maximum load. The central crack is protected by the two others, and therefore experiences a smaller SIF range. As a consequence, its growth is slowed down with respect to the two others that become deeper. The SIF range increases with crack size.

Period II   The opening displacement amplitude at crack mouth 1 and 3 is so great that it inhibits crack 2 opening at its mouth, whereas an open zone is still experienced at its tip. A new equilibrium between cracks corresponding to a new load sharing takes place. Crack 2 still propagates but its SIF range

Case 1	Case 2	Case 3	Case 4
⊤⊤⊤	⊤⊤⊤	⊤⊤⊤	⊤⊤⊤⊤
1 2 3	1 2 3	1 2 3	1 2 3 4 5

Table 1   Two kinds of crack network

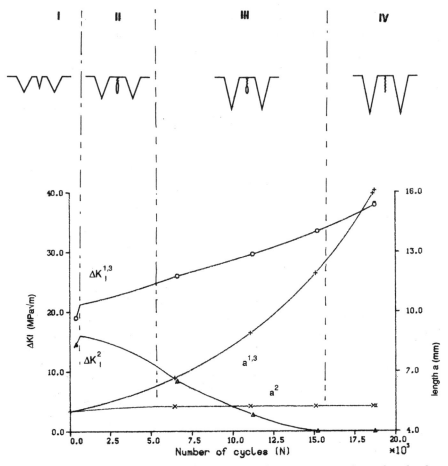

Fig 3   Case 1: variations of crack size and stress intensity factor range versus the number of cycles

decreases against the number of cycles. The SIF range for cracks 1 and 3 continues to increase with crack length. Thus the difference in size becomes more and more pronounced. Furthermore, the deeper cracks 1 and 3 are, the greater the contact stick zone at the mouth of crack 2 is and the smaller its SIF range.

Period III   The SIF range at crack 2 is smaller than the threshold: crack 2 does not propagate any more, but as it still experiences an open zone at its tip, it still takes part in load sharing.

Period IV   Crack 2 is completely trapped; the load is shared between cracks 1 and 3, up to the catastrophic failure.

Despite identical initial conditions in terms of crack length and spatial division, overprotection of the central crack by the exterior ones lies at the origin of the difference in propagation: a lower SIF range is experienced at crack tip

2, and this in turn causes a slower extension, and thus imbalance between cracks.

Figure 4 shows the influence of d on crack propagation: the larger d is, the deeper the non-propagating cracks are, and the greater the number of cycles.

In case 2, the starting size of crack 2 is smaller, and the process described above begins at period II. In case 3, however, an opposite behaviour is obtained: only crack 2 propagates.

Network of five cracks

The next case studied, case 4, allows a generalization of this propagation process. Five cracks are considered, the central one and the two external ones are 6 mm long, and the size of the two others is 5 mm. The distance between two consecutive cracks d is equal to 5 mm. All along the loading process, cracks 2 and 4 never propagate and experience a contact zone at their mouth. The propagation process is summarized in Fig. 5. The increase of the length of the cracks versus the number of cycles is presented in Fig. 6.

Period I As described above, the opening displacement amplitude at the mouth of cracks 1, 3, and 5 is so great that it creates a contact slip zone at the mouth of cracks 2 and 4, and a backward and a forward slip zone for cracks 2 and 4, respectively. Slip is due to the difference in the opening displacement between cracks 1 and 3, and 3 and 5. Crack 3 is overprotected and propagates slower than cracks 1 and 5. Obviously, the deeper the cracks 1, 3, and 5, the greater the contact slip zones at the mouth of cracks 2 and 4. This mechanism goes on until a stick zone appears at the mouth of both cracks. Cycle by cycle, the stick zone becomes more and more important.

Period II The contact zones at the mouth of cracks 2 and 4 are now so great that they inhibit crack 3 from opening. A stick zone is thus observed at its mouth. Nevertheless, crack 3 still propagates.

Period III Crack 3 stops propagating, but still experiences an open zone at its tip.

Period IV Cracks 2 and 4 are completely trapped. The load is shared between cracks 1, 3, and 5 up to the catastrophic failure.

Case 4 is a generalization of the network of three cracks. One crack out of two is trapped at its mouth from the beginning of the propagation process. This goes on until crack 3 is trapped at its mouth too. Finally the network is made of two propagating cracks, cracks 1 and 5, and of three self-arrested cracks of different lengths.

Conclusions

A two-dimensional theory model for the analysis of fatigue crack network propagation is presented. Interactions between cracks are taken into account.

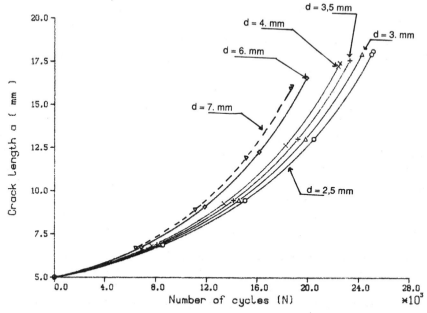

Fig 4(a)   Case 1: influence of *d* on propagation of cracks 1 and 3

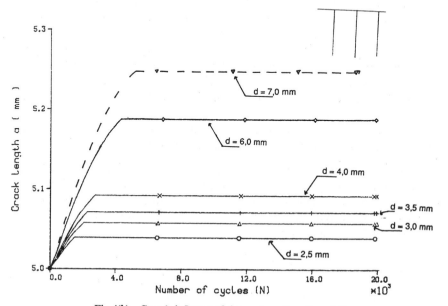

Fig 4(b)   Case 1: influence of *d* on propagation of crack 2

PERIOD I

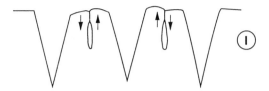

PERIOD II

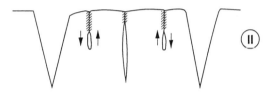

PERIOD III

Fig 5   Case 4: schematization of propagation process

Contact conditions between crack faces including friction are modelled. The
model rests on the dislocation theory and on the contact problem solution
between crack faces as a unilateral contact with friction. Load transfer
between multiple cracks significantly affect their behaviour, depending on the
distance between the cracks, friction between crack faces, and the loading conditions (mode I, II, or mixed). The influence of these interactions on crack
propagation is studied. As a preliminary study, cracks are perpendicular to the
surface and simple constant amplitude tensile/compressive cyclic loading conditions are considered. The first conclusion is that crack network propagation
is considerably slowed down with regard to identical single crack submitted to
identical loading. The second point is the confirmation of experimental observations concerning the network of long and shorter cracks alternatively
located. One crack out of two is trapped by the cracks on its sides. When the
central crack is completely locked, a new equilibrium takes place, reproducing
similar behaviour up to catastrophic failure.

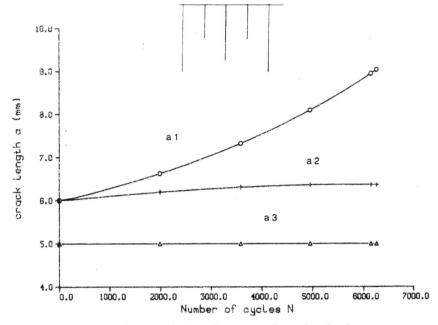

Fig 6   Case 4: crack size evolution versus the number of cycles

References

(1) GODET, M., VINCENT, L., DUBOURG, M. C., LAMACQ, V., CHATEAUMINOIS, A., and REYBET-DEGAT, P. (1993) Programme fretting fatigue, Rapport partial no. 2.

(2) KEER, L. M., BRYANT, M. D., and HARITOS, G. K. (1980) Subsurface cracking and delamination, *Solid contact and lubrication*, Vol. 39, ASME/AMD, pp. 79–95.

(3) KEER, L. M. and BRYANT, M. D. (1983) A pitting model for rolling contact fatigue, *J. Lub. Technol.*, **105**, 198–205.

(4) CHANG, F. K., COMNINOU, M., SHEPPARD, S., and BARBER, J. R. (1984) The subsurface crack under conditions of stick and slip caused by a surface normal force, *J. Appl. Mech.*, **51**, 311–316.

(5) COMNINOU, M. (1977) The interface crack, *J. Appl. Mech.*, **44**, 631–636.

(6) COMNINOU, M. (1978) The interface crack in a shear field, *J. Appl. Mech.*, **45**, 287–290.

(7) COMNINOU, M., and SCHMUESER, D. (1980) Frictional slip between a layer and a substrate caused by a normal load, *Int. J. Engng. Sci.*, **18**, 131–137.

(8) COMNINOU, M., BARBER, J. R., and DUNDURS, J. (1983) Interface slip caused by a surface moving at constant speed, *Int. J. Engng. Sci.*, **25**, 41–46.

(9) HILLS, D. A. and COMNINOU, M. (1985) A normally loaded half plane with an edge cracks, *Int. J. Solids Structures*, **21**, 399–410.

(10) HILLS, D. A. and COMNINOU, M. (1985) An analysis of fretting fatigue cracks during loading phase, *Int. J. Solids Structures*, **21**, 721–730.

(11) NOWELL, D. and HILLS, D. A. (1987) Open cracks at or near free edges, *J. Strain Analysis*, **22**, 177–185.

(12) SHEPPARD, S. D., HILLS, D. A., and BARBER, J. R. (1986) An analysis of fretting cracks. Part 2: unloading and reloading phases', *Int. J. Solids Structures*, **23**, 140–152.

(13) BOWER, A. F. (1988) The influence of crack face friction and trapped fluid on surface initiated rolling contact fatigue cracks, *J. Tribol.*, **110**, 704–711.

(14) DUBOURG, M. C. and VILLECHAISE, B. (1989) Unilateral contact analysis of a crack with friction, *Eur. J. Mech. A/solids*, **8**, 309–319.

(15) DUBOURG, M.-C. and VILLECHAISE, B. (1992) Analysis of multiple fatigue cracks – Part I: theory, *J. Tribol.*, **114**, 455–461.

(16) DUBOURG, M.-C., GODET, M., and VILLECHAISE, B. (1992) Analysis of multiple fatigue cracks. Part II: results, *J. Tribol.*, **114**, 462–468.

(17) DUBOURG, M.-C. and VILLECHAISE, B. (1992) Stress intensity factors in a bent crack: a model, *Eur. J. Mech. A/Solids*, **11**, 169–179.

(18) DUBOURG, M. C., COLIN, F., and VILLECHAISE, B. (1992) Fracture mechanics and small detachment in cyclic Hertzian loading. Theory and computer simulation, *Wear particle: from the cradle to the grave*, (Edited by D. Dowson, C. M. Taylor, and M. Godet), Elsevier Tribology Series Vol. 19, pp. 101–111.

(19) ERDOGAN, F., GUPTA, G. D., and COOK, T. S. (1973) Numerical solution of a singular integral equation, *Method analysis and solution of crack problems*, (Edited by G. C. Sih), Leyden, Noordhoff, pp. 368–425.

(20) KALKER, J. J. (1990) Three-dimensional elastic bodies in rolling contact, (Edited by Kluwer Academic Publishers, Netherlands, p. 314.

(21) DUBOURG, M. C. and KALKER, J. J. (1993) Crack behaviour under rolling contact fatigue, *Rail quality and maintenance for modern railway operation*, (Edited by J. J. Kalker, D. F. Cannon, and O. Orringer), Kluwer Academic Publishers, pp. 373–384.

(22) HOURLIER, F., D'HONDT, H., TRUCHON, M., and PINEAU, A. (1982) Fatigue crack path behaviour under polymodal fatigue, ASTM Conference 'Fatigue sous sollicitations biaxilaes et multiaxiales'.

(23) BATHIAS, C. and BAÏLON, J. P. (1980) *La fatigue des matériaux et des structures*, (Edited by S. A. Maloine), p. 547.

(24) HADJ SASSI, B., BOUSSEAU, M., and LEHR, P. (1987) Cinétique de propagation des fissures de fatigue dans l'alliage de titane TA6V, *Mécanique Matériaux Electricité*, pp. 18–25.

S. Faanes*† and U. S. Fernando*

Life Prediction in Fretting Fatigue using Fracture Mechanics

REFERENCE Faanes, S. and Fernando, U. S., **Life prediction in fretting fatigue using fracture mechanics,** *Fretting Fatigue*, ESIS 18 (Edited by R. B. Waterhouse and T. C. Lindley) 1994, Mechanical Engineering Publications, London, pp. 149–159.

ABSTRACT In fretting fatigue, crack nucleates at a very early stage of loading and the fracture process is dominated by crack growth. Thus fracture mechanics techniques provide a tool in the interpretation of fretting fracture phenomena. In this paper, the fretting fatigue life of BS L65 aluminium alloy under various normal loading and axial loading conditions is considered. An attempt is made to predict the fatigue life using linear elastic fracture mechanics (LEFM) parameters. The elastic stress intensity factors are determined for a crack normal to the contact surface and the effective stress intensity range is evaluated using a simplified crack closure assumption. A model which takes into account the near threshold fatigue crack growth for both short and long cracks is used. Material data are available in the form of plain fatigue properties and as crack growth (da/dN versus ΔK) curves. The data, together with the crack growth model, were used to predict the fretting fatigue life for various loading conditions. The predicted fretting fatigue lives are compared with that of experimental results, and show a good agreement. It has been shown that the modelling of short crack growth behaviour improves the accuracy of the prediction when compared with that obtained by using the conventional, long crack growth models.

Notation

a	Crack length
b	Width of contact zone
d	Crack length correction
Δ	Range of variable
K	Stress intensity factor
N	Number of fatigue cycles
p	Normal stress in contact zone
P	Fretting contact normal force per unit length
q	Tangential stress in contact zone
Q	Fretting contact shear force per unit length
R	Stress ratio
σ	Stress
Y	Geometry factor

Subscripts

ax	Axial load variable
fl	Fatigue limit
max	Maximum of cyclic variable

* SIRIUS, Faculty of Engineering, University of Sheffield, Mappin Street, Sheffield, S1 3JD, UK.
† On leave from Department of Applied Mechanics, Norwegian Institute of Technology, 7034 Trondheim, Norway.

min   Minimum of cyclic variable
th   Threshold value

Introduction

The unfavourable influence of fretting fatigue is well known in a vast number of engineering applications. In structures where 'fail-safe' or 'damage-tolerant' design philosophies are to be followed, the expected life is needed for the component involved. Component life prediction is frequently based on a strain–life or stress–life approach. Material data in this context are rarely based on fretting conditions. In fretting a high stress concentration exists at the contact zone. This concentration accelerates the crack propagation, and hence makes conventional design approaches unconservative.

As the fretting fracture process is dominated by crack growth (1) fracture mechanics methodology is considered to be appropriate in the assessment of fretting fatigue life. In this procedure the fretting forces may be taken directly into account through their influence on the stress field and the stress intensity factors in the fretting zone.

Fretting fatigue life calculations using linear fracture mechanics have previously been performed by Edwards (2) who showed that it is possible to predict life reasonably well by using this approach. Tanaka et al. (3) used the same approach, with success, in fretting fatigue life evaluation of a spring steel and a carbon steel, but short cracks were avoided by 0.5 mm deep slots where the cracks started to grow.

The main deficiency of linear fracture mechanics is its inability to account for the short crack growth behaviour observed in practice. When considering fatigue crack growth in smooth components under uniform loading, this weakness may seriously invalidate the computation results of fatigue life, and often lead to non-conservative life prediction. In fretting environment the presence of a high stress gradient will considerably affect the mechanism of short crack growth, and little work has been done to understand the short crack growth behaviour in fretting.

In this paper fracture mechanics is applied to predict fatigue life for the BS L65 aluminium 4 percent copper alloy, and it is shown how a simple short crack correction can improve the lifetime calculation in fretting fatigue.

Stress intensity factors

The fretting cracks normally nucleate at or near the leading edge of the fretting contact and the crack tip stress intensity factor (SIF) is directly influenced by three forces: the axial load; normal load; and the friction force. In the present analysis it is assumed that the crack is situated at the leading edge of the contact and grows perpendicular to the contact surface, as shown in Fig. 1. The friction force Q and normal load P at the contact patch is assumed to be uniformly distributed. Since the principle of superposition applies in LEFM,

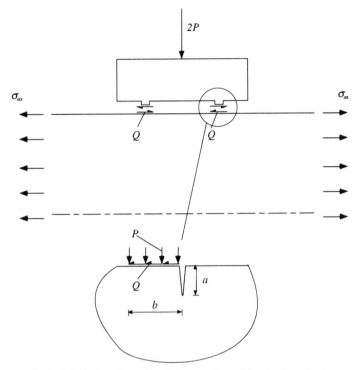

Fig 1   Principal configuration of the experimental fretting investigation

the SIFs due to fretting loads and the bulk load may be evaluated individually
and can be added together to obtain the total SIF.

Stress intensity factors for penetrating edge cracks perpendicular to the
surface are found by using the stress distribution along the crack site in an
uncracked body as described by Hartranft and Sih **(4)**.

$$K_1 = 2 \sqrt{\left(\frac{a}{\pi}\right)} \int_0^1 \frac{\sigma_y(X)\{1 + f(X)\}}{\sqrt{(1 - X^2)}} \, dX \tag{1}$$

The stress distribution due to fretting forces, $\sigma_y(X)$, may be presented as shown
by Faanes and Härkegård **(5)** who also found the SIFs due to uniformly distributed fretting forces based on numerical integration of equation (1). The
accuracy of the SIFs are maintained for very short cracks and may be controlled by the asymptotic solutions given by **(5)**

$$K_P = 2Y \frac{P}{b} \sqrt{\left(\frac{a}{\pi}\right)} \left(\frac{a}{\pi b} - \frac{\pi}{4}\right) \tag{2}$$

$$K_Q = 2Y \frac{Q}{b} \sqrt{\left(\frac{a}{\pi}\right)} \left(\log \frac{2b}{a} - \frac{1}{2}\right) \tag{3}$$

The SIF due to the bulk axial load may be found using equation (1) as well, yielding the approved expression

$$K_{ax} = Y\sigma_{ax}\sqrt{(\pi a)} \qquad (4)$$

where Y is the geometrical factor equal to 1.1215 in all three equations (5).

The total range of the stress intensity factor

In the fretting assembly used here (Fig. 1), it is assumed that through a fatigue cycle, the friction force varies in phase with the axial force with the stress ratio $R = -1$. On the other hand will the contact pressure give a constant negative contribution to the stress intensity factor? This means that the resulting R ratio in fretting will be less than -1 for fully reversed axial loading, and will approach -1 as the crack grows and the influence of the fretting forces fade away.

Under reversed loading, i.e. $R < 0$, the crack opens during tensile loading only and, therefore, the compressive part of the load cycle has a small influence on the crack growth. In such a situation ΔK equals K_{max} which implies that for reversed loading, the crack growth rate and the threshold stress intensity are not affected by R. This is found to be in good agreement with experimental evidences (6) for the material investigated, and is also compatible with recent investigations on crack closure (7)(8).

The total 'effective' stress intensity range in each cycle is calculated based on equations (2)–(4) and equals

$$\Delta K = K_{max} = K_{ax, \, max} + K_{Q, \, max} + K_{P, \, max} \qquad (5)$$

and is used as the load parameter in the crack growth models described below.

The crack growth models

Three different crack growth models have been examined in this paper. Model 1 is Paris' law given by

$$\frac{da}{dN} = C(\Delta K)^n \qquad (6)$$

where $C = 1.5*10^{-11}$ (MPa, m), and $n = 4.08$ are material constants. This model is valid for long cracks with insignificant plastic deformation at the crack tip, and does not take into account the threshold effect found in practice. Model 2 takes account of a threshold stress intensity factor as demonstrated by

$$\frac{da}{dN} = C(\Delta K - \Delta K_{th})^n \qquad (7)$$

where $C = 1.8*10^{-10}$ (MPa, m), $\Delta K_{th} = 1.9$ MPa$\sqrt{m}$ and $n = 3.25$ are material constants, and ΔK is the stress intensity range. The model is represented by the solid line in Fig. 2. The value of ΔK_{th} is assumed to be fixed and independent of crack length and loading. The deficiency of the above model is its inability to account the short crack behaviour.

A third model may be established to take the increased crack growth rate of short cracks into account. The modelling of the short crack growth (model 3) is based on the assumption that the threshold stress intensity is reduced for short cracks. The modified theshold is then used in equation (7). The threshold short crack correction suggested by El. Haddad et al. (9) is used here. This correction is based on the requirement that the crack growth rate must be zero for alternating stresses up to the fatigue limit for smooth components, and may be expressed mathematically as

$$\Delta K_{th}(a) = \Delta K_{th\infty}\sqrt{\left(\frac{a}{a + d}\right)} \qquad (8)$$

where

$$d = \frac{1}{\pi}\left(\frac{\Delta K_{th\infty}}{Y\Delta\sigma_{fl}}\right)^2 - a_0 \qquad (9)$$

and where a_0 is the initial crack length set to the surface roughness, $\Delta\sigma_{fl}$ is the fatigue limit of the material for the given surface roughness, and $\Delta K_{th\infty}$ is the threshold stress intensity factor for a long fatigue crack in the given material. This modification of the threshold stress intensity factor takes care of the observation that the threshold level for commencing crack growth is considerably lower for short cracks.

For sufficiently short cracks, the crack growth threshold vanishes and model 3 gives the crack growth rate plotted with a dotted line in Fig. 2. For comparison, experimental data for short cracks is plotted in the same diagram. When the crack length increases, the threshold stress intensity factor approaches the one for long cracks, and the growth rate reduces and eventually coincides with the solid line given by model 2.

The crack growth behaviour given by model 3 is presented in Fig. 3 in terms of crack growth rate versus crack length. Growth curves are plotted for different bulk load levels and it is seen that the crack growth threshold becomes less pronounced as the stress increases. The crack growth rate according to model 3 for a typical fretting load is depicted in the same figure. It is seen that as the fretting crack grows, the crack growth rate shifts from a growth curve equivalent to a high bulk load towards a low bulk load plain fatigue crack growth rate. This illustrates that the stress gradient in the fretting zone may explain the early accelerated crack growth which is frequently observed in fretting.

The crack growth models described above have been used to predict smooth fatigue endurance behaviour by numerically integrating the crack growth

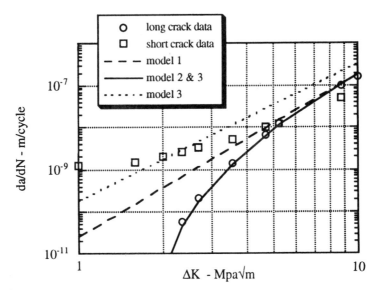

Fig 2   The crack growth rate models plotted versus ΔK and compared with material data available for short and long cracks

models. Figure 4 shows the predicted *S–N* curves which are compared with the experimental endurance data for the material. All data are given for *R* = −1. For model 1 and 3 the initial crack length is assumed to be equal to surface roughness, which is approximately 5 μm for the polished specimens. A

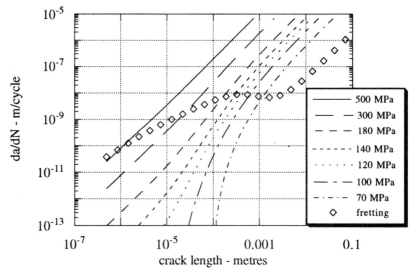

Fig 3   Crack growth rate given by model 3 plotted against crack length for various load levels of uniform loading, and a typical fretting loading

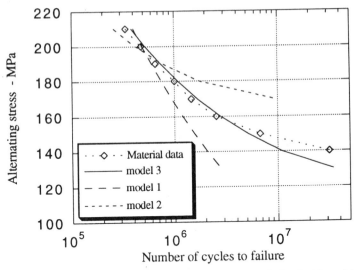

Fig 4 Fatigue lives for plain fatigue under fully reversed loading calculated with models 1–3 and compared with experimental *S–N* data for BS L65

considerably higher initial crack length (30 μm) is required for model 2 to fit the *S–N* material data, which suggests that this model is inadequate for predicting fatigue lives of smooth specimens. It is seen that the *S–N* curve is satisfactorily predicted only when the threshold is modified for short cracks as in model 3. By changing the initial flaw size for model 1 and 2, the corresponding *S–N* curves could be changed to fit the material data for some load values, but the curves will never coincide on all load levels.

Comparison with fretting experiments

Fretting fatigue tests were conducted with various combinations of axial load and normal load. The specimen which has a rectangular cross-section was subjected to cyclic fully reversed axial load. Fretting action was achieved by using two symmetrically placed bridge type pads. The normal load and the axial load were prescribed in the bi-axial testing machine, while the friction force was found by measuring the strain of the fretting pad. Three different pad spans together with nine different load combinations were used to cover both micro and macro slip regimes. A full description of the test facility, details of experiments and the friction, and endurance results are given by Fernando *et al.* **(10)(11)**.

After the performance of the crack growth models in predicting plain fatigue *S–N* curves were monitored, the models were used to predict fretting fatigue lives. The predicted fatigue lives are compared with experimentally obtained results in Figs 5 to 7.

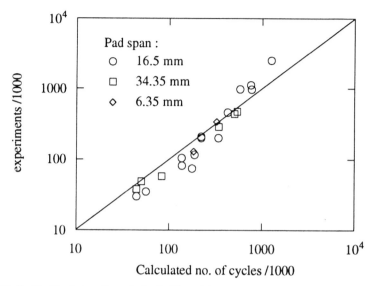

Fig 5 Fretting fatigue lives obtained with model 1 compared with experimental results

Discussion

The encouraging performance of the fatigue life predictions suggests that fracture mechanics is an appropriate tool to evaluate life of components under fretting conditions. This is in agreement with prevous work described in the literature. However, the models give some deviation from the experimental

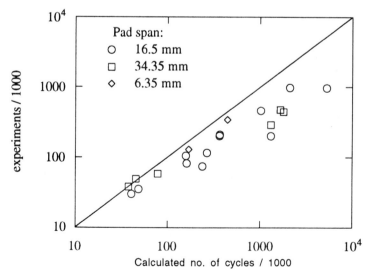

Fig 6 Fretting fatigue lives obtained with model 2 compared with experimental results

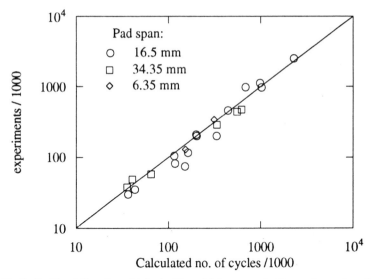

Fig 7   Fretting fatigue lives obtained with model 3 compared with experimental results

results, especially for model 1 and 2 for long fatigue lives. The reasons for this are discussed below.

Influence of short crack behaviour

Despite the fact that life prediction of plain fatigue is poor for model 1 and 2, it is seen that the fretting fatigue life is estimated by these two models with reasonable accuracy. In fretting the initiation and early crack growth is considerably accelerated due to the stress concentration in the fretting zone. Therefore, the ratio of the number of cycles spent while the crack is short to the total number of fatigue cycles may be significantly reduced, and hence any inaccuracy in the description of the short crack growth has a smaller influence on the total predicted life. This may explain the good correlation obtained between experiments and predictions by model 1 and 2, especially for the short fatigue lives. However, the degree of correlation decreases with increasing fatigue life when short fatigue crack growth is ignored. From Figs 5–7 it is clear that the consideration of the anomalous short crack growth behaviour improves the fretting fatigue life prediction, especially for long lives. It should be noted that the short crack correction used here is merely a correction for threshold SIF used in the long crack growth model. A more realistic description that would account for actual short crack growth behaviour may provide even better results.

Crack path and mixed-mode crack growth

An aspect that has been ignored in the analysis is the influence of the crack path in the evaluation of SIF. As observed by many investigations, fretting fatigue cracks are initially found to propagate in an inclined direction (1) beneath the contact zone. In such situations mode II growth mechanisms may intervene and mode I stress intensity factors used here may not be sufficient to describe crack growth. This prolonged growth of fretting cracks in mixed-mode is not yet clearly understood; however, it is expected that the growth rate will be increased due to the presence of a mode II component.

An increased crack growth rate by a mode II component would increase the speed of crack propagation. For crack paths directed nearly perpendicular to the surface, the mode II stress intensity factors are small compared to mode I under fretting loads (10), so large deviations in crack growth should not be expected.

Plastic deformation in the contact zone

The contact forces cause a considerable stress concentration in the fretting zone, and plastic deformation is, therefore, likely. The linear elastic approach using stress intensity factors is invalidated if the plastic zone gets too large. In fretting the proportion of life spent on propagating very short cracks is relatively small as advocated in the previous section. In the loading conditions considered in the present investigation no large scale plastic deformation is observed, and thus the plasticity is expected to have only a minor influence over the predicted results.

The distribution of loads in the contact interface

In this paper the distribution of normal load and friction force are assumed to be uniform under the pad. Great uncertainty is attached to this assumption. An accurate contact load distribution is important for the evaluation of the SIF and hence the crack growth rate, particularly for cracks shorter than the pad width *b*. A large proportion of the fatigue life is spent with such crack lengths and further investigations would be needed.

Effect of the slip amplitude

The pad span affects the slip range and the friction force between the pad and the specimen. In the present analysis the SIF is derived exclusively from the knowledge of loads and, therefore, the *direct influence* of slip is not taken account of. The possible influence of slip *on the friction force* is not discussed, but is indirectly taken account of because the friction force is measured in each experiment. As seen from the Figs 5–7, the pad span seems to have no effect on the accuracy of the life predictions, which indicates that the three load parameters used (axial load, normal load, and friction force) are sufficient to predict the fretting process investigated here. This suggests that the slip amplitude affects fretting fatigue first of all through its influence on the friction force in the fretting zone.

Conclusions

The behaviour of short cracks seems to have less influence on the life time in fretting fatigue compared to plain fatigue. This implies that a simple mode I fracture mechanics analysis may be sufficient to determine fretting fatigue life for high load levels. For low load levels and for long fatigue lives the application of the short crack model improves the life predictions significantly.

The good correlation between experimentally obtained and predicted fatigue lives, irrespective of slip amplitude suggests that the slip influences fretting damage merely through its influence on the friction force.

References

(1) WATERHOUSE, R. B. (1992) Fretting fatigue, *Int. Mater. Rev.*, **37**, 77–97.
(2) EDWARDS, R. P. (1981) The application of fracture mechanics to predicting fretting fatigue, *Fretting fatigue*, (Edited by R. B. Waterhouse), Applied Science Publishers, London.
(3) TANAKA, K., MUTOH, Y., SAKODA, S., LEADBEATER, G. (1985) Fretting fatigue in 0.55C spring steel and 0.45C carbon steel, *Fatigue Fracture Engng Mater. Structure*, **8**, 129–142.
(4) HARTRANFT, R. J. and SIH, G. C. (1973) Alternating method applied to edge and surface crack problems, *Methods of analysis and solutions of crack problems*, (Edited by G. C. Sih), Noordhoff Leyden.
(5) FAANES, S. and HÄRKEGÅRD, G. (1994) Simplified stress intensity factors in fretting fatigue, *Fretting Fatigue*, Mechanical Engineering Publications, London, pp. 73–81. *This volume.*
(6) RAE Technical Report 76115, 1976.
(7) HUDAK, S. J. and DAVIDSON, D. L. (1988) The dependence of crack closure on fatigue loading variables, *ASTM STP 982*, ASTM, Philadelphia, pp. 121–138.
(8) ANDERSON, T. L. (1991) Fracture mechanics: fundamentals and applications, CRC Press.
(9) EL. HADDAD, M. H., SMITH, K. N., and TOPPER, T. H. (1979) Fatigue crack propagation of short cracks, *J. Engng Mater. Technol.*, **101**, 42.
(10) FERNANDO, U. S., FERRAHI, G. H., and BROWN, M. W. (1994) Fretting fatigue behaviour of BS L65 aluminium copper alloy under constant normal load, *Fretting Fatigue*, Mechanical Engineering Publications, London, pp. 183–195. *This volume.*
(11) FERNANDO, U. S., BROWN, M. W., MILLER, K. J., RAYAPROLU, D. B., and COOK, R. Fretting fatigue behaviour of BS L65 aluminium copper alloy under variable normal load, *Fretting Fatigue*, Mechanical Engineering Publications, London, pp. 197–209. *This volume.*

K. Dang Van* and M. H. Maitournam*

Elasto-Plastic Calculations of the Mechanical State in Reciprocating Moving Contacts: Application to Fretting Fatigue

REFERENCE Dang Van, K. and Maitournam, M. H., **Elasto-plastic calculations of the mechanical state in reciprocating moving contacts: application to fretting fatigue,** *Fretting Fatigue*, ESIS 18 (Edited by R. B. Waterhouse and T. C. Lindley) 1994, Mechanical Engineering Publications, London, pp. 161–168.

ABSTRACT A numerical analysis of certain fretting phenomena is presented. It is based on the evaluation of the mechanical quantities with an original computational method for reciprocating moving loads on elasto-plastic structures. This approach, coupled with multiaxial fatigue criteria gives reasonable predictions.

Introduction

Great efforts have been made to understand the different types of damage arising at the contact surfaces of bodies. Experiments carried out on metal to metal contacts have shown the existence of plastic phenomena. In particular, in small amplitude relative oscillatory motions, fretting can appear. This phenomenon cannot be studied by classical engineering methods because the levels of macroscopic stresses and strains are too small to explain the initiation of cracks. A first possible explanation is the existence of microcontacts: nominally flat surfaces in fact present micro-undulations. These imperfections induce local oscillating contacts, and consequently local plastic deformation builds up and damage occurs.

On a micro-scale, relative macroscopic displacements could be quite important compared to the width of the contacts between asperities. To understand the damage mechanism, it is necessary: (a) to evaluate the different mechanical quantities (stresses, strains, plastic strains) and their evolution in connection with such important parameters as the loads, the friction coefficients, and the surface characteristics; (b) to check the results by means of multiaxial damage criteria.

Up to now, the evaluation of residual stresses and strains arising from local plastic deformation has been a difficult problem. Many studies are devoted to the determination of mechanical fields during repeated forward-moving contact and the evaluation of the residual strains and stresses obtained after repeated loadings **(1)–(5)**. The most intensive research in the field has been

* Laboratoire de Mécanique des Solides, CNRS URA 317, École Polytechnique, 91128 Palaiseau cedex, France.

carried out by Johnson and his co-workers (1)–(3)(6)(7). This approach is analytical or semi-analytical, and cannot be used for general kinds of material, geometry, and load distribution. Bhargava et al. (4)(5) proposed a finite element approach of rolling/sliding contacts, based on incremental translations of the loading. This kind of approach is long and expensive from a computational point of view. As far as we know, no analytical or numerical studies on shakedown of structures subjected to forward and backward moving loads are available.

The aims of this paper are:

(a) to present a systematic and reliable computational procedure for the analysis of mechanical quantities in the vicinity of alternating moving contacts;

(b) to compare the obtained results for elastic and plastic shakedown with those evaluated in the case of repeated forward motions as presented in Johnson's studies;

(c) to predict from the stress cycles, the crack initiation features (where it occurs and under which conditions).

Evaluation of the mechanical quantities

The proposed method is based on the assumption that the macroscopic displacements could be quite important compared to the width of the microcontacts. The mechanical quantities in the vicinity of the contact area could be evaluated by considering a half-space subjected to forward and backward moving load. This assumption was used by Dang Van et al. (8) to study the fretting fatigue occurring in a strap. It is also valid in the case of the fretting pads in testing devices used to study the influence of shot peening on fretting fatigue.

The classical two-dimensional rolling contact problem is considered, but the rolling alternates between the forward direction and the backward one. A semi-elliptical pressure distribution (with P_0 the maximum pressure) applied on a width of $2a$, is moving in the x direction on the half-space (defined by $z \leqslant 0$). The semi-infinite body is made of an elasto-plastic linear kinematically hardening von Mises material. The authors' aim was to calculate the stress, residual stress, and plastic strain fields during a finite number of passes, and their asymptotic values in the stabilized state. Elasto-plastic computations could be carried by means of classical existing software. However, for repeated contacts problems, these computations are time-consuming and cumbersome. The incremental translations of the loads used could be prejudicial to the accuracy of the results.

Principle of the method (9)(10)

The basic idea is to make the steady state assumption in a reference frame moving with the loads; for any material quantity A, it leads to the following relation

$$\dot{A} = -\nabla A \cdot V$$

where V is the loading velocity. $V = Ve_1$ for forward motions, and $V = -Ve_1$ for backward motions.

If the material is elasto-plastic, the normality law for the plastic strain rate $(\dot{\varepsilon}^p)$ is written

$$\dot{\varepsilon}^p = \nabla \varepsilon^p \cdot V = \lambda \frac{\partial f}{\partial \sigma}$$

where λ is the plastic multiplier, and $(f = 0)$ defines the elastic domain. The time integration along the loading path, therefore, becomes a space integration along the direction of the motion of the loads. Numerically (using the finite element method), neither the load, nor the structure are translated.

The second idea is to use the stationarity of the limit state (the stress field is periodic, as is the plastic strain in the absence of rachetting) in order to find its value directly.

The full details of the finite elements implementation of this analysis are given in reference (9) where the two numerical procedures which can be used for the determination of the stabilized state are reported.

(1) The first procedure is the pass-by-pass stationary method (PPSM) for the calculation of a single pass. By computing the successive passes by this method, all of the features of the 'stabilized' state can be found (numbers of cycles before reaching it, residual stresses, plastic deformations, ...). This 'stabilized' state can, of course, be rachetting; in such a case, the rachetting rate is immediately deduced as the increment of plastic strains caused by a pass.

(2) The second procedure is the direct stationary method (DSM) for the direct determination of the 'stabilized' state if it is a shakedown (elastic or plastic). The ratchetting is indicated by a non-convergence of the algorithm.

These methods allow for the description of strains and stresses due to cyclicly and alternatively moving contacts, and the quick determination of the nature of the stabilized state (elastic shakedown, plastic shakedown, or ratchetting). For application to fretting, the method is approximate; it works well when the size of local contacts arising from local defects is small relative to the oscillatory distance.

Fatigue analysis

Fatigue analysis is based on the approach developed by Dang Van and co-workers (11). Their models of crack initiation are introduced in a fatigue CAD system (SOLSTICE) by Ballard et al. (12). This software has been systematically used to interpret the results of the previous calculations.

In this paper, the results of fatigue analysis are presented in the Dang Van's diagram. This diagram is the representation of the loading path in the τ (microscopic shear), p (hydrostatic pressure) space. Fatigue failure does not occur if the loading path is inside the safety domain, which is delimited by the experimentally-determined straight lines. By our method, for 'each' depth, the loading path in the stabilized state is directly obtained as the evolution of the mechanical quantities when x_1 varies from $+\infty$ (far ahead) to $-\infty$ and vice versa.

Numerical results

For the numerical calculations, a finite sized mesh represents the semi-infinite body. The material parameters are: E (Young's modulus) = 207 GPa; v (Poisson's ratio) = 0.3; k (shear resistance) = 236.7 MPa and C (hardening modulus) = 15.9 GPa.

Examples of results

As an illustration of the computational method, let us consider the load parameter $P_0/k = 3.21$ and the friction coefficient $\mu = 0.2$. The stabilized state obtained in the case of forward and backward motions is an elastic shake-down. The evolution of the longitudinal stress σ_{11} in the superficial mesh layer is represented in Fig. 1. The longitudinal residual stress patterns are shown in Fig. 2. These stresses are higher than those obtained in the case of repeated forward motion. This could be the reason why the shakedown limits obtained

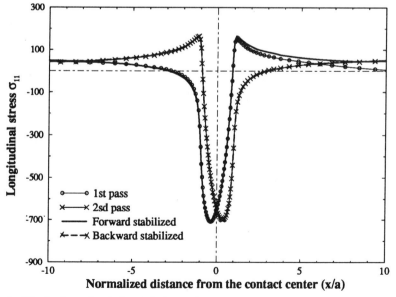

Fig 1   Distribution and evolution of the longitudinal stress near the surface $P_0/k = 3.21$; $\mu = 0.2$

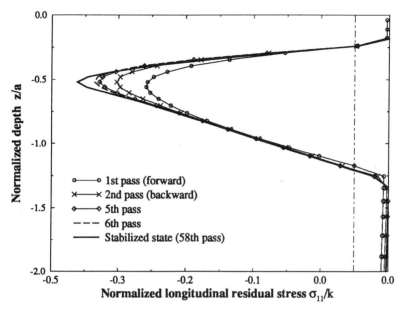

Fig 2   Distribution and evolution of the longitudinal stress near the surface $P_0/k = 3.21$; $\mu = 0.2$

for reciprocating motion (reported below) are lower than those corresponding to repeated forward motion. The whole stress tensor cycle is obtained for every point of the structure. Then fatigue analysis is performed using SOL-STICE software. Figure 3 shows the stress loading paths in the Dang Van's fatigue diagram (11)(12). Crack initiation can occur at or beneath the surface, depending on the material fatigue characteristics (the slope of the fatigue limit in the Dang Van's diagram). For a higher friction coefficient, cracks initiate at the surface. For instance for $P_0/k = 2.8$ and $\mu = 0.4$, the stabilized state is plastic shakedown; the loading paths represented in Fig. 4 show that the critical layer is at the surface.

Shakedown limits

Shakedown limits are studied for reciprocating motions as carried out for the repeated forward motions by K. L. Johnson *et al.* (13), and numerically by Dang Van and Maitournam (9). In this study, the PPSM is used to find the shakedown limits. A stabilized state is reached if, after a loading cycle (forward and backward motion), the plastic strains found are the same, with a precision of 0.01, as those before the cycle. For instance, sixteen forward and backward motions is sufficient to reach shakedown for $P_0/k = 4$. and $\mu = 0$. The shakedown limits obtained (Fig. 5) in the case of reciprocating motions are lower than those obtained for forward motions. This fact can explain the particular *nocivity* of alternating moving contacts as encountered in fretting phenomena.

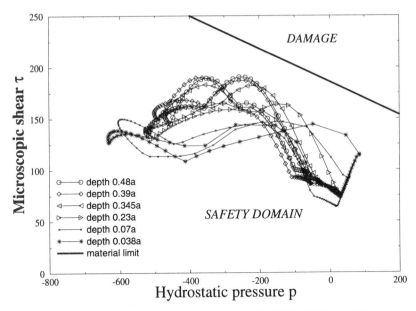

Fig 3   Loading paths in Dang Van's representation $P_0/k = 3.21$; $\mu = 0.2$

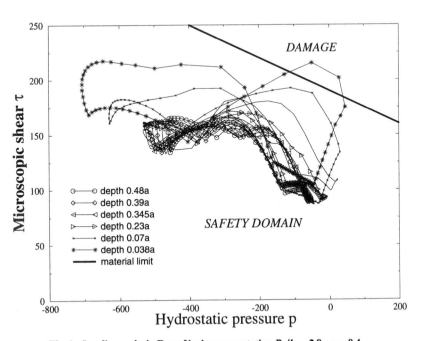

Fig 4   Loading paths in Dang Van's representation $P_0/k = 2.8$; $\mu = 0.4$

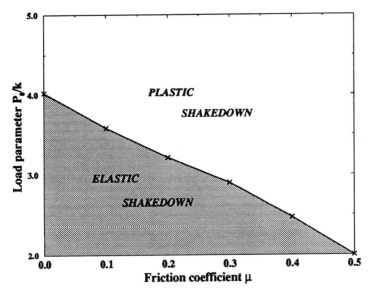

Fig 5   Shakedown limits for line contact for linear kinematic hardening material

Conclusion

A quantitative analysis of fretting phenomena is presented. This relies upon an evaluation of the mechanical quantities at the vicinity of the contacts, based on a powerful and suitable computational method for forward and backward loads moving on elasto–plastic structures. This approach is coupled with multiaxial fatigue criteria. The combination of both methods provides a good tool for reasonable quantitative analysis of fretting phenomena.

References

(1) BOWER, A. F. and JOHNSON, K. L. (1989) The influence of strain hardening on cumulative plastic deformation in rolling and sliding line contact, *J. Mech. Phys. Solids*, **37**, 471–493.
(2) HEARLE, A. D. and JOHNSON, K. L. (1987) Cumulative plastic flow in rolling and sliding line contact, *J. appl. Mech.*, **54**, 1–5.
(3) JOHNSON, K. L. (1986) Plastic flow, residual stress and shakedown in rolling contact, Proceedings of Second International Symposium on Contact Mechanics and Wear of Rail/Wheel Systems, pp. 83–97.
(4) BHARGAVA, V., HAHN, G. T., and RUBIN, C. (1985) An elastic–plastic finite element model of rolling contact – Part 1, *J. appl. Mech.*, **52**, 67–74.
(5) BHARGAVA, V., HAHN, G. T., and RUBIN, C. (1985) An elastic–plastic finite element model of rolling contact – part 2, *J. appl. Mech.*, **52**, 75–82.
(6) JOHNSON, K. L. and JEFFERIS, J. A. (1963) Plastic flow and residual stress in rolling and sliding contact, Institution of Mechanical Engineers Symposium on Rolling Contact Fatigue, pp. 50–61.
(7) MERWIN, J. E. and JOHNSON, K. L. (1963) An analysis of plastic deformation in rolling contact, *Proc. Inst. Mech. Engrs*, **177**, 676–685.
(8) DANG VAN, K., BATHIAS, C., and BERNARD, P. (1988) 'Sur une nouvelle approche du phénomène de fretting, *Arch. Mech.*, **40**, 543–555.

(9) DANG VAN, K. and MAITOURNAM, H. (1993) Steady-state flow in classical plasticity: application to repeated rolling and sliding contact, *J. Mech. Phys. Solids*, **41**, 1691–1710.

(10) DANG VAN, K., INGLEBERT, G., and PROIX, J. M. (1985) Sur un nouvel algorithme de calcul de structure élastoplastique en régime stationnaire, 3ème Colloque Tendances actuelles en calcul de structures.

(11) DANG VAN, K. (1992) On structural integrity assessment for multiaxial loading paths, *Theoretical concepts and numerical analysis of fatigue* (Edited by C. J. Beevers and A. F. Blom).

(12) BALLARD, P., DANG VAN K., DEPERROIS, A., PAPADOPOULOS, Y. (1992) High cycle fatigue and finite element analysis, Euromech 297, Fatigue Fracture Engng Mater. Structures.

(13) JOHNSON, K. L. (1985) *Contact mechanics*, Cambridge University Press.

EXPERIMENTAL METHODS

D. A. Hills* and D. Nowell*

A Critical Analysis of Fretting Fatigue Experiments

REFERENCE Hills, D. A. and Nowell, D., **A critical analysis of fretting fatigue experiments,** *Fretting Fatigue*, ESIS 18 (Edited by R. B. Waterhouse and T. C. Lindley) 1994, Mechanical Engineering Publications, London, pp. 171–182.

ABSTRACT A review of fretting fatigue experiments is carried out and a distinction made between tests to rank a material's resistance to fretting fatigue, and those made to gain understanding of the process of fretting fatigue itself. Of those in the latter category, a critical appraisal of the effect of the geometry of the contact is undertaken, including the methods of analysis available to describe the state of stress and degree of slip. The adoption of tests utilizing a convex specimen geometry in partial slip is recommended and some of the difficulties of realizing and analysing such configurations are discussed.

Introduction

Fretting fatigue is a surface phenomenon. Unlike plain fatigue, where the initiation phase of crack development is normally associated with the presence of some pre-existing macroscopic defect, or free initiation from some surface irregularity, fretting fatigue is essentially a process involving the interaction of two bodies. However, once a crack has grown to a length which is comparable with the characteristic dimension of the contact, the influence of the contact stress field itself will have diminished to a low level, and the crack might equally be one associated with plain fatigue. The fretting process itself is controlled by the bulk geometry of the contacting bodies, their surface finish, the interfacial coefficient of friction, the physical properties of the bodies (particularly their elastic properties and yield stress), and the applied loading history. These quantities are sufficient to enable a complete description of the contact problem to be specified, both on a bulk and an asperity scale, and hence a knowledge of the state of stress, strain, and other relevant quantities to be deduced. By themselves, these do no more than describe one aspect of the environment in which cracks initiate, and in which other processes such as wear and plastic flow occur. The initiation process will also be affected by the presence of any corrosive agents, and the mechanics of initiation may well be different from those occurring in plain fatigue.

These comments are intended to emphasize that when the term 'fretting fatigue' is used, it encompasses a combination of a contact problem, together with a consideration of the environment and conditions for crack initiation and early growth. Thus, when fretting fatigue tests are carried out, they actually characterize a combination of surface loading history, resulting from *geometric and loading* properties, and crack initiation/growth, which are

---

* Department of Engineering Science, University of Oxford, Parks Road, Oxford, OX1 3PJ, UK.

material properties. Most fretting fatigue tests actually assess these attributes in tandem but it is important to recognize that they might, at least in principle, be separated out, so the material property results inferred may be applied to other geometries. Fretting fatigue tests may be divided into two broad categories: those which seek to rank materials' resistance to fretting fatigue (defined in different ways), and those which seek to understand and quantify the effects of the severe contact stress fields (at both the bulk and asperity scale) on crack development in order to gain a more fundamental understanding of the process itself.

Classes of test used

Traditionally, the 'materials ranking' type of fatigue test has sought to apply a controlled and repeatable fretting history to a standard type of fatigue specimen. There are many ways in which this may be done, such as by using a rotating-bending specimen with clamped pads, or a similar geometry under push–pull conditions (Fig. 1); there are several other variations on this basic theme. Each of these geometries possess certain features which influence the fretting problem. The first is that the loading is normally applied by imposing a relative displacement between specimen and pads. This means that the shearing force associated with fretting is derived from a mixture of the compliance of the specimen and the compliance of the contact itself (which in turn depends on the coefficient of friction); it is not possible independently to vary the shearing force, which then becomes dependent on the geometry of the apparatus. It is also quite difficult to estimate the shearing force, either analytically or experimentally, and as it is a function of the coefficient of interfacial friction, which almost certainly changes during the test, so may the shearing

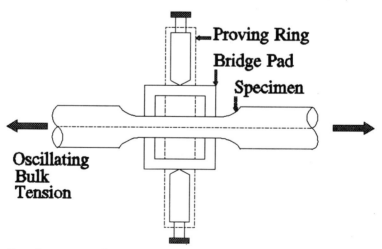

Fig 1   General appearance of tests using bridge-type specimens. Overall loading may be in push–pull (as here) or rotating bending

force itself. Secondly, it is commonplace in this kind of test to use fretting pads which have flat faces. These fall into the class of contacts referred to as 'complete' and the use of such a geometry has a profound influence on many attributes of the contact.

Fretting tests aimed at obtaining a greater understanding of the process itself have generally used more carefully controlled geometries of the 'incomplete' type, and are often run in the partial slip regime. It is essential to understand the reasons for these choices when considering the design of any fretting fatigue experiment.

Complete or incomplete contact?

It is important to distinguish between the two general classes of contact which may be used. These are shown schematically in Fig. 2. The first is the so-called complete contact, exemplified by a flat-ended punch. A complete contact is one in which the contact size is independent of the contact load, which is not, by itself, undesirable, but it does have several consequences. The first is that it is extremely difficult to determine the interfacial contact pressure distribution (1)(2) and secondly the presence of any kind of burr or other geometric imperfection, such as misalignment, will have a profound influence on the contact pressure. Incomplete contacts, exemplified by the classic Hertzian configuration, have a completely different set of characteristics; the size of the contact increases as the applied load is increased, they are relatively insensitive to minor surface imperfections and, because for many non-conforming contacts (specifically, those in which the contact size is small compared with the radius of both bodies) each body may be represented by a half-plane, closed form analytical solutions are known.

Quite apart from a distinction in both the form and clarity of the contact pressure distribution, the two classes of contact produce qualitatively different stick–slip regimes. Let us assume, for the time being, that one of the contacting

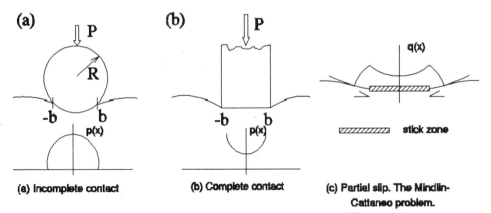

(a) Incomplete contact   (b) Complete contact   (c) Partial slip. The Mindlin-
                                                              Cattaneo problem.

Fig 2   Classes of contact and the partial slip problem

bodies is flat and elastic, whilst the other is rigid. The contact pressure distribution (neglecting the influence of elastic mis-match on surface normal displacement) will be given by $p(x) = (1 - x^2)^s$, where $s = 1/2$ if the contact is Hertzian, and $s = -1/2$ if a flat-ended punch is used. If a monotonically increasing shear force is now applied to the Hertzian geometry the classic Mindlin–Cattaneo stick–slip regime results, whereby the initial condition of stick everywhere gives way to a regime with a central stick zone bordered by two slip zones of the same sign, Fig. 2. By contrast, in the case of the flat punch, if we assume that *no* slip takes place, the shear traction required is given by $q(x) = (1 - x^2)^{-1/2}$. This is of precisely the same form as the contact pressure distribution and hence no slip will take place anywhere while $q(x) < -fp(x)$, i.e. $Q < fP$. However, as soon as an equality is achieved, instantaneous sliding will ensue, i.e. there is no partial slip regime. This suggests that for problems involving complete contact under imposed displacement conditions (such as in the fretting tests in Fig. 1), the displacement amplitude is controlled exclusively by the overall geometry of the problem and not by the contact itself. It should be noted that in real cases, where neither body is rigid, a more complex traction distribution pattern will exist, and that it is possible for partial slip to occur; however, the traction distributions obtained are extremely complex, and, for the reasons cited, are likely to be imperfectly known, so that the stick–slip pattern will be capricious and probably vary dynamically as wear and changes in the coefficient of friction occur. For these reasons test geometries involving flat-ended pads are very much a second choice when designing a test apparatus, although these do have the merit that they are easy to manufacture. Even if the test is to be used to carry out no more than a materials ranking procedure, their sensitivity to manufacturing imperfections means that repeatability will have to be sacrificed.

Partial slip or sliding?

The question of how the shearing force might be imposed is now addressed. As discussed above, in the tests shown in Fig. 1 the clamped contact probably experiences full slip under imposed displacement conditions but contacts of the incomplete type can conveniently be run in the partial slip regime where there is relative slip over only part of the contact area. The resulting distributions of shear traction and micro-slip cannot easily be determined directly from experimental measurements but may be predicted from measurements of the load history together with a knowledge of the coefficient of friction, f. Unfortunately f depends on the surface condition of the contacting bodies and this is frequently modified during the course of an experiment so that f varies with time. At the start of a test, assuming that the surfaces have been machined uniformly, a constant (uniform) coefficient of friction will be exhibited, that is the ratio of the shear to the direct traction (q/p) obtained if the contact is slid will be equal to some constant value, f_0 at all points. This pointwise definition will be consistent with the macroscopically measurable quantity defined as the ratio of Q/P when the contact undergoes gross sliding. However, as the test proceeds relative motion of surface particles within the slip zone will cause a modification of surface topography, increasing the intimacy of the contact, and this in turn will result in a higher coefficient of friction, *within the slip region.* A higher proportion of the applied load will now be carried in the slip zones and if the same total load is to be applied regions near the margins of the slip zones will give way to stick. Furthermore, it will not, in general, be possible to deduce the local coefficient of friction from some average value measured by sliding the entire contact and evaluating the ratio Q/P. One way around this difficulty is to use a well-defined, plane contact where it is possible to obtain explicit expressions for the stick and slip zone distribution using, for example, the Mindlin–Cattaneo model for Hertzian contact (3)(4). From this, assuming that surface modification occurs uniformly within a slip region, i.e., that all points within the slip zone(s) have the same coefficient of friction at a given time, it is possible to produce an explicit equation relating the mean, measured coefficient of friction with the true value (5). Once a value for f in the slip zones has been determined the distributions of shear traction and relative slip may readily be evaluated. In a partial slip problem the relative displacement of surface particles within the slip zone is related directly to the geometry of the problem and the loading history experienced. For the Mindlin–Cattaneo problem cited, for example, the surface displacements within the slip zones are given in closed form, albeit approximately, by Johnson (6).

It is also perfectly possible to consider conducting tests in which the amplitude of displacement is imposed at a very large value (exceeding the compliance of the components), so that gross sliding occurs. Note that using a force controlled feedback circuit is not a practical proposition for conducting sliding tests, as the displacement will increase without limit once sliding ensues, unless additional precautions are taken. Also, it should be borne in mind that surface modification will occur as described above, so that tests may well be carried out which are initially under sliding conditions but where an increase in the coefficient of friction subsequently causes the partial slip regime to be entered. Sliding tests, however, suffer from two significant drawbacks. First, appreciable wear may well occur which will alter the geometry of the contact and render the prediction of the contact pressure inaccurate as well as affecting the initiation of fretting cracks. Secondly, if the contact experiences sliding the slip amplitudes are independent of load and it will be necessary to employ careful experimental techniques to measure the degree of relative slip occurring at the contact. At first sight it may not be clear why the relative displacement of surface particles should influence fretting at all, and indeed, in the context of a macroscopic (bulk) contact it does not. However, if we re-focus attention at the scale of an asperity it is clear that for each loading cycle of the overall contact any single asperity within the contact may undergo several cycles of loading, as several asperities on the contacting surface may move over the point in question (7). It is interesting to reflect that this means that different areas of the surface experience different numbers of individual loading cycles at any point during the life of the component.

A further consideration affecting the choice between full and partial slip is the sophistication of the test machine available. Partial slip tests require careful control of the fretting force and this is most conveniently achieved by use of a second actuator, controlled independently from that applying a bulk stress to the specimen. It is, however, perfectly possible to produce a closely controlled fretting force and a specimen stress from a single actuator machine provided that the two are applied in phase (e.g. (7)).

Towards an ideal test geometry

The observations and deductions made above lead us towards certain ideals to seek in a test machine designed to generate a greater understanding of the fretting process. The first is that the contact should be incomplete, i.e., convex, and where an increase in load leads to an increase in contact patch size. This will produce a well-defined state of stress reasonably insensitive to the influence of any minor manufacturing imperfections, and free of stress singularities. Secondly, the tests should normally be run in a partial slip regime in order to minimize the difficulties associated with wear and with measurement of relative slip. Further, it needs to be decided whether an essentially plane (two-dimensional) configuration should be adopted or a three-dimensional geometry.

Two- or three-dimensional contacts?

Possible incomplete contact geometry which might be considered are wedge or cylindrical contacts and their three-dimensional counterparts the cone and sphere. Difficulties occurring with the first of each of these pairs are that the central point of contact is singular, thereby producing some localized plasticity, and the consequent influence this has on accuracy of the size of the contact patch predicted. Hertzian contacts would, therefore, seem to be an ideal choice, particularly as the state of stress is fully defined (8), together with the displacement field and compliance.

Two-dimensional Hertzian contacts have the advantage that the stress state is simpler to obtain, and the analysis of any two-dimensional crack which might need to be modelled is also straightforward. Furthermore, the determination of the stick slip regime is exact by the Mindlin–Cattaneo method (3)(4), and it is feasible to incorporate various relaxations of the standard Hertzian assumptions into the model in a closed form, e.g., the influence of a material mis-match (9), the influence of a superimposed bulk tension (10) and the influence of finite thickness of the tensile specimen (11). These advantages in analysing the problem are to a large extent outweighed by the practical difficulties in realizing a true two-dimensional contact; first, there is always a transition from plane stress towards plane strain conditions which produces some three-dimensional effects in the contact pressure distribution (12), with implications for the stress state induced. More serious, however, are the corner effects which inevitably arise at the ends of the fretting pads (13). The best compromise is to arrange for the pads and specimen to be the same width (Fig. 3), but unless this is perfectly achieved and the axis of the pad is truly transverse to the longitudinal axis of the main specimen corner, singularities will inevitably arise, and these will initiate cracks. Although remote from the corners the state of stress rapidly becomes uniform the cracks themselves will grow into a complex shape, and hence be difficult to analyse. Further, it is impossible to probe questions of crack initiation criteria as the state of stress adjacent to the corner is ill-defined. To some extent these effects can be reduced by careful setting up of the test, and adjusting the pads using pressure-sensitive paper to obtain a near-uniform transverse pressure distribution. Also, any peaks in the contact pressure arising at the very corners of the specimen will be relieved by local plasticity; this will in turn have a profound influence on crack initiation conditions.

These practical problems lend weight to the argument that a contact having convexity in both planes should be used. One possibility might be to use bar-relled rollers, but this would increase the complexity of the contact problem to be solved, and hence an axi-symmetric contact is probably the best choice of indenter to produce fretting damage. The states of stress induced by a static or sliding spherical contact are known in closed form (14)(15) and are reasonably straightforward to evaluate, particularly on the plane of symmetry where the stress is most severe, and hence cracks are most likely to initiate. Edge effects are, of course, completely absent, but it is difficult to extend the contact solutions in the ways mentioned above, and it should also be mentioned that the

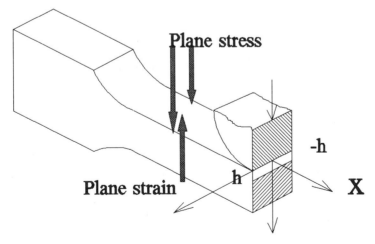

Fig 3   General appearance of a test specimen. The transition from a plane stress to plane strain

Mindlin–Cattaneo solution normally cited as providing a correct description
of the stick–slip zone geometry is imprecise insofar as the transverse slip component is neglected. This means that the predicted slip direction is not precisely aligned with the shear traction direction, which is physically incorrect. A
numerical solution to this problem has recently been carried out (16), and this
shows that Mindlin's solution is correct when Poisson's ratio is zero, and that
the maximum discrepancy occurs when the contacting bodies are incompress-
ible. The true slip zone was also found, and it was shown that the frictional
energy expenditure in the slip zones is localized in two crescent shaped regions
either side the plane of symmetry. In the same paper the influence of material
mis-match is also addressed, within the limits of the Goodman approximation
(17), i.e., that the contact pressure remains Hertzian. This gives rise to a
complex pattern of stick and slip which shakes down to a steady state within a
few cycles of loading. It is very difficult to carry out the other relaxations of
the Hertzian assumptions mentioned above in relation to the two-dimensional
problem, i.e., taking into account bulk tension and finite specimen thickness,
without recourse to purely numerical methods.

Crack detection

A consequence of using a three-dimensional fretting contact is that cracks
usually initiate on the plane of symmetry, where the severest stresses occur.
This makes for geometrically simple cracks growing in a well-defined stress
state, but also makes them difficult to observe and measure. In an idealized
two-dimensional contact problem, and ignoring end-effects, a through-
thickness crack with an approximately straight front might be expected to
develop, and this has the big advantage that it may be observed optically at
the free ends. In reality, however, for the reasons stated above, cracks grown
under nominally plane conditions do not always have such a simple shape and
in any case optical measurements made at the side of the specimen are not
necessarily representative of conditions towards the middle. Three-dimensional
spherical contacts tend to generate thumbnail-shaped cracks which cannot be
observed at all until they have grown to very large dimensions. In such cases
non-optical techniques must be relied upon, and the following would seem to
be possible methods.

Potential drop methods

A constant-current generator is used to impose a known current across the
plane of the crack, and the potential developed across the crack mouth is
measured. The technique may be made much more sensitive by using alternating current (at a frequency of about 7 kHz, but the optimum choice is
material-dependent), which exploits the skin effect, and hence concentrates the
current flux near the surface. Unfortunately the presence of the fretting pads
may well affect the current distribution and also hinders access for the PD
probes. In fretting it is usual for the crack to initiate within the contact patch

itself and it is, therefore, impossible to position a probe close to the crack mouth.

Beach marking

As cracks grow it is often observed that, from time to time, a line delineating the shape of the crack front is laid down. The precise reason for the existence of such marks is imperfectly understood, but they are thought to occur when the crack front advances rapidly during a small number of crack cycles. In some materials the effect can be promoted by changing the frequency of the applied loading; therefore, by changing the frequency at known points during the life of the test beach marks can be laid down at known numbers of cycles. Unfortunately the process is not completely reliable, but is useful for certain materials.

Heat tinting

Some materials such as titanium alloys respond well to the heat-tinting process. In this the specimen is removed from the test fixture at controlled intervals and placed in an oven. This produces a permanent tint on the exposed surface of the crack. By decreasing the temperature used for successive bakes a set of bands can be developed showing the crack front at any particular time. The principal objection to using this technique is that the heat treatment will relax out many of the residual stresses present, whether on the scale of the crack-tip plasticity zone, or more widespread residual stresses, and hence the fatigue performance may not be representative of an untreated component. In fretting fatigue it is desirable to avoid disturbing the contact during such a process and the heating may well need to be applied in situ. Even if this is done, the increased temperature may well cause surface modification and a consequent change in the coefficient of friction.

Hence, although spherical contacts are preferable from a contact mechanics standpoint, there are real problems associated with the measurement of any cracks generated. These difficulties restrict the usefulness of such geometries in investigating crack growth rates in fretting fatigue.

Analysis of cracks

It was stated at the outset that fretting fatigue research should concern itself with surface effects, as that is the region where differences from plain fatigue are likely to be most pronounced. The initiation phase is something that is difficult to describe analytically using continuum mechanics, although a theory is gradually being developed for the development of persistent slip bands and their evolution into true embryo cracks **(18)**. It is, however, likely to be some time before the gestation period of a crack can be predicted, and even then many imponderables are likely to make the predictions imprecise. Of course we might look for initiation criteria experimentally, using crude correlations of probable controlling physical quantities, such as the expenditure of frictional energy (19), possibly combined with the influence of in-plane tension (20). But these hypothesis will do no more than to enable the results of a set of tests carried out using one geometry to be carried over to another using the same materials, and with the same surface texture and cleanliness.

A more fruitful area for the analysis of fretting fatigue data might be in the description of embryo cracks which, though fully formed, are still no bigger than the smallest characteristic contact dimension, and hence are still propelled significantly by the contact stress field itself. At this stage the crack is almost certainly experiencing a decreasing stress field as it grows away from the contact, but the stress intensity factor itself may be increasing as the crack length increases; these two effects clearly act in opposite directions. If the crack tip stress intensity actually falls below the threshold level for finite crack growth it is certainly possible for self arrest to occur. It has been shown that if evaluation of the crack tip stress intensity factor is carried out on the scale of the bulk contact it is very probable that its value will increase monotonically (21). On the other hand, if such a calculation is performed on the scale of an *asperity* it is quite possible for the self arrest phenomenon to occur, although some caution is required in ensuring that the absolute size of the crack is such that it can be considered to be fully developed, existing in a continuum, and hence controlled by the stress intensity factor.

We now turn to questions of evaluating the stress intensity factor itself for cracks propelled by a fretting contact. As with the analysis of contacts, a decision has to be made whether the model used should be two- or three-dimensional. The former has the advantage that great flexibility can be built into the solution fairly readily, and the influence of crack closure, kinking and even the very steep stress gradients found near the contact may easily be incorporated, providing a suitable method of analysis is chosen. The method of distributed singularities or dislocations, pioneered at Sheffield and Cambridge in the early 1960s in the context of crack tip plasticity was taken up by Comninou and others a decade later to analyse crack tip closure in layered bodies (22)(23). This process of distributing dislocations along the line of the crack to render it traction-free, results in a singular integral equation which may be inverted very efficiently numerically, and it copes extremely well with all of the requirements described above. It is equally well suited to the analysis of cracks in a tensile environment (24), and its computational efficiency would seem to render approximate methods unnecessary.

In reality all cracks are, to a degree at least, three-dimensional in character. When the crack front is curved, using a plane (two-dimensional) model for the behaviour near the centre of the crack is valid only if the curvature of the crack front is slight, so that the stress gradients along the crack front are small. In cases where this does not hold, the use of an Eigenstrain method, described in an accompanying paper (25), has much to recommend it. It may be viewed as a three-dimensional analogue of the dislocation method, although it does differ fundamentally insofar as the domain of the crack is split up into a finite number of regions, within which the relative crack face displacements are modelled in a prescribed way. The principal advantages of the Eigenstrain method over finite element techniques is that cracks at (or near) the free surface may be modelled much more efficiently: the presence of the semi-infinite region and the free surface are built into the model, and so only the crack faces need to be cleared of traction, and whatever loading conditions are to be imposed satisfied. This produces a very much smaller matrix of equations needed to solve a particular problem to a given accuracy.

Conclusions

In this paper many aspects of the fretting fatigue test and its analysis have been reviewed. In order to gain a greater understanding of the fretting process it is essential to devise tests in which there is a well-defined state of contact stress. This in turn strongly suggests that incomplete (convex) geometries should be preferred to flat-ended contacts. Further, the most precise way in which relative displacement can be controlled within the contact is to operate the apparatus in the partial slip regime so that the stick zone provides a firm reference point and the compliance of the apparatus is unimportant.

Two- and three-dimensional geometries can both be employed. The former is easy to analyse but difficult to realize in practice. The latter is now becoming amenable to analysis with advances in techniques for dealing with three-dimensional cracks but further work on the measurement of crack dimensions in situ is required before the potential advantages of this geometry can be fully exploited.

References

(1) KHADEM, R. and O'CONNOR, J. J. (1969) Adhesive or frictionless compression of an elastic rectangle between two identical elastic half-spaces, *Int. J. Engng Sci.*, 7, 153–168.
(2) SATO, K. (1992) Determination and control of contact pressure distribution in fretting fatigue, *Standardization of fretting fatigue test methods and equipment, ASTM STP* 1159, (Edited by M. H. Attia and R. B. Waterhouse), ASTM, Philadelphia, pp. 88–100.
(3) MINDLIN, R. D. (1949) Compliance of elastic bodies in contact, *J. appl. Mech.*, 16, 259–268.
(4) CATTANEO, C. (1938) Sul Contatto di due corpi elastici: distribuzione locale degli sforzi, *Atti della Reale Accademia Nazionale Dei Lincei*, 27, 342–348, 434–436, 474–478.
(5) HILLS, D. A., NOWELL, D., and SACKFIELD, A. (1988) Surface fatigue considerations in fretting, *Interface dynamics*, (Edited by D. Dowson, C. M. Taylor, M. Godet, and D. Berthe), Elsevier, Amsterdam.
(6) JOHNSON, K. L. (1955) Surface interaction between elastically loaded bodies under tangential forces, *Proc. Roy. Soc. Lond., Ser A*, 230, 531–548.
(7) NOWELL, D. and HILLS, D. A. (1990) Crack initiation criteria in fretting fatigue, *Wear*, 136, 329–343.
(8) HILLS, D. A., NOWELL, D., and SACKFIELD, A. (1993) *Mechanics of elastic contacts*, Butterworth-Heinemann, Oxford.
(9) NOWELL, D., HILLS, D. A., and SACKFIELD, A. (1988) Contact of dissimilar elastic cylinders under normal and tangential loading, *J. Mech. Phys. Solids*, 36, 29–75.
(10) NOWELL, D. and HILLS, D. A. (1987) Mechanics of fretting fatigue tests, *Int. J. Mech. Sci.*, 29, 355–365.

(11) NOWELL, D. and HILLS, D. A. (1988) Contact problems incorporating layers, *Int. J. Solids Structures*, **24**, 105–115.
(12) NAYAK, L. and JOHNSON, K. L. (1979) Pressure between elastic bodies having a slender area of contact and arbitrary profiles, *Int. J. Mech. Sci.*, **21**, 237–247.
(13) JOHNSON, K. L. (1985) *Contact mechanics*, Cambridge University Press.
(14) HAMILTON, G. E. and GOODMAN, L. E. (1966) The stress-field created by a circular sliding contact, *J. appl. Mech.*, **88**, 371–376.
(15) SACKFIELD, A. and HILLS, D. A. (1983) A note on the Hertz contact problem – a correlation of standard formulae, *J. Strain Analysis*, **18**, 195–197.
(16) MUNISAMY, R. L., HILLS, D. A., and NOWELL, D. (1992) Contact of similar and dissimilar elastic spheres under tangential loading. *Proceedings of Contact Mechanics International Symposium*, Presses Polytechniques et Universitaires Romandes, Lausanne, pp. 447–461.
(17) GOODMAN, L. E. (1962) Contact stress analysis of normally loaded rough spheres, *J. appl. Mech.*, **29**, 515–522.
(18) MURA, T. and NAKASONE, Y. (1990) A theory of fatigue crack initiation, *J. appl. Mech.*, **57**, 1–6.
(19) RUIZ, C., BODDINGTON, P. H. B., and CHEN, K. C. (1984) An investigation of fatigue and fretting in a dovetail joint, *Expl. Mech.*, **24**, 208–217.
(20) KUNO, M., WATERHOUSE, R. B., NOWELL, D., and HILLS, D. A. (1989) Initiation and growth of fretting fatigue cracks in the partial slip regime, *Fatigue Fracture Engng Mater. Structures*, **12**, 387–398.
(21) MILLER, G. R., KEER, L. M., and CHENG, H. S. (1985) On the mechanics of fatigue crack growth due to contact loading, *Proc. Roy. Soc., Lond., Ser. A*, **397**, 197–209.
(22) COMNINOU, M. and BARBER, J. R. (1983) Frictional slip between a layer and a substrate due to a periodic tangential surface force, *Int. J. Solids Structures*, **19**, 533–539.
(23) SCHMUESER, D., COMNINOU, M., and DUNDURS, J. (1980) Separation and slip between a layer and a substrate caused by a tensile load, *Int. J. Engng Sci.*, **18**, 1149–1155.
(24) NOWELL, D. and HILLS, D. A. (1987) Open cracks at or near free edges, *J. Strain Analysis*, **22**, 355–365.
(25) DAI, D. N., HILLS, D. A., and NOWELL, D. (1994) Stress intensity factors for three-dimensional fretting fatigue cracks, *Fretting Fatigue*, Mechanical Engineering Publications, London, pp. 59–71. *This volume.*

U. S. Fernando*, G. H. Farrahi*, and M. W. Brown*

Fretting Fatigue Crack Growth Behaviour of BS L65 4 percent Copper Aluminium Alloy under Constant Normal Load

REFERENCE Fernando, U. S., Farrahi, G. H., and Brown, M. W., **Fretting fatigue crack growth behaviour of BS L65 4 percent copper aluminium alloy under constant normal load,** *Fretting Fatigue*, ESIS 18 (Edited by R. B. Waterhouse and T. C. Lindley) 1994, Mechanical Engineering Publications, London, pp. 183–195.

ABSTRACT Fretting fatigue crack growth behaviour of BS L65 copper aluminium alloy is investigated under various combinations of axial cyclic loading and constant normal load. In this paper the observed friction behaviour, crack propagation response, and fretting fatigue life results are presented. Crack extension is monitored using a DC potential drop technique, and it is shown that cracks can be detected as early as 10–20 percent of the fretting fatigue life, which clearly indicates that the fretting fracture process is dominated by progressive stable growth of cracks.

The friction results indicate that the stable friction amplitude for fully developed macroslip is dependent on the applied normal load. Higher coefficients of friction are found at lower normal load, and the value of coefficient of friction is found to decrease as the normal load is increased. The influence of normal load and the friction force on the crack growth rate and the fretting fatigue life are discussed. The early stage of propagation is significantly faster than the speed of equivalent mode I cracks.

Introduction

Fretting is caused by the microscopic rubbing of a pair of surfaces of two components that are in contact and pressed together. Under fretting the severe stress concentration due to contact forces, together with the surface degradation caused by rubbing, leads to rapid nucleation of microcracks. In situations where one or both of the components are subjected to cyclic loading of sufficient intensity, the microscopic surface cracks generated by fretting can then grow to cause catastrophic fatigue failure. Fretting-initiated fatigue failure is frequently observed in a wide range of engineering assemblies (**1**); fir-tree roots of turbine blades and their mating slots, collars of discs shrunk on to rotating shafts, keyways of transmission assemblies, and riveted or clamped joints are common examples.

Unlike fatigue failure of plain specimens at low stress levels, in fretting fatigue the presence of an intense cyclic surface friction force virtually eliminates the crack initiation period and considerably accelerates the early phase of microstructural-dependent crack growth. Thus the major part of fretting fatigue lifetime is considered to be spent on the propagation of fatigue cracks of the order of one grain size or more in length (**2**)(**3**). The contact loadings, i.e., contact pressure and the friction force, their distribution over the contact area, and the stress gradients at the contact zone (which are distinct from the

* SIRIUS, Faculty of Engineering, University of Sheffield, PO Box 600, Mappin Street, Sheffield S1 3JD, UK.

cyclic body stresses created by the external applied forces) have a significant influence on the fretting crack growth process. It follows that a study of the growth of fretting cracks and an understanding of the influence of the above factors in crack growth is essential for an accurate fatigue evaluation of close-fitting assemblies.

Although it is a widely recognized fact that the fracture process in fretting fatigue is dominated by crack growth, to date only a few studies have been performed to investigate crack growth behaviour in a fretting environment (3)–(5). A limited amount of work has been done to identify the influence of normal pressure and friction force on the fretting crack growth. In this paper an attempt is made to partly fill the gap in our understanding in this area.

Experimental procedure

A new servo-hydraulic testing facility was developed to perform fretting fatigue experiments. The loading arrangement is schematically illustrated in Fig. 1. The facility was primarily developed for the purpose of performing variable normal load-fretting tests and is described in detail elsewhere (6).

The specimen, as shown in Fig. 2(a) was of rectangular cross-section, providing a wide area for unrestricted crack growth. The normal load was applied via two symmetrically placed bridge pads, each with two feet, providing flat contacts to create a fretting action. The geometry of the pad is given in Fig. 2(b). The position of the applied loads on the fretting pads and their stiffness were selected to minimize bending of the pad under normal load and friction

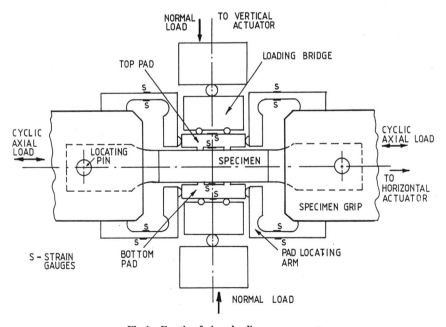

Fig 1   Fretting fatigue loading arrangement

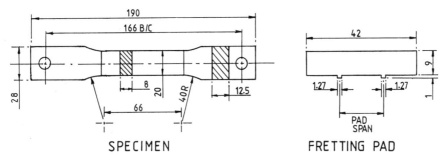

Fig 2   Specimen and fretting pad (all dimensions in mm)

force, respectively. This was considered necessary in the case of flat contact since the biting effect caused by bending of the pad can significantly influence the distribution of contact stress.

The friction forces on the contact area were measured by strain gauging the pads. Each pad was kept in position, relative to the specimen, by using two symmetrically placed locating arms that were attached to the grips. The arms were strain gauged and the deflection of the arms was used to measure the slip associated with each contact point. Both axial load of the specimen and the normal load of the pad were applied via servo controlled actuators. In addition, some tests were performed using a proving ring to create normal load. The fretting crack growth was monitored using a direct current potential drop (PD) technique. A personal computer based data acquisition system was developed and used to record axial load and friction hysteresis loops and PD measurements during tests.

The material investigated was a fully artificially-aged 4 percent copper aluminium alloy, a general purpose material widely used in aerospace/aircraft structures and components. The fretting pads were made of BS S98 steel. The material combination has been extensively used in fretting experiments (2)(7). The specimen and the pad sections of the current investigation were selected such that the slip conditions under a given loading were the same as those of previous investigations, so that the results could be compared. In all tests, fully reversed cyclic axial load was applied with a sinusoidal waveform of 15 Hz frequency.

Three pad spans, namely 6.35 mm, 16.5 mm, and 34.35 mm, were investigated, together with three axial load amplitudes, 70 MPa, 100 MPa, and 125 MPa. Tests were performed with various values of normal pressure covering the range 20 MPa–120 MPa. These values were chosen to include both micro- and macroslip regimes.

Crack measurement

In the fretting test arrangement used, four contact locations were involved in each specimen tested. Since all these locations were equally favoured for crack generation, the possibility of measuring four growing cracks in each test was envisaged. Several techniques were considered for measuring cracks. Optical measurement using a travelling microscope was considered simple and inexpensive, although it could not give sufficient sensitivity and accuracy when monitoring all four locations. The acetate replica method could not be used as it would involve disturbing the test and the contact between the pad and the specimen. Both these methods require regular operator involvement and could not be automated. Moreover, they would have provided only a superficial aspect of the crack development, which was considered to be unrepresentative of the behaviour in the interior. The fact that fretting cracks almost always nucleate at or near the leading edge of the contact patch enabled the use of the electric potential method for crack measurement. This not only permitted the automation in the crack measurements, but it also allowed several crack measurements to be conducted at different locations without much difficulty. However, a development of an accurate crack length calibration from PD measurements was required. Since there is no added advantage in using an alternating current method in non-magnetic aluminium alloy, in the present study a pulsed direct current method was used.

Crack measurement was performed using a Howden direct current potential drop crack monitor. Pulsed current with a stable peak value of 20 Amps was passed through the specimen, and potential differentials were measured from leads spot welded on to front and back faces of the specimen. At each crack location two measurements, a direct reading and a reference reading, were taken from each face, giving a total of eight pairs of readings. The leads were placed as shown schematically in Fig. 3. The leads for direct readings were placed 1 mm from the contact surface and 4 mm apart, and the reference leads were placed 2 mm from the contact surface and 20 mm apart. Both pairs of leads were placed equidistant from the leading edge of the contact patch. The potential readings were taken at regular intervals during the test.

The crack shapes observed in fretting tests were found to vary considerably. Broadly these could be categorized into three groups, namely semi-elliptical

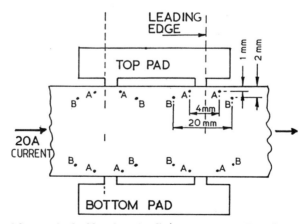

Fig 3   Potential measuring lead locations. A – direct measurement; B – reference measurement

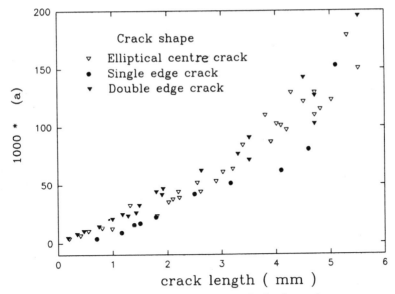

Fig 4   **Potential drop calibration curve for fretting cracks**

centre cracks, single edge cracks, and double edge cracks. The crack length calibration from PD measurements was obtained experimentally for these three crack shapes, Fig. 4. The way in which the calibration was conducted is as follows. Fretting cracks were initiated in specimens and these cracks were then allowed to grow under cyclic bending. The PD measurements were taken at regular intervals and the crack length was also measured using an optical microscope, and by marking the fracture surface through application of periodic overloads. The crack length is evaluated using a parameter $\chi(a)$ defined as

$$\chi(a) = \left\{\frac{\Phi_A}{\Phi_B}\right\} - \left\{\frac{\Phi_A}{\Phi_B}\right\}_0 \tag{1}$$

where a is the crack length, Φ_A and Φ_B are direct and reference readings of the potential differentials. The subscript 0 denotes the value of the potential ratio at the beginning of the test when no crack was present. The form of $\chi(a)$ was chosen in such a way as to cancel the error due to any change in current, and to minimize the error in calibration for short cracks due to slight mis-positioning of welding leads. The recorded data for potential differentials Φ_A and Φ_B together with the crack calibration curves shown in Fig. 4 were used to evaluate crack length in the fretting tests.

Friction results

The evolution of the friction force during tests indicated that a rapid increase in friction amplitude occurs in the early part of the test. In the tests where sufficient slip was encountered, the friction force versus axial load hysteresis loops showed a distinct microslip and macroslip behaviour at this early stage. The highest friction amplitude was observed at approximately 5–10 percent of the lifetime when the fretting debris began to appear at the contact interface. During subsequent loading the friction amplitude was found to be more or less constant, and the micro/macroslip behaviour became less apparent in the hysteresis loops. Towards the end of the test, a considerable reduction in friction amplitude was observed, which may be related to the growth of fretting cracks and the alteration in slip characteristics. Approximately the same friction behaviour was observed for both top and bottom pads.

The dependence of the friction force amplitudes on the applied normal load (N) for various values of axial load and for three pad spans is shown in Fig. 5. It can be seen that for a given value of normal load, the friction amplitude is significantly affected by the axial load and the pad span. For a given pad span and axial load, a higher coefficient of friction, μ, is observed at lower normal loads where possibility of macro-slip behaviour is reduced. The friction coefficient falls gradually as the normal load is increased. The results indicate that there is an upper bound value of coefficient of friction μ^* that can be achieved under a given normal load. This limiting value is highest at low normal loads, being approximately equal to 2 for the material combination used.

The value of μ^* was considerably reduced as the normal load was increased. The friction force amplitude was found to be related to the amplitude of slip. It was apparent that the coefficient of friction approached the value of μ^* only

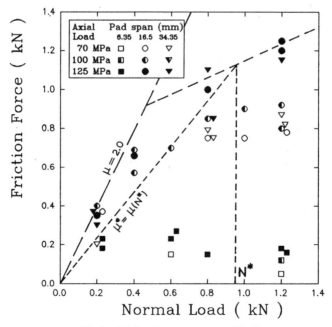

Fig 5　Friction force versus normal load

when a fully-developed macroslip condition was achieved. Lower values of friction were achieved for underdeveloped macroslip conditions. In general, higher slip amplitude is expected for longer pad spans and higher axial loads, while normal load and friction have a secondary influence. The influence of slip was clearly visible from the reduction in the friction amplitude at lower axial load amplitudes and at smaller pad spans. In situations where micro-slip prevails, an increase in the normal load reduces the amount of micro-slip: the influence of friction can be clearly seen from the results from the 6.35 mm pad span.

The observed friction behaviour can be explained by considering the involvement of surface asperities and intersurface debris in the creation of friction force in fretting. The friction force is considered to be created by either direct interlocking of surface asperities, or by trapping oxide debris in between the surface asperities. At the beginning of the test the surfaces were polished and the friction was created by direct interlocking of the asperities. This was apparent by clear micro/macro behaviour in the hysteresis loop. At low normal loads, the value of the friction force was insufficient to break the surface asperities and, therefore, lead to higher μ. At higher normal loads sufficient friction was generated to plastically deform and break some of the surface asperities to create oxide debris particles which in turn reduce locking-up effects. It is postulated that μ^* is primarily dependent on the number of the surface asperities and the strength of the interface debris particles.

In general the mean value of the friction force in the tests was found to be nearly zero. However, in some tests it was noticed that a mean friction force was induced during the test assembly. In tests with higher slip amplitude the value of the mean friction gradually diminished during the initial phase of the tests; in tests that involved very small slip amplitude, however, some mean friction was observed to persist until the end of the test. In such tests the peak value of the friction force was found to be considerably higher than the amplitude of friction.

Crack growth results

The PD measurements recorded during the progress of tests clearly indicated that the growth of fretting fatigue cracks could sometimes be detected as early as 10–20 percent of the fatigue life. Typical results of crack length measurements are presented in Fig. 6, in terms of number of cycles, n, normalized by fatigue life, N_f. Due to extremely low sensitivity of the measuring technique at crack lengths below 0.1 mm, accurate detection of small cracks was difficult, as the change in potential signals for such cracks is of the same order of the scatter observed in the PD readings. It was also possible to detect the exact point of crack nucleation and make confident measurement of a very small crack. Furthermore it was realized that the smallest crack length that could be confidently measured by the present technique was limited to approximately 80–100 μm.

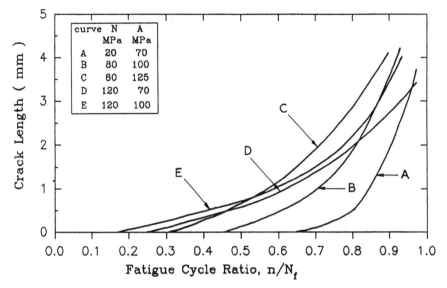

Fig 6   Typical results of crack length versus fatigue cycle ratio

The crack growth results reflect the significant influence of friction force and normal load on the crack growth rate. Typical results are presented in terms of crack growth rate (da/dn) versus crack length (a), in Figs. 7–10. For the case of a 16.5 mm pad span the growth rates for three axial stress amplitudes are compared, for three different values of normal pressure, 20 MPa, 80 MPa, and 120 MPa, in Figs. 7(a) to 7(c), respectively. The growth curves for 100 MPa axial stress and for various values of normal stress are given in Fig. 8. Some results for a 34.55 mm pad span are shown in Fig. 9. The linear elastic fracture mechanics (LEFM) prediction of the crack growth curve for the material, corresponding to the respective values of axial load, is also shown in the figures. It can be seen that the measured growth curves approach the value of LEFM predictions at larger crack lengths, approximately above 4 mm. No account was taken for closure of these short cracks in compression, when making LEFM prediction of propagation.

In every case, significant increases in growth rate are evident at smaller crack lengths in fretting; however, the magnitude of friction force must also be taken into account when interpreting crack growth results. Comparing Fig. 8 with the friction behaviour shown in Fig. 5, it is clear that the friction force may have a marked influence on the crack growth rate in the early stage of crack growth. Acceleration of short crack growth rates is clearly evident in the tests where the axial stress is 125 MPa and 100 MPa, Fig. 7, in particular with 80 MPa and 120 MPa normal pressure. The acceleration is less marked in the tests with 20 MPa normal pressure, Fig. 7(a). The friction forces measured in the latter tests are much less compared to those observed in other tests. It can

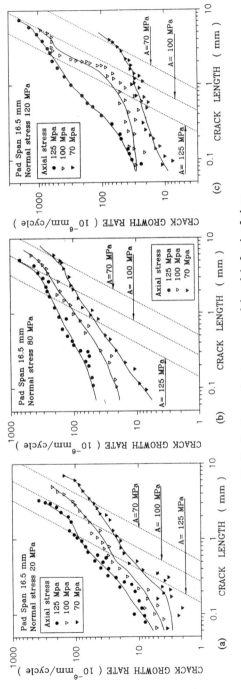

Fig 7 Crack growth rate versus crack length in fretting fatigue

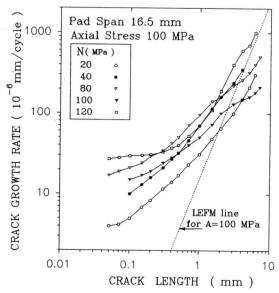

Fig 8   The influence of normal stress on fretting fatigue crack growth rate

also be seen that the influence of friction diminishes as the crack length is increased.

The influence of the normal load on da/dn is clearly evident from Fig. 7(c) and Fig. 8. A considerable retardation in growth rate is observed under 120

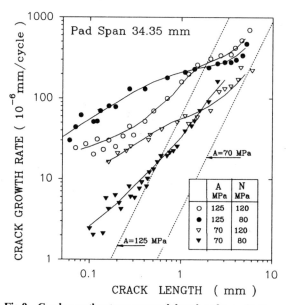

Fig 9   Crack growth rate versus crack length pad span 34.35 mm

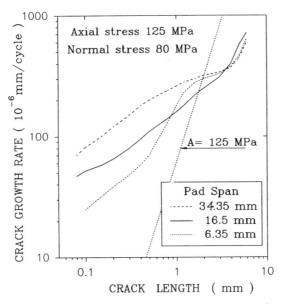

Fig 10 **The influence of pad span on fretting fatigue crack growth rate**

MPa normal pressure, in particular for 70 MPa and 100 MPa axial stress amplitudes. This retardation effect is less apparent at 125 MPa axial load. It can also be seen that the influence of the normal load is more pronounced at medium crack lengths, indicating that the retardation effect is probably caused by crack closure due to high compressive normal load.

The da/dn results for three pad spans are compared in Fig. 10, for the same values of axial stress and normal pressure. A higher growth rate is observed for a larger span; however, as the friction force for a larger span is much higher than that of the smaller span, it is not clear whether there is any direct influence of pad span on the crack growth process.

Fretting fatigue life

Fatigue life results are presented in Figs 11 and 12 for all tests. For the case of 16.5 mm 34.35 mm pad spans, Fig. 11 clearly indicates the considerable reduction in fatigue strength due to fretting. The clear influence of normal load is apparent from Fig. 12. It should be noted that, as in the case of crack growth rates, the magnitude of friction force must be taken into account when interpreting fatigue life data. Similar fatigue life behaviour is observed under both pad spans. It is thought that for a given value of coefficient of friction, the fatigue life is reduced to an increase in the contact pressure up to a critical value of the normal load. Above this normal load further increase in the normal pressure tends to increase fatigue life.

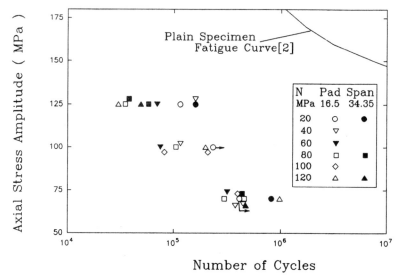

Fig 11   Axial stress amplitude versus fretting fatigue life

As clearly indicated in Fig. 12, two dominant regions of fretting behaviour can be identified, depending upon the magnitude of the normal load. In region A an increase of the normal load tends to decrease the life while in region B the life is enhanced by increasing the normal load. Therefore, a critical value of normal load exists where the fretting fatigue life attains a minimum value. It is postulated that the retardation effect produced by crack closure due to high

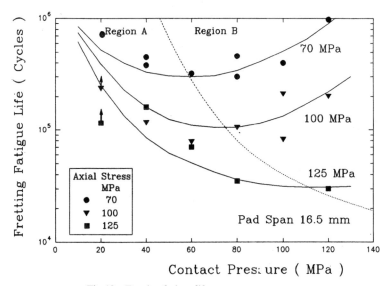

Fig 12   Fretting fatigue life versus contact pressure

compressive normal load, as observed in the crack growth measurements, is the reason for this increase in the life in region B.

Conclusion

A systematic investigation of the effects of the three controlled variables, normal load, axial load amplitude, and pad span, permitted a study of the evolution of friction and the propagation of fretting cracks. It was found that fretting cracks of the order of grain size could be detected at a very early stage of loading, indicating that the fracture process is dominated by crack growth, and the microstructural-dependent crack growth phase is comparatively small in fretting. It was also shown that friction force and normal load have a significant influence over crack growth. The influence of the former is at the very early stages of the growth while the retarding influence of the latter is more pronounced at medium crack length. Both fatigue life and crack growth data seemed to indicate that crack closure due to high compressive normal load increases fatigue life. It can also be concluded that the friction force in fretting is governed by the number and size of surface asperities and the strength of the interface debris particles.

Acknowledgements

The authors gratefully acknowledge MOD (RAE, Farnborough) and the Science and Engineering Research Council for funding this investigation. Special thanks are due to R. Cook, Dr D. Rayaprolu and Dr R. B. Waterhouse for their help, advice, and valuable discussions throughout the work. The authors would like to thank the Director of SIRIUS for providing the test facilities, and also the University of Sheffield for providing financial support (USF).

References

(1) FORSYTH, P. J. E. (1981) Occurrence of fretting fatigue failure in practice, *Fretting Fatigue*, (Edited by R. B. Waterhouse), Applied Science, London.
(2) EDWARDS, P. R. and RYMAN, R. J. (1975) Studies in fretting fatigue under variable amplitude loading conditions, RAE Technical Report No. 75132.
(3) ENDO, K. (1981) Practical observation of initiation and propagation of fretting fatigue cracks, *Fretting Fatigue*, (Edited by R. B. Waterhouse), Applied Science, London.
(4) TAKEUCHI, M., WATERHOUSE, R. B., MUTOH, Y., and SATOH, T. (1991) The behaviour of fatigue crack growth in the fretting-corrosion-fatigue of high tensile roping steel in air and seawater, *Fatigue Fracture Engng Mater. Structures*, **14**, 69–77.
(5) SATO, K., FUJII, H., and KODAMA, S. (1986) Crack propagation behaviour in fretting fatigue of S45C steel, *Bull. JSME*, **29**, 3253–3258.
(6) FERNANDO, U. S., BROWN, M. W., MILLER, K. J., COOK, R., and RAYAPROLU, D. (1994) Fretting fatigue behaviour of BS L65 aluminium alloy under variable normal load, *Fretting Fatigue*, Mechanical Engineering Publications, London, pp. 197–209. *This volume.*
(7) EDWARDS, P. R. and COOK, R. (1978) Friction force measurements on fretted specimens under constant normal loading, RAE Technical Report 78019.

U. S. Fernando*, *M. W. Brown**, *K. J. Miller**, *R. Cook*†, *and*
D. Rayaprolu†

Fretting Fatigue Behaviour of BS L65
4 percent Copper Aluminium Alloy under
Variable Normal Load

REFERENCE Fernando, U. S., Brown, M. W., Miller, K. J., Cook, R., and Rayaprolu, D., **Fretting fatigue behaviour of BS L65 4 percent copper aluminium alloy under variable normal load,** *Fretting Fatigue*, ESIS 18 (Edited by R. B. Waterhouse and T. C. Lindley) 1994, Mechanical Engineering Publications, London, pp. 197–209.

ABSTRACT A new fretting fatigue testing facility is developed which enables control of the normal load on the contact area. The normal load can be varied in-phase or out-of-phase with the specimen axial load, and also with any desired waveform, thus allowing simulation of a wide range of fretting load patterns. The facility is briefly described and results are presented for a variable normal load test programme on BS L65 4 percent copper aluminium alloy.

Tests are performed with two variable normal loading patterns, which are representative of loads typically experienced by industrial fasteners such as bolted joints and lugs. Friction behaviour and fatigue life for the two cases of loading are compared with those from constant normal load tests. It is shown that the friction force response in fretting can be markedly affected due to the normal load waveform. In-phase normal loading patterns used are found to reduce peak values of friction at higher normal load. However, the influence of the high normal load on fatigue life is less decisive, due to the dual influence of normal load on fretting crack growth. For medium values of normal load, the variable normal load waveforms are found to enhance the resistance of the material in fretting fatigue.

The results of tests with changing phase between the axial and normal load indicate that the friction response is significantly affected by the phase of loading. The versatility of the testing facility is highlighted in relation to its ability to control crack driving forces in fretting.

Introduction

Fretting occurs in situations where two components are in constant contact and a small oscillatory movement exists between them. Often these oscillating movements are caused by the cyclic loads exerted in either one or both of the components. The fretting cracks that nucleate due to stress intensification at the contact zone will subsequently grow under this cyclic loading action, resulting in fretting fatigue failure. In component assemblies where fretting fatigue failure is likely to occur **(1)(2)** such as riveted bolted joints, lug assemblies, press-fit joints, pinned connections, and blade fir-tree root fastenings in turbine discs, the contact geometry is distorted by the application of external load, and the contact stress at the interface is, therefore, a variable that depends on the bulk body stress. So it is important to understand fretting behaviour where the normal load varies with bulk body stress.

Multiaxial loading often exists in complex structures such as aircraft wings. For fretting in such situations, normal loading patterns may follow waveforms that are in-phase or out-of-phase with the bulk body stresses. The latter are

* SIRIUS, Faculty of Engineering, University of Sheffield, PO Box 600, Mappin Street, Sheffield, S1 3JD, UK.
† Materials and Structures Division, DRA (RAE), Farnborough, Hants, UK.

more likely where multiaxial applied stresses arise from different load sources, which are out-of-phase with each other. In order that the structural integrity of such components can be properly assessed, it is important that the fretting fatigue behaviour should be investigated by experiments which extend beyond constant normal load conditions, to those typical of more general loading encountered in service. Moreover, an ability to control the normal load in fretting tests is expected to help our understanding of the mechanisms of the fretting fatigue phenomenon by identifying and isolating the critical factors that cause fretting damage. The work presented here is primarily aimed at fulfilling these objectives.

There are many factors (3) affecting fretting fatigue damage processes, of which the influence of mechanical variables such as bulk stress, contact pressure, amplitude of slip, frequency of loading, loading history, nature of the surface, and residual stress are of significant importance from an engineering view point. The ability to control these variables in a desired fashion is an important requirement of any proper experimental set-up.

In conventional fretting test arrangements a specimen of suitable shape is subjected to cyclic axial load, and fretting action is provided by some form of fretting pad. Bridge pads consisting of either one or two contacting feet are commonly used. The amount of slip is varied by suitably changing the length of the bridge (pad span) and the amplitude of axial load. In conventional machines a pair of fretting pads are used which are clamped on to the specimen test area by using a proving ring or a similar arrangement, which allows adjustment of the clamping load. Such an arrangement, however, can provide only a static or periodically changed contact load and will not permit the application of a dynamic normal load on the contact area that can be varied within a loading cycle. From the literature it was noticed that none of the available test facilities were capable of fulfilling the latter requirement, and a new facility had to be designed and developed to fulfil these objectives.

In the new facility the normal load can be varied in-phase or out-of-phase with the specimen axial load, with any desired waveform, allowing simulation of a wide range of fretting load patterns. The facility is briefly described in the paper. In addition, results of a variable normal load test programme on 4 percent copper aluminium alloy (BS L65) specimens with steel fretting pads are presented. The tests were performed with two types of variable normal load patterns, which were representative of loads typically experienced by bolted joints and lugs. Friction behaviour and fatigue life for the two loading patterns are compared with those from static normal load tests. The influence of the normal load waveform and the effect of changing phase between the axial and normal load on the friction response are discussed.

Fretting fatigue testing facility

The primary objective of the present investigation was to conduct fretting fatigue tests that enables study of the specific influence of a varying normal load on both the induced friction response and the fretting damage. As none of the available test methods provide the means for imposing dynamic normal load on the fretting pad, a new facility was developed to satisfy this requirement.

The loading frame of the new facility is a Mayes biaxial machine which provides both the normal load on the fretting pad and the specimen axial load. The machine has a rigid vertical frame with four servo-controlled hydraulic actuators; two on the vertical axis and two on the horizontal axis. Both actuators on a given axis are controlled by a single servo valve so that symmetrical loading can be applied. The loading arrangement is shown in detail in Fig. 1. The specimen is fixed between two horizontal actuators using identical grips. The grips are permanently attached to the actuators so that the specimen can be placed parallel to the actuator axis. Alignment of the specimen is obtained by using two pins which locate the specimen on each grip. The specimen is clamped using four screws at each end, which are sufficiently preloaded to create the necessary friction to apply a maximum cyclic load of 80 kN. The grips have been specifically designed to eliminate all backlash during cyclic loading. Guide plates and displacement limiters are set on each horizontal actuator so that the specimens always stay in the vertical plane and the centre of the specimen cannot move more than 0.5 mm away from the vertical actuator axis. The specimen is electrically insulated from the machine actuators and the frame using insulating pieces inserted at appropriate places.

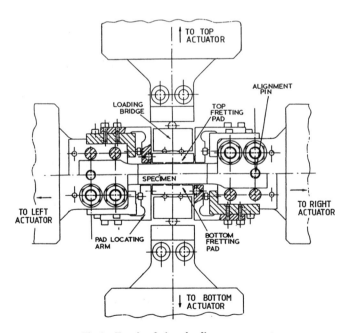

Fig 1   Fretting fatigue loading arrangement

Each fretting pad is positioned symmetrically on the specimen using two locating arms. The locating arms are fastened to the specimen grips and can be adjusted to create the required compression and to maintain the proper contact between the arms and the pad at all times during the loading cycle. The signals from the strain gauges bonded to the locating arms are used to measure the slip at each pad contact. The normal load is derived from the vertical actuators. The grips of these actuators are identical to those of the horizontal axis and carry two flat blocks. The flat block compresses the respective fretting pad via a steel ball and a loading pad. Two rollers placed between the fretting pad and the loading pads allow a symmetrical distribution of the vertical load on to each pad contact. The rollers can be located according to the pad span so that they are exactly aligned to the pad contacts, thus avoiding any bending of the fretting bridge due to the normal load.

The basic dimensions of the specimen and the fretting bridge are given in Fig. 2. The specimen consists of a reduced test area with blended fillets at each end. Two reamed holes are provided at each end for locating the specimen on the machine. Specimens were manufactured from flat bar stock, and during manufacture care was taken to align the holes to the test area with a high degree of accuracy. Before the test the surface of the test area was polished with emery paper and cleaned with acetone.

The pads, made of S98 steel, consisted of two contact feet each 1.27 mm in width and 1.00 mm in height. The pad span was selected to obtain the desired slip amplitude. The bridge section was designed to have sufficient rigidity so as to minimize bending due to friction force. The signals from the strain gauges bonded to the fretting bridge were used to measure friction force. Each bridge was individually calibrated for friction force sensitivity. The same fretting pad was used for several tests, and the pad contact area was carefully polished with emery paper and cleaned with acetone to remove fretting debris before it was reused. The contact surfaces of the pads were periodically (after approximately five tests) ground to maintain the proper contact.

The load on each axis was measured from the Mayes control system. A six-channel strain gauge bridge amplifier was used to measure the friction loads on the two fretting pads and the displacement signals of the four locating arms. The growth of fretting cracks was measured using a Howden direct

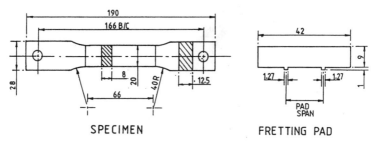

SPECIMEN   FRETTING PAD

Fig 2 Specimen and fretting pad (all dimensions in mm)

current potential drop crack monitor. The details of the crack measurement technique are given elsewhere **(4)**. A personal computer based digital data acquisition system was used to record all relevant information during the test. This was comprised of periodic recording of variables, i.e, axial load, normal load, top and bottom friction force and displacement signals for a complete loading cycle, and periodic recording of several potential readings that are necessary to evaluate crack growth from each contact location. A digital waveform generator combined with a digital-to-analogue converter unit was incorporated to generate the desired command signal waveforms for both the axial and the normal load control. The command signals were appropriately modified electronically to obtain maximum control accuracy during loading cycle. A description of the command control and data acquisition system is given elsewhere **(5)**.

Variable normal load test programme

In the current investigation, the fretting tests were performed using a constant amplitude fully reversed axial load, with a sinusoidal waveform of 15 Hz frequency. Three pad spans were investigated, but only the results for the 16.5 mm pad span are presented in this paper. For normal load, two waveforms were considered namely, A and B as shown in Fig. 3, together with static normal load (denoted as waveform C). The normal load values quoted in the figures correspond to the maximum load acting on each contact foot.

Under waveform A the compressive normal load was varied proportionally to the magnitude of axial load, so that the highest normal load was exerted at

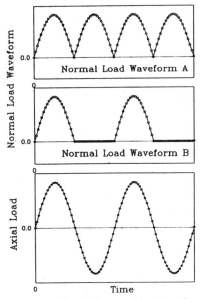

Fig 3   **Types of variable normal load waveforms**

both tensile and compressive peaks of the axial load cycle. The normal load approached zero at the instant of zero axial load. Waveform B is identical to waveform A but the normal load was kept at zero during whole compressive part of the axial load cycle. During tests, however, a small compressive normal load necessary to maintain the contact between the pad and the specimen was used instead of zero normal load.

It is not difficult to understand that the type B waveform represents the typical loading patterns exerted in many fretting situations such as in closely fitted pin loaded lugs, keyways of transmission assemblies, turbine blade/disc dovetail fixings, etc. In such assemblies a small clearance may exist between the mating parts; consequently the mating surfaces move apart during compressive parts of the applied load cycle, resulting in zero normal load. The situation in which waveform A is applicable is more difficult to visualize; however, it can be shown that such loading may occur in joints that are simultaneously subjected to two or more external forces.

Friction response

Friction force plays a critical role in fretting crack nucleation and the early phase of crack growth. Therefore, the understanding of friction behaviour in fretting is paramount for the development of models necessary to describe fretting fatigue damage mechanisms. The friction force versus axial load hysteresis loops recorded in tests for the three normal load cases investigated are compared in Fig. 4 for various combinations of axial stress amplitude and normal load. It is clearly visible that the loop shape is significantly influenced by the normal load waveform. It should be highlighted that the nominal slip amplitude (the amount of slip derived from axial load and pad span) is the same for all three loading cases compared. The regions of micro- and macro-slip can be clearly recognized in the case of static normal load, for all values of axial and normal load. During the increasing and decreasing parts of the axial load waveform, the partial slip behaviour was observed at the beginning, just after the load reversal, which gradually converted into total slip behaviour as the load was increased. This was manifested by the reduction in the gradient of the curve. More complex slip behaviour was observed under both waveforms A and B.

Under waveform A, during both increasing and decreasing parts of axial load, partial slip was apparent near the axial load peak values, and total slip behaviour was visible as the axial load approaches zero. For the case of waveform B, little partial slip behaviour was visible for the compressive part of the axial load cycle, in particular at higher normal load values and lower axial load amplitudes. It can be seen that both waveforms A and B showed similar slip behaviour as the slip amplitude was increased. From Fig. 4, it can be seen that the change in slip behaviour mentioned above did not significantly influence the peak values of friction force, for the cases of static and waveform A

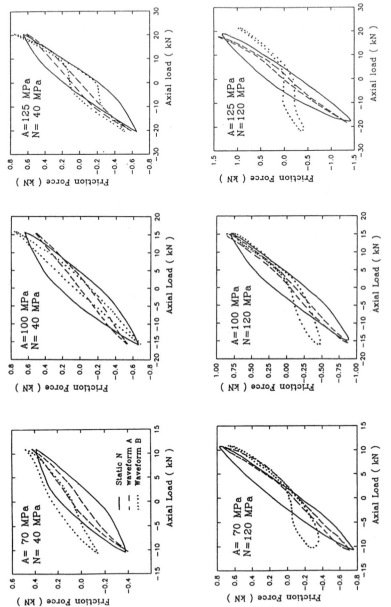

Fig 4 Friction force versus axial load hysteresis loops for static and variable normal load

normal loads. The peak friction force under waveform B was found to be higher than that for the static case at lower normal load, while it was found to be lower than the static case at higher normal loads. However, there is con- siderable difference if the friction forces are compared on the basis of root- mean-square values, which takes account the time-dependent damage influence of the friction force.

The evolution of friction force during the life of each test showed that the friction force increased rapidly at the start of the test. This can be attributed to the removal of the surface oxide layer and subsequent metal-to-metal contact which increases the coefficient of friction. Stable friction behaviour was observed during the major part of the lifetime. The mean friction force of the loading cycle for both the static and waveform A normal loads was found to be nearly zero. Under waveform B, however, a positive mean friction force was observed, as was expected. In many tests with waveform B a progressive increase in friction force amplitude was observed during lifetime. The mean friction force was also found to change periodically in these tests. This may be related to changes in slip amplitude, as the slip displacements can change from one contact foot to the other due to slight difference in friction resistance (or wear) in each contact foot.

For all three normal load cases, the friction force corresponding to tensile peak of axial load was plotted against the normal load, in Fig. 5. As seen the coefficient of friction, defined using the friction amplitude for fully developed macro-slip, was not significantly affected by the waveform used. At low normal loads, the waveform B results appear to be closer to the fully developed

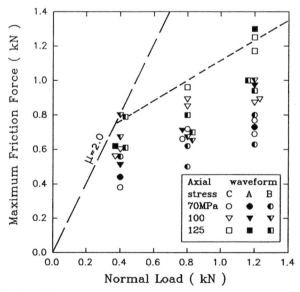

Fig 5   **Maximum friction force versus normal load for static and variable normal load**

macroslip when compared with the other two cases. At higher normal loads the opposite is true. It should be noted that the nominal slip amplitude primarily depends on the magnitude of axial load and the pad span and, therefore, is same for the three loading cases compared. The difference in the peak friction force is, therefore, merely a result of different loading paths associated with individual waveforms.

Fatigue life

Fatigue life results for all three cases of normal loading are presented in Fig. 6. The constant normal load results are found to be in overall agreement with those of previous data on the same material combination (6). A clear influence of the normal load waveform on the fatigue life is apparent from this figure; however, the magnitude of friction force in these tests must also be taken into account when interpreting the scatter in fatigue life data. The same results are presented in terms of the normal load versus fatigue life, in Fig. 7.

From Figs 6 and 7, it is apparent that under variable normal load the fatigue lives are enhanced when compared with static normal load test data. For the case of 0.8 kN normal load a clear enhancement of fatigue life, by about a factor of three or more, was observed under both waveforms A and B. However, for 0.4 kN and 1.2 kN normal loads the effect of waveform on the lifetime was decisive. This was considered to be due to the dual influence of the normal load on the fatigue crack growth. For a given slip amplitude the higher normal load result in higher friction, which considerably accelerates the early phase of fretting crack growth (4). On the other hand the increase in the normal load causes retardation in the crack growth by increasing crack

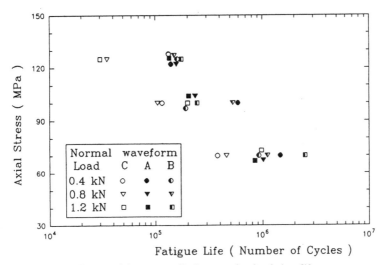

Fig 6   Axial stress amplitude versus fretting fatigue life

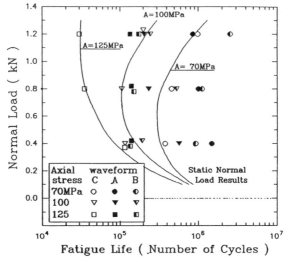

Fig 7   The influence of normal load on fretting fatigue life

closure effects. The change in slip behaviour due to variable normal load may considerably complicate this dual influence of the normal load on fretting damage.

Influence of the phase of loading on friction response

The results presented so far correspond to in-phase control of normal load with respect to specimen axial load. Some out-of-phase tests were performed with waveform A and with changing phase of the waveform as shown in Fig. 8, for the case of 100 MPa axial stress amplitude and 120 MPa nominal contact pressure (assuming a uniform distribution). The influence of the phase between the axial load and the normal load on the friction response is shown in Fig. 9. As can be seen, the friction force was found to be significantly affected due to the change of phase. More importantly the peak friction and the instant where the peak friction occurs in the loading cycle depended on the phase of loading. It should be emphasized that although different peak friction was observed, the nominal slip amplitude is the same for all these tests. The ability to control the friction response by this method, without altering slip amplitude, is thought to be useful in investigating models to describe dynamic friction behaviour. Moreover such a method would provide an effective means of isolating the influence of slip and friction on fretting damage and allow understanding the critical influence of friction on fretting crack growth.

Fretting fatigue fracture is known to be dominated by stable crack growth, and fracture mechanics is considered to be a useful tool in evaluating fretting fatigue life. In situations where plasticity is highly localized, linear elastic fracture mechanics (LEFM) is applicable and stress intensity factor is considered

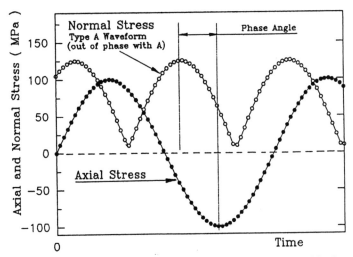

Fig 8 Variable normal load waveform *A* out-of-phase with axial load

to be a useful parameter in the description of crack driving force. In fretting environments the stress intensity factor is affected primarily by three loads: bulk stress; contact normal load; and friction force. The varying influence of these loads on the stress intensity factor at different values of fretting crack length has been identified (7). The ability to control friction force independent of the axial stress amplitude is thought to be useful as it enables control of the range of the peak value of the stress intensity during loading cycle without altering the slip amplitude or the amplitude of axial load. This also provides a means of controlling the mixed-mode loading behaviour in fretting crack growth in any desired manner, thus enabling assessment of the applicability of LEFM in the description of fretting crack growth processes.

Conclusion

A new fretting fatigue testing facility was developed which enabled control of the normal load on the contact area. The normal load can be varied in-phase or out-of-phase with the specimen axial load, with any desired waveform, allowing simulation of a wide range of fretting load patterns encountered in component assemblies. Fretting fatigue behaviour of BS L65 4 percent copper aluminium alloy was presented for two normal load waveforms. The results of variable normal load tests were compared with those of static normal load data. It was shown that the friction behaviour and fatigue life are markedly affected by the normal load waveform.

The versatility of the testing facility has been highlighted in relation to its ability to control crack driving forces in fretting. Possible ways of assessing the critical influence of slip amplitude, and the applicability of LEFM in evaluation of fretting fatigue life have been discussed.

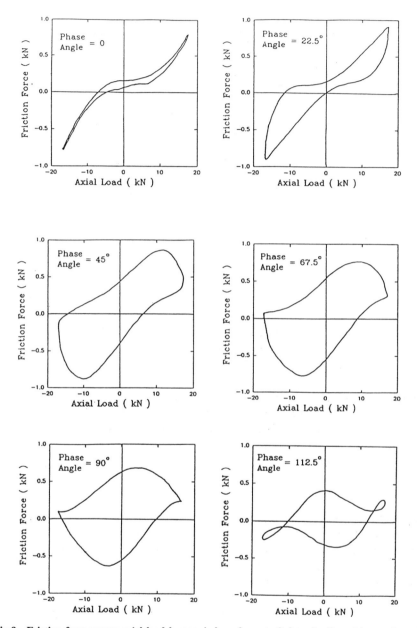

Fig 9 Friction force versus axial load hysteresis loop for out-of-phase loading with waveform A.
      $A = 100$ **MPa**; $N = 125$ **MPa**

References

(1) FORSYTH, P. J. E. (1981) Occurrence of fretting fatigue failure in practice, *Fretting fatigue*, (Edited by R. B. Waterhouse), Applied Science, London.

(2) WATERHOUSE, R. B. (1992) The problems of fretting fatigue testing, *Standardisation of fretting fatigue test methods and equipment, ASTM STP 1159*, (Edited by M. H. Attia and R. B. Waterhouse), ASTM, Philadelphia, pp. 13–19.

(3) DOBOROMIRSKI, J. M. (1992) Variables of fretting process: are there 50 of them? *Standardisation of fretting fatigue test methods and equipment, ASTM STP 1159*, (Edited by M. H. Attia and R. B. Waterhouse), ASTM, Philadelphia, pp. 160–166.

(4) FERNANDO, U. S., FARRAHI, G. H., and BROWN, M. W. (1994) Fretting fatigue crack growth behaviour of BS L65 4 percent copper aluminium alloy under constant normal load, *Fretting Fatigue*, (Edited by R. B. Waterhouse and T. C. Lindley) (Mechanical Engineering Publications, London, pp. 183–195. *This Volume.*

(5) FERNANDO, U. S., FARRAHI, G. H., and SHEIKH, M. A. (1992) Fretting fatigue behaviour of BS L65 aluminium alloy, Progress report, SIRIUS, University of Sheffield, Sheffield, UK.

(6) EDWARDS, P. R., and RYMAN, R. J. (1975) Studies in fretting fatigue under variable amplitude loading conditions, RAE Technical report No. 75132. RAE, Farnborough, UK.

(7) SHEIKH, M. A., FERNANDO, U. S., BROWN, M. W., and MILLER, K. J. (1994), Elastic stress intensity factors for fretting cracks using the finite element method, *Fretting Fatigue*, (Edited by R. B. Waterhouse and T. C. Lindley), Mechanical Engineering Publications, London, pp. 83–101. *This volume.*

C. B. Elliott III* and D. W. Hoeppner*

A Fretting Fatigue System useable in a Scanning Electron Microscope

REFERENCE Elliott, C. B. III and Hoeppner, D. W., **A fretting fatigue system useable in a scanning electron microscope,** *Fretting Fatigue*, ESIS 18 (Edited by R. B. Waterhouse and T. C. Lindley) 1994, Mechanical Engineering Publications, London, pp. 211–217.

ABSTRACT Research in the Quality and Integrity Design Engineering Center (QIDEC) of the University of Utah is currently being conducted in an attempt to clearly elucidate the role of corrosion, specifically oxidation in ambient air, on fretting fatigue. To perform the testing associated with this research, a system capable of performing fretting fatigue tests in a laboratory air environment or in a scanning electron microscope (SEM) vacuum environment has been developed by QIDEC. This paper discusses the fretting fatigue test system and an example of early results obtained with its use.

Introduction

The presence of fretting has been shown to modify the fatigue response of many structural materials. Fatigue life at a given stress or strain level generally decreases, when compared to unfretted fatigue life under similar material, loading, and environmental conditions. Many investigations have been conducted to identify the failure mechanisms which cause fretting fatigue degradation. There is general agreement on several concepts. For example, investigators believe that fretting causes early crack nucleation, but has less effect on subsequent crack propagation **(1)**–**(4)**. They also tend to believe that the debris, which results when fretting wear particles become trapped in the fretted area, affects the process due to its composition, characteristics, and rate of accumulation **(4)**–**(6)**.

However, there are several factors upon which investigators disagree. An example is the role of corrosion. In some cases, results from specific experiments have led investigators to conclude that fretting fatigue can be due predominantly to corrosion **(4)(7)**–**(10)**. Others contend that corrosion can have only a minor role in fretting fatigue **(3)(6)(9)**.

The purpose of this paper is to describe the fretting fatigue system which has been developed by QIDEC. This system allows fretting fatigue testing to be conducted in an SEM. The test can be interrupted at any time, the fretting pads removed from the specimen, and the specimen moved under the electron beam of the SEM for observation, without the vacuum of the SEM being disturbed. This allows observation of the fretted surface without air, or any environment other than the vacuum, ever being introduced to the specimen during or following the fretting fatigue process. It is hoped that comparison of the fretted surfaces of specimens that have been tested in air with those tested

* Quality and Integrity Design Engineering Center, 3209 MEB University of Utah, Salt Lake City, Utah 84112, USA.

only in a vacuum will provide insights into the effects of corrosion, specifically oxidation, on the fretting fatigue process.

System overview

The total system includes the control, loading, and data acquisition subsystems. It allows monitoring of the applied fatigue loads, the residual loads at the far end of the specimen from the actuator, and the normal load applied to the fretting pads. Frequency, normal load and applied fatigue loads are controlled for each test. Although all components of the total system are necessary, most are standard MTS or Hewlett Packard components and are not of significance to this paper. Therefore, the remainder of this paper will deal with the fretting fatigue load frame (referred to as the system) that was developed by QIDEC.

The system consists of the following four subsystems, which are described in subsequent sections of this paper:

(1) The *external fatigue loading subsystem* which is outside of the SEM allows the fatigue load to be transmitted to the specimen inside the SEM. It also enables the specimen to be moved further into the SEM chamber for viewing.
(2) The *internal fatigue loading subsystem* applies the fatigue loading to the specimen.
(3) The *internal fretting loading subsystem* applies the normal loading to the specimen through the fretting pads.
(4) The *vacuum lid subsystem* replaces the front door of the vacuum chamber. All interfaces between the outside and the inside of the vacuum chamber are in the vacuum lid. The vacuum lid subsystem also provides the ability to move the specimen left, right, and up within the SEM chamber.

A top view drawing of the system and its four subsystems is shown in Fig. 1.

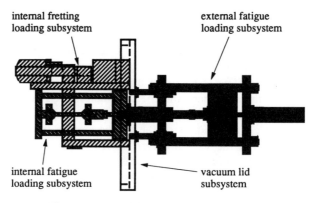

Fig 1   Top view of the fretting fatigue load frame

The external fatigue loading subsystem

The top view drawing of the external fatigue loading subsystem, shown in Fig. 2, should be referred to while reading the following.

The near load cell measures the load applied by the actuator. This load, which is used for control, is transmitted by the load rod through the bellows, exit cylinder, and other components into the vacuum chamber.

Three support rods provide support to the external fatigue loading subsystem, two of these are shown in Fig. 2. The third rod, which has been removed from this figure for clarity, goes over the centre of the subsystem. These rods are attached to the near and far disks.

The exit cylinder passes through the vacuum lid subsystem and is threaded into the far load cell, which is a part of the internal fatigue loading subsystem.

The components described above can be moved towards or away from the vacuum chamber by adjusting nuts on three movement rods, two of which are shown in Fig. 2. The third rod, which has been removed from this figure for clarity, goes under the centre of the subsystem. By adjusting the nuts, the far disk is moved, which causes the other components, and the internal fatigue loading subsystem, to move. This procedure allows the specimen to be moved under the electron beam of the SEM for viewing. It may also allow the specimen to be repositioned between the fretting pads for further testing, but this capability has not been validated.

The internal fatigue loading subsystem

The top view drawing of the internal fatigue loading subsystem, shown in Fig. 3, should be referred to while reading the following.

Fatigue loading enters the SEM by the load rod, which was also shown in Fig. 2. This loading is transmitted through the specimen and grips to the load train end piece, and back through the two side rods to the far load cell. When no normal (fretting) load is applied to the specimen, the far load cell measures

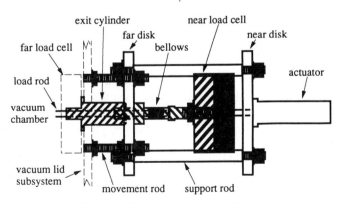

Fig 2   Top view of the external fatigue loading subsystem

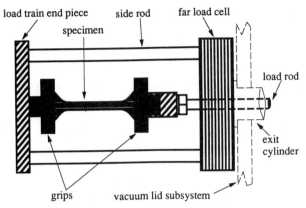

Fig 3   Top view of the internal fatigue loading subsystem

the same loading as the near load cell in the external fatigue loading subsystem. When fretting is applied by the internal fretting loading subsystem, the
far load cell measures the applied load as modified by the frictional loading
that results from the fretting process.

As discussed previously, the exit cylinder of the external fatigue loading subsystem is threaded into the far load cell of the internal fatigue loading subsystem. This provides the link for supporting the fatigue loading. It is also the
interface necessary to allow the internal fatigue loading subsystem to be
moved into the SEM chamber, where the specimen may be viewed, by adjusting the nuts on the three movement rods of the external fatigue loading subsystem.

The internal fretting loading subsystem

Figure 4 shows a top view drawing of the internal fretting loading subsystem.
The three figures identified by this drawing should be referred to while reading
the following.

A back view drawing of the internal fretting loading subsystem is shown in
Fig. 5. The fretting pads (test material on top and a phenolic material on the
bottom) are clamped against the top and bottom of the specimen by the pivoting action of the two fretting bars. The fulcrum for this pivoting action is the
left support bar which is shown in Fig. 5 and also in the left view of the
internal fretting loading subsystem in Fig. 6. Two stops and an end plate, seen
in Figs 5 and 6, keep the fretting bars in place. The spring, seen in Fig. 5,
causes the fretting bars to open when the specimen is to be moved further into
the SEM for viewing.

The fretting bars are clamped on the left side of Fig. 5 and held in place by
the two hydraulic rams that can be seen in Fig. 7. The vertical ram operates
two clamping bars which close on the ends of the fretting bars. A load cell
between the top clamping bar and the top fretting bar measures the applied

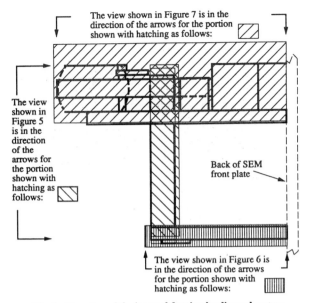

The view shown in Figure 7 is in the direction of the arrows for the portion shown with hatching as follows:

The view shown in Figure 5 is in the direction of the arrows for the portion shown with hatching as follows:

Back of SEM front plate

The view shown in Figure 6 is in the direction of the arrows for the portion shown with hatching as follows:

Fig 4   Top view of the internal fretting loading subsystem

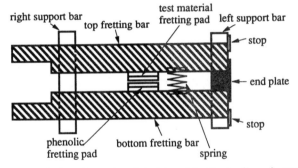

right support bar
top fretting bar
test material fretting pad
left support bar
stop
end plate
stop
phenolic fretting pad
bottom fretting bar
spring

Fig 5   Back view of a portion of the internal fretting loading subsystem

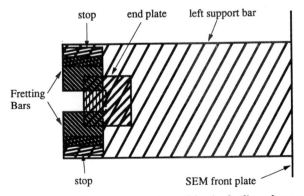

stop
end plate
left support bar
Fretting Bars
stop
SEM front plate

Fig 6   Left view of a portion of the internal fretting loading subsystem

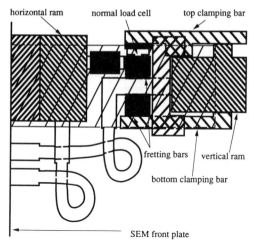

Fig 7   Right view of a portion of the internal fretting loading subsystem

force, from which the normal fretting force is computed. The horizontal ram holds the top fretting bar in place so that it does not move as a result of the frictional loading forces.

The vacuum lid subsystem

A front view drawing of the vacuum lid subsystem is shown in Fig. 8. This subsystem replaces the front door of the SEM vacuum chamber. All interfaces between the outside and inside of the vacuum chamber are in the vacuum lid subsystem. This is so that the system can be removed from the SEM and placed in another fixture for testing in air, thereby allowing the SEM to remain multipurpose instead of requiring it to be dedicated.

Bars mounted behind the vacuum lid on the sides and top allow the system to be moved left, right or up within the vacuum chamber. This is accomplished

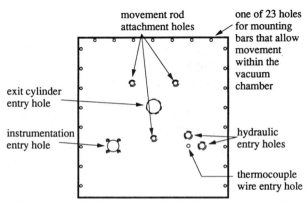

Fig 8   Front view of vacuum lid subsystem

by turning screws (not shown in Fig. 8) that are fitted into threaded holes in the bars.

Concluding remarks

This paper has discussed the fretting fatigue test system designed by the Quality and Integrity Design Engineering Center of the University of Utah. This system is currently being used in an experimental programme to conduct fretting fatigue tests in air and a scanning electron microscope. Preliminary analysis of the test data indicates that the system is effective. For example, under one set of material and loading conditions, the average life of three 7075–T7351 aluminium alloy specimens was approximately 80 000 cycles. The one comparable specimen tested under the same loading conditions, but in a vacuum, had a fretting fatigue life of over 1 600 000 cycles.

Acknowledgements

The authors are grateful to the University of Utah for providing the facilities and the scanning electron microscope used in this programme, and to FASIDE International Incorporated which funded the building of the fretting fatigue system and provided the support equipment being used in this programme. Appreciation is expressed to Alcoa which provided 8090–T73 aluminium alloy in support of this QIDEC research.

References

(1) HOEPPNER, D. W. and GOSS, G. L. (1974) A fretting-fatigue damage threshold concept, *Wear*, **27**, 61–70.
(2) WATERHOUSE, R. B. (1981) Introduction to fatigue, *Fretting Fatigue*, Applied Sciences Publishers, London, pp. 1–22.
(3) KANTIMATHI, A. and ALIC, J. A. (1981) The effects of periodic high loads on fretting fatigue, *J. Engng Mater. Technol.*, **103**, 223–228.
(4) BILL, R. C. (1983) Fretting wear and fretting fatigue – How are they related?, *Trans. ASME*, **105**, 230–238.
(5) COLOMBIE, CH., BERTHIER, Y., FLOQUET, A., VINCENT, L., and GODET, M. (1984) Fretting: load carrying capacity of wear debris, *Trans. ASME*, **106**, 194–201.
(6) BERTHIER, Y., VINCENT, L., and GODET, M. (1989) Fretting fatigue and fretting wear, *Tribol. Int.*, **22**, 235–242.
(7) POON, C. J. and HOEPPNER, D. W. (1981) A statistically based investigation of the environmental and cyclic stress effects on fretting fatigue, *Trans. ASME*, **103**, 218–222.
(8) WATERHOUSE, R. B. (1981) Fretting fatigue in aqueous electrolytes, *Fretting fatigue*, Applied Sciences Publishers, London, pp. 159–176.
(9) HOEPPNER, D. W. (1981) Environmental effects in fretting fatigue, *Fretting fatigue*, Applied Sciences Publishers, London, pp. 143–158.
(10) KOVALEVSKII, V. V. (1981) The mechanism of fretting fatigue in metals, *Wear*, **67**, 271–285.

A. Del Puglia*, F. Pratesi*, and G. Zonfrillo*

Experimental Procedure and Parameters Involved in Fretting Fatigue Tests

REFERENCE Del Puglia, A., Pratesi, F., and Zonfrillo, G., **Experimental procedure and parameters involved in fretting fatigue tests**, *Fretting Fatigue*, ESIS 18 (Edited by R. B. Waterhouse and T. C. Lindley) 1994, Mechanical Engineering Publications, London, pp. 219–238.

ABSTRACT From the research activity on fretting fatigue carried out in the DMTI Department of the University of Florence, a consistent set of fretting fatigue results is reported. The experimental conditions and some metallurgical evidence are described. The representative variables are defined by introducing a tribological model, which is then calibrated using experimental data. The model is used to represent and interpret the observed fretting behaviour and to provide relationships between the relevant variables, in particular between the parameters controlling the test and the associated local conditions which determine fretting action in the contact area. In addition, the resulting reduction in life is discussed within the framework of a general interpretation of the experimental results.

Notation

Ah	Contact area at start of test, evaluated with Hertz theory
Am	Contact area derived from measurements on the pad at end of the test
Cd	Coefficient of sliding friction, limit value of Ct
Ct	Maximum in a cycle of the ratio $\Delta T/Fn$
Fn	Fretting load (normal to specimen surface)
Hd	$1/Kd$
Hp	Coefficient used in the intermediate slip stage
Hs	$1/Ks$
Kd	Stiffness of device
Ks	Stiffness per unitary area of specimen part between contact and device clamp
N	Number of cycles at rupture
Nf	Number of cycles at rupture in plain fatigue tests
Nff	Number of cycles at rupture in fretting fatigue tests
Ph	Mean contact pressure at start of test ($Ph = Fn/Ah$)
Pm	Mean contact pressure at end of test ($Pm = Fn/Am$)
R	Fretting fatigue stress parameter, percentage ratio between fretting fatigue and plain fatigue stresses for $Nff = Nf$
S	Slip in a cycle
Sr	Slip without the slide contribution
T	Tangential force due to friction

* Dipartimento di Meccanica e Tecnologie Industriali, University of Florence, Via S. Marta 3, 50139 Firenze, Italy.

Ud Device displacement
Us Specimen displacement
Wd Work dissipated by tangential force T during a cycle
Wh Specific work dissipated per unitary area at start of test during a cycle ($Wh = Wd/Ah$)
Wm Specific work dissipated per unitary area at end of test during a cycle ($Wm = Wd/Am$)
Wr Work dissipated during a cycle, without the slide contribution
$\Delta\sigma$ Fatigue stress amplitude
ΔT Tangential force amplitude in a cycle
ΔTd Value of ΔT in slide condition (equal to $Cd*Fn$)
ΔTs Value of ΔT, limit of the stick condition
ΔUd Device displacement amplitude in a cycle
ΔUs Specimen displacement amplitude in a cycle
σ Applied stress

Introduction

Fretting fatigue phenomena may become life-determining for some components in actual working conditions; therefore, much testing and research activity has been carried out in this field in recent years. At present, the main requirement for most testing devices and procedures is that they reproduce fretting situations that occur in actual practice. The variety of practical fretting conditions is so large that many experimental approaches have been suggested to cover their diversity. Because an exhaustive simulation of actual working conditions is very complex, different simplifications are used to accomplish each particular aim, and to provide achievable control conditions for the tests. Thus, it is still very difficult to compare the results obtained by different laboratories and standard testing procedures are lacking.

On reviewing the literature one finds that there are not only differences in experimental conditions selected for carrying out the tests, but there are also differences in the meaning assigned to some parameters. For instance, 'slip' can refer to different quantities involved in fretting. In the case of a partial stick, slip values from zero to a maximum are present in the contact area. This maximum – corresponding to the displacement of the outer parts of the nominal contact zone – is the quantity generally measured and defined as slip. In other cases, slip is evaluated as a global relative displacement of the two bodies in forced contact; in this situation, slip may be large and accompanied by more or less consistent wear. A similar situation exists for the friction coefficient. At times it is globally considered, and thus it is practically a mean value; at times it is taken as the value locally present in a specific zone between the two surfaces held in forced contact. Consequently, if one compares results from different works, it is essential to know whether they are based on the same definitions and thus have the same meaning.

The main mechanisms controlling fretting fatigue effects are now well understood, at least from a qualitative point of view (1). Several investigations with variation of a single parameter have already been carried out in the past and some fundamental conclusions have been reached (2). However, test results depend not only on specific materials, but also on the fretting experimental equipment and on testing conditions. It is thus important to obtain further information, with special attention to reciprocal effects among parameters corresponding to the test situation and in particular to specific devices and procedures.

A complementary approach has been that of observing the kind of damage which has occurred on the surfaces in forced contact and relating this damage to parameters representing the tribological situation (3). Recently, much attention has been devoted in the literature to applications of fracture mechanics techniques to both propagating and non-propagating fretting cracks (see for instance (4)). The evidence of a predominant effect of fretting damage on the nucleation stage has led to the use of a two-stage fatigue life prediction method to quantitatively evaluate the life reduction occurring in fretting fatigue tests (5).

A consistent set of fretting fatigue results and its context of experimental conditions are reported. Initially, the fretting device behaviour is described, some metallurgical evidence is considered, and the representative variables are defined. The selected variables can be divided into two classes: 'external' and 'local'. The former are fixed before the test and controlled during it; the latter change within the fretting process, depending on the values of the external variables. For a better understanding of the global experimental conditions, a tribological model is introduced, representing the specific fretting behaviour and providing relations among variables. Lastly, a general interpretation of experimental results is given, with particular reference to life reduction caused by fretting.

Experimental equipment and procedure

The experimental apparatus is based on an MTS hydraulic testing machine (100 kN maximum applied load) used for both plain fatigue and fretting fatigue tests. The specimen of rectangular section (8 mm × 2.5 mm) was sized according to ASTM recommendations for fatigue. When the fretting fatigue tests were carried out, the specimen was fatigue loaded and a suitable device, directly mounted on it, provided the required fretting conditions, Fig. 1(a). This device was originally developed at BAM, Berlin (6)(7).

The device is composed of two parts which are clamped together at both sides of the specimen in a section near its end. On either side of the specimen at the mid-section, are a pad and a rolling bearing, which are pushed against the specimen itself by a force Fn, normal to the surface, provided by an external elastic ring. Fretting conditions are created by the pad, owing to the movement of the specimen under fatigue load. The tangential force T resulting from

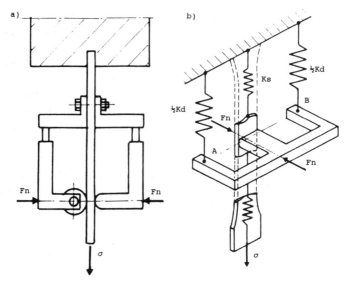

Fig 1 Schematic representation (a) and mechanical scheme (b) of the fretting device

these conditions is measured by strain gauges suitably located on the device (8).

The cylindrical pad used (radius 6 mm) is in linear contact with the surface of the specimen and has its axis normal to the fatigue loading direction. In practice, the contact has a finite area; its value at the beginning of the test Ah, may be evaluated with the Hertz theory. During the test, wear occurs and the area of contact at the cylindrical pad increases. For given stress conditions and specimen, the increase in the contact area obviously depends on the pad material and can be greatly reduced by using a very hard alloy or a ceramic material. There are, however, some reasons for choosing the same alloy for the pad as for the specimen, i.e., to simulate cases occurring in actual practice and to establish a reference case for chemical–mechanical interactions between the two surfaces. In these experiments, the pad has always been manufactured using the same alloy as the specimen.

Several steels have been tested (9); the data reported here were obtained on the steel with specification 38NCD4K. Sinusoidal fatigue loads (frequency 10 Hz) were applied around a mean value so that the specimen was never under compression. The mean value of the stress was kept constant at 500 MPa in all the tests. Thus the tests were carried out by fixing two parameters, the normal force Fn, and the oscillation amplitude of fatigue loads (or the equivalent amplitude of fatigue stress $\Delta\sigma$).

Measurements were recorded step-wise during the test. During acquisition, the tangential force T and the actual applied stress σ were simultaneously recorded as function of time; typical examples are shown in Fig. 2. In addition to this representation, it was possible to obtain diagrams similar to hysteresis

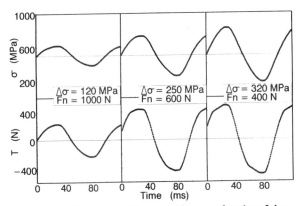

Fig 2 Applied stress and tangential forces as function of time

cycles, by plotting T as a function of σ. Depending on different loading conditions, the shape of the cycles was apparently quite different (Fig. 3) as discussed below.

The mechanical scheme of the fretting device is shown in Fig. 1(b). The global elastic stiffness (of the device and of the part of the specimen included between contact and device clamp) is represented by the corresponding springs with constants Kd and Ks. The arms linking the pad to the points A and B (end of the springs representing the device) are assumed to be perfectly rigid. With reference to this scheme, the vertical displacements Us and Ud can be

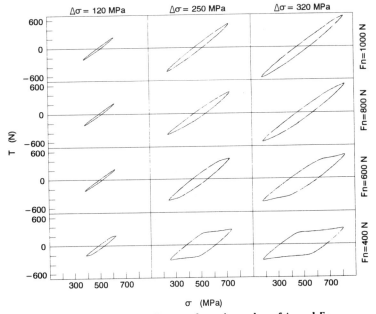

Fig 3 Hysteresis-like diagrams for various values of $\Delta\sigma$ and Fn

associated with given values of σ and T. These are, respectively, the displacement of the initial ideal contact line on the specimen, and of the line joining A and B points, representing the pad position. The maximum ranges of these displacements in a cycle are termed ΔUs and ΔUd and the slip S represents the quantity that balances their difference.

The most significant experimental parameters which can be derived from the hysteresis-like cycles are the slip S, and the tangential force amplitude during a cycle ΔT. ΔT is taken as half the difference between the highest and the lowest values reached by T during the cycle. Experimentally, S is obtained as the ratio to Ks of the width of the cycle (distance along the σ axis between its two oblique lateral branches).

As reported in the literature (e.g. (10)), in experimental conditions of this kind, the hysteresis-like cycles change with time. The main reason for this is modifications occurring in the contact area. In particular, remarkable changes are observed in the very initial phase. Subsequently, a practically stabilized cycle is observed during the main part of the test. Therefore, all cycles shown in Fig. 3 have been obtained after a value of about 5000 cycles, taken as reference. Significant changes in the stabilized hysteresis-like cycles occur later only in the case of tribological conditions leading to strong wear phenomena, see Fig. 4.

After the test was completed (rupture or survival after four million cycles), the final contact area Am was determined on many specimens, by measuring the width of the track left on the worn pad zone.

Fractographic analysis

In order to obtain further information, fractographic observations were carried out by SEM after rupture of the specimens. The investigation considered both the lateral part of the surface near the rupture and the fracture surface itself. The lateral surface is very indicative of fretting damage since

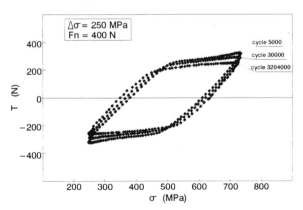

Fig 4   Evolution during the test of the hysteresis-like cycle

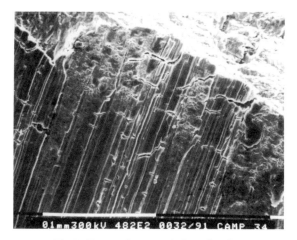

Fig 5 Damaged area on the lateral surface

after plain fatigue tests, it shows the grooves in the loading direction deriving from mechanical machining; when fretting is present, this zone is more or less worn. Usually, toward the border line with the undamaged outer part of the lateral surface, debris and scales are observed which have been removed from the surface and pushed back onto the surface at a different location. At times these fragments appear to be heavily oxidized. Moreover, in this zone several cracks can be observed, parallel to the fracture surface, in addition to these leading to rupture. Generally they are rather small, closed, and not connected with each other, Fig. 5.

On the fracture surface, the difference between samples broken under plain fatigue and under fretting fatigue conditions is mainly related to the nucleation stage. Whereas the fatigue propagation surface generally derives from a single crack, fretting failure appears to derive from the confluence of several cracks propagating on parallel planes, Fig. 6. It may occur that most of these cracks are practically propagating on the same plane; in this case the shell-like morphology of the fretting fatigue zone is much more open than the almost semi-circular one typical of plain fatigue. In conclusion, the main effect of the fretting action consists of the nucleation of several neighbouring cracks. Apart from the different number of cracks, the propagation of any crack and the rupture process practically occur with the same morphology as for plain fatigue.

Test results

The main aim of the reported fretting fatigue tests has been to derive Wöhler curves for various values of fretting load Fn. With respect to the Wöhler curve obtained in plain fatigue conditions, the fretting fatigue curves in general show a lower strength and a different behaviour for different Fn, Fig. 7. In particu-

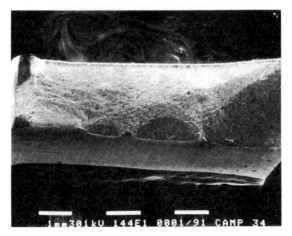

Fig 6   **Typical fracture surface**

lar, for $Fn = 400$ N the fretting fatigue curve practically coincides with the
plain fatigue one. For $Fn = 600$ N the behaviour is more complex: with
respect to plain fatigue, at high stresses there is a small deviation, which
increases with decreasing $\Delta\sigma$ up to a fretting fatigue limit lower than in the
case of greater Fn. The curves for $Fn = 800$ N and for $Fn = 1000$ N are very
near to each other and show an almost constant reduction in strength for any
life value.

A conceptual scheme of fretting fatigue behaviour, as compared with plain
fatigue, is shown in Fig. 8. In the case of plain fatigue, the applied stress amplitude $\Delta\sigma$, together with the material and test characteristics, determines the
number of cycles at rupture Nf. In the case of fretting fatigue, the number of

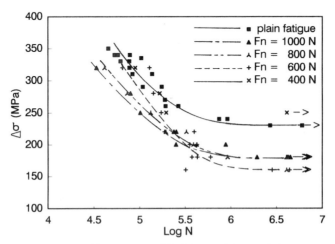

Fig 7   **Wöhler curves for plain fatigue and fretting fatigue**

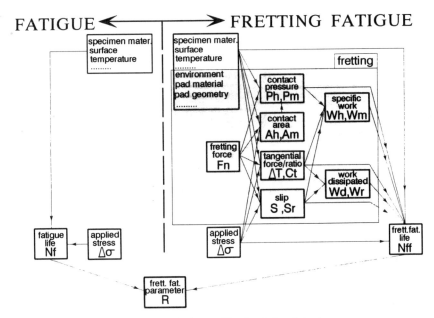

Fig 8   Conceptual scheme of fretting fatigue behaviour

cycles at rupture Nff is determined directly by the same variables and by the system of fretting parameters, which for the moment can be considered globally as a single box. In addition to plain fatigue variables these parameters are controlled by the fretting load Fn and other factors linked to the pad (material and shape) and to the test conditions (in particular, environment).

An evaluation of the decrease of fretting fatigue life as compared to plain fatigue life can be obtained in terms of the reduction in $\Delta\sigma$ that leads the specimen to failure at the same number of cycles. Thus the fretting fatigue stress parameter, R, will be used. It is the percentage ratio between $\Delta\sigma$ leading to rupture in the presence of fretting and $\Delta\sigma$ leading to plain fatigue rupture in the same number of cycles. The complement to 100 of R represents the percentage reduction in stress due to the fretting action. Owing to the usual scatter of fatigue test results, some experimental points can occur at values higher than 100 in the following diagrams.

In order to determine relations between experimental results and fretting parameters, the directly obtained experimental values ΔT and S can be plotted against Nff, Figs 9 and 10. In the first case, the dependence is similar to that observed in a Wöhler diagram, but with some deviations to shorter lifetimes. The diagram for S shows more dispersed points, strongly dependent on Fn.

The same quantities ΔT and S can be plotted against life reduction, expressed as R, in order to point out the damaging effect of fretting with respect to plain fatigue. These diagrams are reported in Figs 11 and 12. While

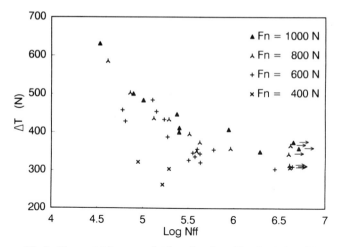

Fig 9   **Tangential force amplitude as function of fretting fatigue life**

Fig. 11 does not show any evident dependence, Fig. 12 indicates that a consistent life reduction occurs only for low values of slip S.

In the specific experimental conditions, neither ΔT nor S can be fixed independently, as they derive from various externally fixed quantities. Therefore, different values of other life-determining parameters may correspond to the same value of ΔT or S. Hence the study was directed towards defining the characteristics of the device and describing them by introducing a tribological model, which provides a consistent set of relations among the various parameters.

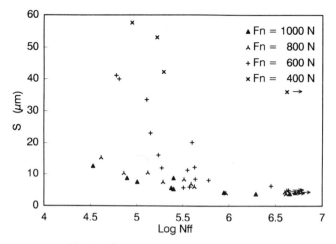

Fig 10   **Slip as function of fretting fatigue life**

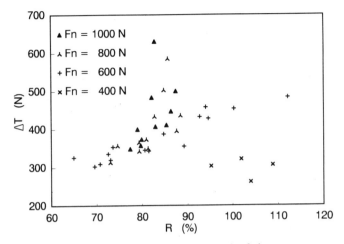

Fig 11   Tangential force amplitude versus fretting fatigue parameter

Tribological model

From the experimental diagrams of Figs 2 and 3, it appears that the tribological behaviour can differ greatly depending on the actual values of
applied stress $\Delta\sigma$ and fretting load Fn. The general trend, however, is evident
from the hysteresis-like cycles reported in Fig. 3. Specifically, in correspondence to the lowest $\Delta\sigma$ values and the highest Fn ones, the dependence of the
tangential force T on σ is very near to linearity and the slip is practically zero.
Increasing $\Delta\sigma$ and/or lowering Fn, the slip becomes greater, as may be seen by
the appearance of wider hysteresis-like cycles. In these cycles, after a phase in
which it is proportional to σ, T increases at a progressively lower rate. After

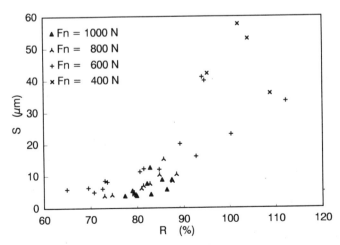

Fig 12   Slip versus fretting fatigue parameter

these two phases, for sufficiently high $\Delta\sigma$ and sufficiently low Fn, T tends to an almost constant value. In the extreme cases, the cycle appears as a parallelogram, with height depending on Fn. Similar trends are found in analysing the values of T and σ as a function of time. In the three typical cases shown in Fig. 2, a function T, that progressively deviates from the sinus shape with decreasing Fn and increasing $\Delta\sigma$, corresponds to the almost sinusoidal applied function σ. Moreover, the function T is cut off at an Fn-dependent typical value and a time correlation between T and σ becomes evident, as shown in the figure.

The large slip corresponding to the stage with almost constant T can be easily associated to the global relative displacement of the pad with respect to the specimen ('slide' condition). The slip corresponding to the intermediate stage without proportionality between T and σ can be associated with various physical phenomena in the contact zone. In particular, on the basis of the actual distribution of the stresses (which may be derived from the Mindlin model taking the tangential force T into account (11)), a cyclic variation of the stick-zone location within the contact area is expected, which contributes in the experimental procedure to the measured slip S. Further contributions to S may derive from plastic deformations of subsurface zones of both specimen and pad (sometimes connected with local formation of microscopic welded joints and some transfer of material from one surface to another (1)).

In line with other research (12), the reported behaviour has been interpreted by a simple tribological model relating the externally fixed parameters (such as $\Delta\sigma$, Fn, elastic characteristics of specimen and device) to the parameters controlling the fretting fatigue phenomena (such as S, ΔT, and derived quantities). The analytical relations of the model are reported in the Appendix. The diagrams of Fig. 13(a) and (b) – where ΔT and S are shown as function of $\Delta\sigma$ for three different values of Fn – are useful for illustrating the behaviour in the various stages.

The main assumption of the model is that the actual situation in the contact area can be described at global level by one of three different stages – A, B, C. They can be seen as generally occurring one after another when the absolute value of σ increases, although the intermediate stage may be absent for low Fn. In particular, the proposed relations are defined in terms of the variables that characterize a cycle; from these relations, the current value of the associated variables σ and T within the cycle can be derived, in the same way as in LCF studies for relating the hysteresis cycles to the cyclic stress–strain law.

In stage A it is assumed that both surfaces are held together by Fn and show no slip. This corresponds to the physical assumption that existing local microslips can be neglected, ('stick' condition). With reference to Fig. 1(b), the displacement of the test section on the specimen is the same as the displacement of the pad, so that the resulting S is zero. Therefore, ΔT is proportional to $\Delta\sigma$ with a constant depending on Ks and Kd. This behaviour is maintained as long as a 'stick' condition ($\Delta T < \Delta Tp$) is present.

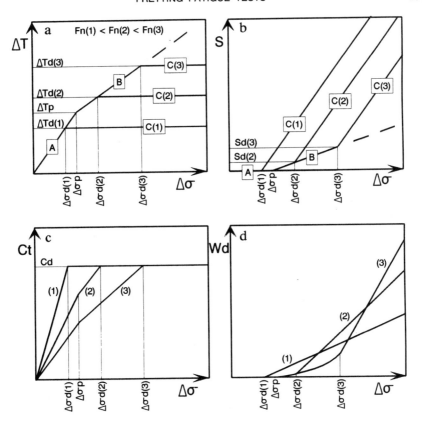

Fig 13   Relationships according to the model for three different fretting loads

Stage B is intermediate: it allows for some slip occurring due to small relative displacements of the surface layers. The displacement of the ideal contact line on the specimen is equal to the sum of the elastic displacement of the pad and the slip. This slip may have various physical causes and ΔT dependences. As a first approximation, it is assumed to be proportional (with coefficient Hp) to the difference between the values of ΔT, the current value and the one corresponding to the end of the stick condition, ΔTs. In terms of ΔT, the behaviour is still linear and similar to that described for stage A, although with a lower proportionality constant.

Stage C corresponds to 'slide' conditions; free slip occurs between the two surfaces after the sharp change in contact conditions occurring when ΔT reaches the specific value ΔTd (defined for each fixed Fn and given by the product of Fn and the sliding friction coefficient Cd, typical of the tribological condition and independent of Fn). ΔT is independent of further increases of $\Delta\sigma$; relative slips between the two surfaces in forced contact balance further displacements of the ideal contact line on the specimen.

In the model, ΔT and S are determined, respectively, by the lowest and the highest value among those calculated for given $\Delta\sigma$ and Fn with the different expressions proposed for the three stages. Therefore, when all three stages are present, as in cases (2) and (3), Fig. 13, the following behaviour is observed for increasing $\Delta\sigma$. The tangential force amplitude ΔT shows an initial proportional phase (A or stick stage), a change in proportionality constant when ΔT reaches ΔTp (B or intermediate stage), and a constant value when ΔT reaches ΔTd (C or slide stage), see Fig. 13(a). The slip S is null in stage A, then it linearly increases during stage B, independent of Fn values; as soon as slide occurs (stage C), the slope changes to a new value, independent of Fn, so that the representative lines for various Fn remain parallel, see Fig. 13(b). A different case may occur, with the absence of stage B, if ΔTd is lower than ΔTp, as represented by (1) in Fig. 13.

The constants of the model – Kd, Hp, ΔTp, Cd – have been determined by fitting them to the total set of experimental points, in terms of both ΔT and S. Ks has been obtained by analytical calculations. In Fig. 14 quantitative comparisons between the experimental values and those obtained from the model are outlined. The agreement indicates that the model suitably represents the various behaviours corresponding to a large range of external parameters and of tribological situations.

From the relations obtained, the values of Ct, defined as the ratio between ΔT and Fn, can be directly derived. In Fig. 13(c), Ct is reported as a function of $\Delta\sigma$. Moreover, a further parameter may be considered – the work dissipated during a cycle, Wd (roughly evaluated as twice the product of ΔT and S). The behaviour of this parameter according to the model is shown in Fig. 13(d).

In conclusion, although the representation of the intermediate stage with linear dependence of ΔT and S on $\Delta\sigma$ is apparently schematic, the model provides an essential aid to interpreting the experimental data, because it satis-

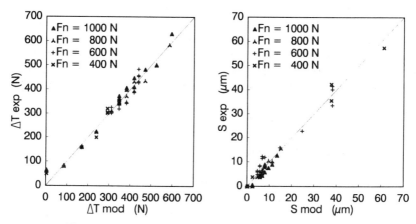

Fig 14   Comparison between experimental data and model values

factorily describes the various relations existing between externally applied parameters and local conditions of contact.

Study of parameters

The tribological model is helpful in understanding the shape of the diagrams of Figs 9 and 10 and in explaining why they did not satisfactorily account for life reduction in terms of the selected parameters ΔT and S. The main reason is their dependence on $\Delta\sigma$ and the strong effect of $\Delta\sigma$ itself on rupture. Actually, if stage C is not reached during the cycle, a single relation exists between $\Delta\sigma$ and either ΔT or S, independent of the value of Fn: as a consequence, in Figs 9 and 10, the experimental points are mainly located near a single curve. If stage C is reached, ΔT becomes constant for any $\Delta\sigma$ value, while S becomes large, and thus the corresponding points are found below the main line in Fig. 9 and mostly above it in Fig. 10.

The main contents of the 'box', used in Fig. 8 to schematically represent the complex phenomena of fretting effect can be described by introducing suitable parameters, besides S and ΔT already considered. It should be noted that this representation is related to the experimental device and procedure used, and that some of these relations have already been expressed in the tribological model.

Useful parameters can be: the areas Ah and Am, as defined in describing the experimental equipment and procedure; Ct, as introduced in describing the model; and mean contact pressure, considered here as the mean value on the whole contact surface, which may be Ph or Pm depending on whether Ah or Am is used as contact surface. Note that Ph can be evaluated before carrying out a specific test, because it is linked to pad shape, material, and applied Fn; Pm is a typical experimental quantity, measured after the test.

Other parameters can be derived from those already considered, such as specific work Wh and Wm. The last two parameters represent the specific work (i.e., dissipated per unitary contact surface), evaluated on the basis of Ah or Am. Because work is dissipated in stages B and C, and the latter stage is accompanied by large slip and wear, the contribution of the C stage to the work can be omitted and a reduced work Wr can be introduced. This requires the definition of a reduced slip Sr, corresponding to slip occurring only in stage B.

The set of functional relations of these parameters with external data (for example pad characteristics), with applied quantities (as $\Delta\sigma$ and Fn) is outlined in Fig. 8. Once the tribological model has been fitted to the experimental data, some of these relations may be quantitatively obtained from it and from the Hertz theory. Others can only be obtained aposteriori from a specific test. In a previous work (13) the possible relations between Nff or R and these parameters have been systematically studied. The most significant diagrams are reviewed below.

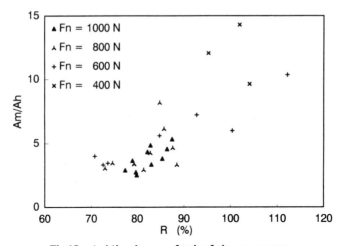

Fig 15 *Am/Ah* ratio versus fretting fatigue parameter

By plotting the ratio *Am/Ah* versus the fretting fatigue stress parameter *R*, Fig. 15, it can be noted that *R* is lower (corresponding to greater damage) for lower values of this ratio; that is, for conditions limited to A and B stages in the model and thus to low slips and small wear. The highest *Am/Ah* ratios correspond to the presence of stage C and to consistent wear, and therefore show the well-known effect of wear in reducing fretting damage (2). This diagram, although very indicative of the phenomena, is not particularly useful since it requires the experimental value of *Am*, available only at the end of the test.

In Fig. 16, where the dissipated work *Wd* is reported as a function of *R*, high values of work correspond to low *R* and the presence of life reduction

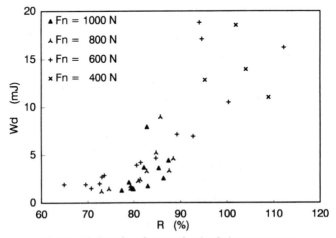

Fig 16 Dissipated work versus fretting fatigue parameter

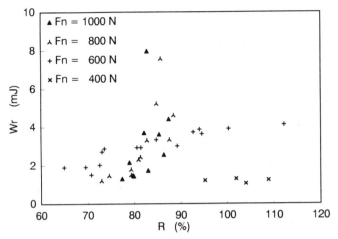

Fig 17   Reduced work versus fretting fatigue parameter

($R < 90$ percent) can be associated to Wd values below about 5 mJ. This behaviour may be explained by the fact that large slips give high values of work, spent, however, mainly in wear.

Figure 17 shows that the reduced work Wr, although it does not include the work dissipated as wear in stage C, does not provide a better correlation with R and an evaluation of life reduction on this basis seems to be impossible. This derives from the fact that the work dissipated in wear is not only lost for fretting damage, but also contributes directly to shortening the already formed cracks.

In Fig. 18, the specific work Wh is plotted versus R. The diagram obtained is similar to that of Wd (Fig. 16) owing to the strong connection existing

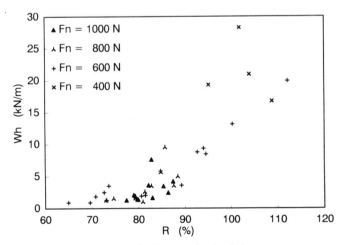

Fig 18   Specific work Wh versus fretting fatigue parameter

between the represented quantities, but the experimental points are less scattered. The life-reducing effect of fretting becomes insignificant only above a given value of this parameter (about 10 kN/m) and it tends to increase when Wh decreases. Note that Wh may be deduced before carrying out a specific test, by using the model for deducing Wd and applying Hertz theory for Ah. Once the external parameters are set, the value of Wh can indicate whether or not a consistent life reduction will occur.

Conclusions

Results reported by different research groups are generally difficult to compare, as they are obtained with different testing procedures and devices. For deriving useful information from any set of experimental results, it is important to have a clear understanding of the many relations (linked to the equipment used) among fixed test parameters and local conditions in the contact area, as well as between these and fretting damage.

While carrying out a set of experimental tests, these relations between parameters have been investigated and collected to construct a tribological model. The model, fitted to the experimental results, provides a useful insight into actual dependences to be expected among external and local parameters, although it does not describe in detail the phenomena occurring in the contact.

Although the proposed tribological model is directly connected to the experimental device and test procedure used, it can represent a common framework for a large class of devices, to which the model can be applied with simple modifications. In some cases, the behaviour of the experimental equipment can be directly represented by the proposed model, requiring only the values of the five quantities Ks, Kd, Hp, ΔTp, and Cd, which can be derived from a few suitable tests.

Various parameters derived from the model can be related to the life reduction due to fretting. In particular, a useful parameter for deducing the conditions leading to early ruptures has been found to be Wh, the work dissipated by tangential force per unitary area of contact surface.

There is agreement between results obtained from the model and fractographical observations. In both cases, the result is that a large slip in stage C does not contribute to life reduction. It is evident that the corresponding wear is stronger than in other stages. Formation of a great amount of debris, however, does not correspond to formation or propagation of cracks; it may instead lead to the opposite effect of shortening existing cracks.

To achieve consistent progress it would be useful to design new devices where, for instance, experimental parameters could be individually modified. However, the characterization of existing devices and comparison of the results obtained with them would benefit from a previous analysis of corresponding parameter relationships, as described here for the specific device used.

Acknowledgements

This work has been carried out with partial financial support by MURST.

APPENDIX – Analytical relations of the model

The elastic displacement amplitudes, see Fig. 1(b), of the initial contact line on specimen (under the action of $\Delta\sigma$) and on device (under the action of ΔT), with respect to the section of device clamp on specimen, are given by

$$\Delta Us = Hs*\Delta\sigma \quad \Delta Ud = Hd*\Delta T$$

Any difference between ΔUs and ΔUd, when present, is balanced by slip S

$$\Delta Us = \Delta Ud + S$$

Stage A

The slip is zero, therefore

$$S = 0 \quad \Delta T = \Delta\sigma*Hs/Hd$$

Stage B

The slip is assumed to be proportional to the part of ΔT greater than the stick limit ΔTp, with a coefficient Hp; therefore,

$$S = \Delta Us - \Delta Ud = (\Delta T - \Delta Tp)*Hp$$

$$\Delta T = [\Delta\sigma*Hs + \Delta Tp*Hp]/(Hd + Hp)$$

$$S = Hp*(\Delta\sigma*Hs - \Delta Tp*Hd)/(Hd + Hp)$$

Stage C

In the slide condition the tangential force ΔT is independent of $\Delta\sigma$ and depends on Fn only, therefore

$$\Delta T = Cd*Fn \quad S = \Delta\sigma*Hs - Cd*Fn*Hd$$

Notes

Since ΔT is very small with respect to the force amplitude on the specimen corresponding to $\Delta\sigma$, it has been assumed that ΔUs depends only on $\Delta\sigma$.

During the fitting of the model to test results, the experimental values of S have been corrected by a constant quantity (4 μm), taking clearances in the experimental equipment into account.

References

(1) WATERHOUSE, R. B. (1981) *Fretting fatigue*, Applied Science, London.
(2) NISHIOKA, K. and HIRAKAWA, K. (1968)–(1972) Fundamental investigations of fretting fatigue, Parts 1–6, *Bull. JSME*, **11**, 437–445, **12**, 180–187, 397–407, 408–414, 692–697, **15**, 135–144.

(3) VINGSBO, O. and SÖDERBERG, S. (1988) On fretting maps, *Wear Maters*, 885–894.

(4) HATTORI, T., NAKAMURA, M., SAKATA, H., and WATANABE, T. (1988) Fretting fatigue analysis using fracture mechanics, *JSME Int. J.*, **31**, 100–107.

(5) ZONFRILLO, G., PRATESI, F., and DEL PUGLIA, A. (1989) Application of multistage life prediction methods to fretting fatigue interactions, *Proceedings of the 10th Conference SMiRT*, (Edited by A. H. Hadjian), American Association for Structural Mechanics in Reactor Technology, Vol. L, pp. 165–170.

(6) KLAFFKE, D. (1983) Einfluß von Oberflächenschichten auf das Schwingungs-verschleißverhalten von Stählen, Vortrag auf VDI.

(7) SANDER, H. (1986) Multitechnique studies on fretting fatigue: influence of surface treatment, Report AD R + D 4038–AN European Research Office of the US Army, London.

(8) DEL PUGLIA, A., PRATESI, F., and ZONFRILLO, G. (1989) Fretting-fatica: problematica e prove su un acciaio da cementazione, *Organi di Trasmissione*, **11**, 60–66.

(9) ZONFRILLO, G., PRATESI, F., and DEL PUGLIA, A. (1991) Resistenza a fatica per varie condizioni di fretting, Atti 20° Conf. AIAS, pp. 223–232.

(10) BERTHIER, Y., VINCENT, L., and GODET, M. (1988) Velocity accommodation in fretting, *Wear*, **125**, 39–52.

(11) HILLS, D. A., NOWELL, D., and O'CONNOR, J. J. (1988) On the mechanics of fretting fatigue, *Wear*, **125**, 129–146.

(12) ROOKE, D. P. and COURTNEY, T. J. (1989) The effect of final friction coefficient on fretting fatigue waveforms, *Fatigue Fracture Engng Mater. Structures*, **12**, 227–236.

(13) ZONFRILLO, G., PRATESI, F., and DEL PUGLIA, A. (1992) Relazioni tra parametri locali e resistenza in prove di fretting-fatica, Atti 6° Conv. Naz. AIM Tribologia, pp. 65–76.

T. C. Lindley* and K. J. Nix†

An Appraisal of the Factors which Influence the Initiation of Cracking by Fretting Fatigue in Power Plant

REFERENCE Lindley, T. C. and Nix, K. J., **An appraisal of the factors which influence the initiation of cracking by fretting fatigue in power plant**, *Fretting Fatigue*, ESIS 18 (Edited by R. B. Waterhouse and T. C. Lindley), 1994, Mechanical Engineering Publications, London, pp. 239–256.

ABSTRACT Fretting fatigue is a mechanism which has resulted in a number of failures in turbine and generator rotors. A review is, therefore, made of the factors which influence the initiation of cracking by fretting fatigue in a component. A number of factors are addressed, including applied stresses, contacting materials and the magnitude of relative slip between components. In addition, the methods available for the assessment of fretting fatigue are discussed and the preferred methods indicated. Any further developments required in fretting fatigue assessment procedures are also discussed.

Introduction

Low pressure turbine and generator rotors are subject to very large numbers of fatigue cycles ($> 10^{10}$) during their operating life. Rotors are designed to withstand such fatigue cycling and failure by fatigue is extremely unlikely unless a small surface defect can form by some means. A number of possible mechanisms are available for the formation of such defects, including surface machining damage and corrosion pitting, but one of the most common is that of fretting fatigue. Fretting fatigue has resulted in the failure of a number of turbine and generator rotors in recent years. Various methods are available to estimate the likelihood of cracking due to fretting fatigue and one method, employing fracture mechanics, has been developed into an assessment procedure **(1)** which has been applied to turbogenerator plant. However, some uncertainties remain in the assessment of fretting fatigue. The purpose of this paper is to review the factors which influence cracking by fretting fatigue and the methods of assessment, highlighting any current deficiencies.

The factors which influence fretting fatigue

Fretting is defined as repeated oscillatory sliding motion between two contacting surfaces. The distance over which the two components slide relative to one another or 'relative slip range' is small in fretting, i.e., 10–100 μm. Fretting is potentially a risk to the structural integrity of a component by one of two processes: the first is fretting wear, in which relative slip results in material

* Ecole des Mines de Paris, Evry, France. Now at Department of Materials, Imperial College of Science and Technology, London, SW7 2BP, UK.
† National Power, Swindon, Wilts, UK.

removal from one or both components and the second, which may occur concurrently with wear, is fretting fatigue – a mechanism by which small cracks initiate and grow from the region of fretting contact between the two components. Fretting fatigue tends to occur most commonly when components are firmly clamped together, such as clamped joints and shrink fitted components. In these situations, the wear due to fretting is often very little and the fretting fatigue crack formation process dominates.

A large number of experimental and analytical studies have been performed in order to identify the factors which might influence the occurrence of fretting fatigue. The likely influencing factors are:

(1) the range of relative slip,
(2) the normal contact (clamping) pressure between the two components,
(3) the chemical composition and mechanical properties of the contacting materials,
(4) the magnitude of any fatigue loading present in the component at risk from cracking.

The importance of each of these factors is discussed below.

Relative slip range

Some experimental studies of fretting fatigue have attempted to investigate the influence of relative slip range on fretting fatigue (2)–(4). Most experiments employed a test assembly similar to that shown in Fig. 1. Fatigue loading was

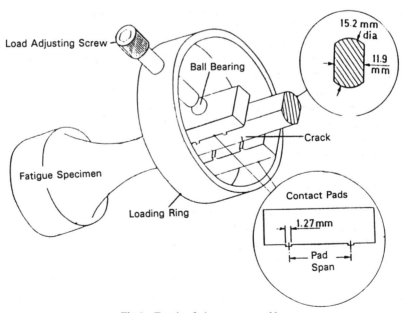

Fig 1 Fretting fatigue test assembly

applied to the specimen with fretting pads clamped to machined flats on the specimen side surfaces. A pre-determined contact pressure was applied between pads and specimen. Under fatigue loading, relative slip occurred between pads and specimen and the magnitude of this slip was measured using a variety of techniques. In each study, the fatigue strength under fretting conditions (defined as applied stress amplitude for 10^8 life) was determined for various testing conditions which imposed different magnitudes of relative slip: these fatigue strengths are plotted as a function of the relative slip range measured at applied stresses equal to the fatigue strength in Fig. 2. Each set of data refer to a different grade of steel tested under different test conditions and hence the fretting fatigue strengths themselves are highly variable. Despite this, all data show a similar trend in that the fretting fatigue strength reduces with increasing slip, achieving a minimum at a relative slip range of between about 10 to 30 μm. In some cases the fretting fatigue strength then increases with further increasing slip. Similar experiments have been performed in which the minimum fretting fatigue life has been determined at a particular level of applied stress (5)(6). These studies gave similar results in that the minimum fatigue life was found to occur at relative slip ranges of around 30 to 50 μm.

The data presented above indicate a clear relationship between relative slip range and fretting fatigue strength, and would suggest a 'critical' range of slip of between 10 and about 30 μm for minimum fatigue strength. Despite this observed experimental relationship between relative slip and fretting fatigue strength, most studies have concluded that fretting fatigue strength is not in fact controlled by relative slip per se. The magnitude of relative slip between two components is determined by the geometry of the two components, the

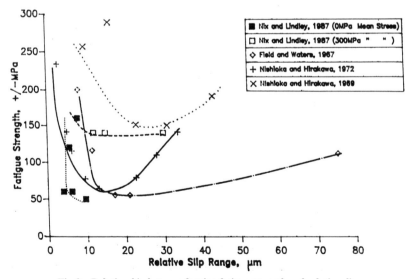

Fig 2   Relationship between fretting fatigue strength and relative slip

fatigue stresses, and the stresses at the contact interface. In a similar way, these factors also control the frictional forces between two components, and it is possible that these forces control fretting fatigue strength, since they determine the magnitude of shear stresses at the fretting location.

The relationship between the frictional forces between two components in fretting contact and the fatigue stress amplitude applied to one component is shown schematically in Fig. 3. At low applied stress, the two components are essentially 'locked' together and there is no significant relative slip. The magnitude of the frictional force is controlled solely by the geometry of the two components. As applied stress levels increase, relative slip is apparent and slight surface fretting damage is apparent as a 'scar'. This form of relative slip is often referred to as 'micro-slip' by Johnson and O'Connor (7) in that the two components are not bodily sliding over one another and a proportion of the relative slip occurring is taken up by shearing of the surface layers of the materials rather than by sliding. The measured relative slip range δ would

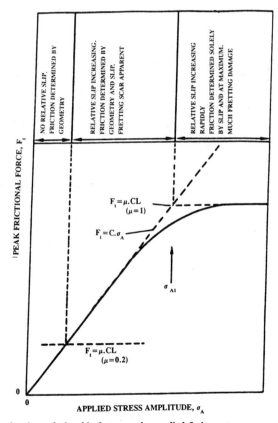

Fig 3   Schematic showing relationship between the applied fatigue stress on a component and the frictional forces at a fretting contact

typically be less than 15 μm in this regime. As applied fatigue stresses increase further, the frictional force achieves a limiting value which does not increase further with increasing stress. This regime is termed the 'macro-slip' regime in that the components are sliding over one another and considerable surface fretting damage is apparent in the form of loose oxide etc. The limiting value of the frictional force is determined by the product of the normal contact load (CL) and the coefficient of friction μ. The relative slip range in this regime would typically be greater than 15 μm and would increase rapidly with increasing applied fatigue stress.

It is clear from the above that relative slip range and frictional forces are intimately related over a range of applied stress conditions and hence the observed relationship between fretting fatigue strength and slip range does not necessarily imply a direct influence of relative slip over fretting fatigue strength. A number of experimental studies have demonstrated that frictional forces rather than relative slip range control fretting fatigue strength. For example, Nishioka and Hirakawa (3) determined the relationship between relative slip and frictional force for a particular fretting fatigue test assembly. They then determined the total cyclic stress at the fretting contact, including both the externally applied fatigue stress and the tangential surface stresses resulting from the frictional forces. It was then possible to show that the fretting fatigue strength of the assembly was related to this total stress.

If relative slip were to have any influence over fretting fatigue strength then it would be over the early stages of crack initiation. Nix and Lindley (4) investigated this aspect by performing fretting fatigue tests at stresses just below the fretting fatigue limit. Such testing resulted in the formation of small non-propagating cracks. The size of these cracks was then related both to the frictional forces at the fretting contact and to relative slip range. No relationship with relative slip range was observed, whereas the size of cracks was directly related to the frictional force, Fig. 4. It was, therefore, concluded that for the range of relative slip investigated, from about 5 to 30 μm, relative slip range had no influence over fretting crack initiation and hence over fretting fatigue strength.

The conclusions reached from the experimental studies described above are fully supported by analytical studies which have been carried out: for a spherical or cylindrical surface in contact with a flat surface, the stresses at the contact are described using Hertzian theory. This theory has been applied to estimate the stresses at such a contact under fretting conditions (8)(9). The stresses calculated in this way have then been used to predict the fretting fatigue strength of a test assembly employing such spherical or cylindrical contacts. The results demonstrate that the fretting fatigue strength is related to the total cyclic stress at the location of fretting.

The investigations summarized above which demonstrate no effect of relative slip range all refer to fretting conditions in which fretting wear was relatively little. If fretting wear is severe then it is possible that small cracks

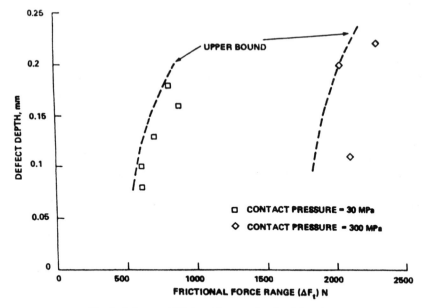

Fig 4   Influence of frictional force on fretting defect size

initiated by the fretting fatigue mechanism can be worn away due to fretting wear. This then results in a gradual increase in fretting fatigue strength as the amount of wear increases. The total volume of material removed by fretting is related to the stresses at the fretting interface and to the relative slip range. Hence, at high slip ranges, the fretting fatigue strength may be controlled in part by the relative slip range. Clear examples of such behaviour are shown in Fig. 2 where, for some of the data sets, the fretting fatigue limit increases under high slip conditions. It must be emphasized, however, that such conditions which result in large amount of fretting wear have not been observed in most incidences of fretting fatigue which have occurred in turbogenerator rotors.

A second situation under which relative slip range may influence fretting fatigue, occurs where the fretting fatigue strength of a component is being improved by the imposition of a surface treatment, such as glass bead peening, or a surface coating, such as a solid lubricant (10). Since fretting is very much a surface related phenomenon, such treatments can result in large improvements in the fretting fatigue strength of a component. However, a relatively small amount of fretting wear could remove the surface treated layer, thus resulting in a large reduction in fretting fatigue strength.

In summary, it is apparent that fretting fatigue strength is related to surface stresses generated by fretting and that the magnitude of relative slip is of secondary importance, unless it is high enough to result in significant wear, or the fretting fatigue strength of a component is enhanced by a surface treatment or coating.

Normal contact pressure

The effect on fretting fatigue of the normal contact pressure between pad and specimen in a fretting test assembly has been explored experimentally by a number of workers. Results for a number of materials are shown in Fig. 5 compiled by Waterhouse (11). These show that fretting fatigue strength initially decreases with increasing contact pressure but at pressures of greater than some critical value (about 50 MPa in the case of a steel) the fretting fatigue strength is approximately constant. A study of the effect of contact pressure on the fretting fatigue strength of 3.5NiCrMoV rotor steel (12) showed no difference between the fretting fatigue strengths at contact pressures of 30 and 300 MPa, Fig. 6.

The above discussion covering the effects of relative slip indicates that fretting fatigue strength is determined by the stresses at the fretting contact resulting from friction between components. Since frictional forces are governed by the contact pressure, via the coefficient of friction μ, it is entirely consistent that contact pressure influences fretting fatigue strength. However, it is perhaps surprising that beyond a minimum value, contact pressure is observed to have no further effect. The rationale for this is that as contact pressure increases, an increasingly large constant compressive stress is imposed in the surface of the component. This compressive stress tends to act in opposition to the cyclic shear stresses resulting from friction: whilst the cyclic shear stresses promote the growth of small cracks into the component, the compressive stress tends to close any such cracks, restricting their growth. The overall

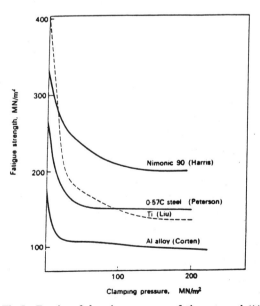

Fig 5   **Fretting of clamping pressure on fatigue strength** (11)

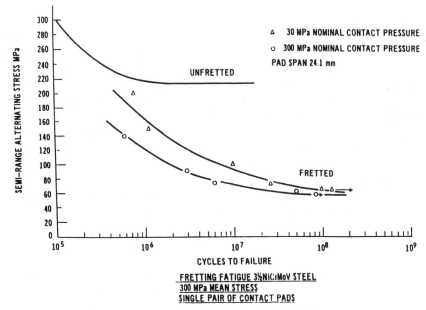

Fig 6   Effect of clamping pressure on fretting fatigue

effect, which can be modelled using fracture mechanics **(1)(13)(14)**, is that fretting fatigue strength is relatively independent of contact pressure.

Chemical composition and mechanical properties of contacting components

A number of experimental studies **(4)(6)(15)(16)** have investigated the effects of contact materials on fretting fatigue strength. In these studies, the test specimen material was kept constant and the contact material varied. Spink **(6)** employed a 2.5NiCrMoV LP rotor steel as the test specimen material. The contacting materials employed were 3CrMoV disc steel, 8Mn8Ni4Cr end ring steel, and mild steel. Different degrees of surface damage were associated with the different contact materials, but no differences were apparent in fretting fatigue strength. Wharton *et al.* **(16)** investigated the fretting fatigue strength of an aluminium alloy, in contact with aluminium, copper, brass, and steel. Except in the cases of copper and brass pads, where severe pad wear occurred, the fretting fatigue strengths were independent of pad material. Nix and Lindley **(4)** employed 3.5NiCrMoV rotor steel specimens in contact with 1CrMo steel and aluminium alloy 2014A. Small differences in fretting fatigue strength were observed between the tests using the two contact materials, but frictional force measurements showed that these differences were due to lower frictional forces (and hence stresses) being imposed on the specimen by the more compliant aluminium alloy fretting pads.

The results described above demonstrate conclusively that for the materials studied, there is no effect of contact material itself on fretting fatigue strength.

All results can be explained on the basis of the stresses generated at the fretting contact. In general it is clear that modifying the material making up the contacting component would not, therefore, give any improvement in fretting fatigue strength. The exception to this would be the case where modifying the contacting material gives a reduction in the frictional forces generated by fretting. An example of this would be the replacement of a component with one of similar dimensions manufactured from a material of lower Young's Modulus (e.g., a change from steel to aluminium). Such a change would give a contacting component of lower stiffness, which would tend to reduce the frictional forces generated by fretting and hence improve fretting fatigue strength.

Magnitude of fatigue loading

It is clear from the discussions presented above that fretting fatigue strength is controlled by the stresses generated at the point of fretting contact between two components. These stresses are the sum of stresses from two sources: (1) the fatigue stresses present in the component at risk from cracking and (2) the stresses generated by fretting at the contact due to normal pressure and frictional loading.

The relative sizes of the stresses from the two sources are highly dependent on the geometry and external loading present. In many situations where fretting is observed in service the fatigue stresses in the component might be very low, or zero, with stresses generated at the fretting contact being the only stresses present. It is pertinent to consider whether fretting fatigue cracking can occur in such a situation.

A relatively small number of experimental studies have sought to investigate separately the effects of remotely applied and fretting generated stresses. This is because, in most conventional fretting test assemblies, the size of the applied fatigue loading controls the range of relative slip and frictional force; sizeable frictional forces, and hence fretting generated stresses, cannot be generated in the absence of applied fatigue stresses. Wright and O'Connor (17) devised a test apparatus which enabled fretting to occur on a specimen surface in the absence of any fatigue loading, but under a range of applied contact stresses. Surface fretting damage was observed but no specimen failures occurred. In addition, subsequent fatigue testing of fretting damaged specimens showed that this damage had not resulted in any reduction in fatigue strength. The conclusion from the work was that some level of fatigue stress was required to enable the initiation and growth of cracks by fretting fatigue.

Using a conventional fretting fatigue test assembly (Fig. 1), Nix and Lindley (4) were able to vary the relative sizes of the fretting cracks generated and applied stresses. This was achieved by employing fretting contact pads of various sizes. Tests carried out at stresses below the fretting fatigue strength resulted in the formation of small non-propagating cracks. The sizes of these cracks were found to be directly related to frictional force at the fretting contact (and hence to the fretting generated component of stress): the sizes were not controlled by the externally applied stress. The conclusion reached here was that in the initiation and early crack growth phase of fretting fatigue it is the fretting generated stress which is of primary importance.

The fretting generated stress must reduce with depth below the surface and hence, in the absence of sufficient applied fatigue stress, crack arrest occurs at a shallow depth (<0.3 mm for the test apparatus used). The results of this study suggest that an externally applied stress might not be necessary to initiate fretting fatigue cracks. This is possibly in contradiction to the studies of Wright and O'Connor (17) where no reduction in fatigue strength was observed for specimens previously tested under fretting stresses only: if cracks had initiated in the fretting test, a reduction in fatigue strength might be expected in subsequent fatigue testing. It is not possible to clarify this point since no metallographic observations were made by Wright and O'Connor (17). However, it is equally possible that in the fretting test, frictional forces were not sufficiently high to initiate cracks.

In summary, both components of stress generated at the fretting contact and externally applied, are necessary for failure to occur by fretting fatigue. The indication is that fretting generated stresses are of primary importance in the initiation phase with externally applied stresses being required for continued crack growth. It is also possible that in the complete absence of any externally applied fatigue stress, small cracks might initiate and grow to some depth before arresting. These cracks would not then grow further by fretting fatigue but could act as sites of initiation for other crack growth mechanisms such as stress corrosion, in for example steam turbine discs.

Methods of assessing fretting fatigue

The ESDU method

The Engineering Sciences Data Unit (ESDU) data items 67012 and 68005 (18)(19) were produced to provide a means of carrying out a fretting fatigue assessment. The ESDU route is very simple and employs a fretting fatigue strength reduction factor K_{ff} = Plain Specimen Fatigue Strength divided by Fretting Fatigue Strength. The size of K_{ff} is assumed to depend on the relative slip range δ. The route specifies that if the relative slip range $\delta \leqslant 5$ μm then $K_{ff} = 2$ at low mean stress and $K_{ff} = 4$ at high mean stress. If $\delta > 8$ μm then the route specifies that $K_{ff} = 5$ at low mean stress and $K_{ff} = 10$ at high mean stress. In the case of shrunk on discs or collars the further proviso is made that fretting damage is only apparent if $\sigma/UTS \geqslant 0.06$ (σ_a is the externally applied fatigue stress and UTS is the Ultimate Tensile Strength of the material, both in MPa). Clearly this route must be regarded as suspect on scientific grounds, since most studies have concluded that fretting fatigue strength is not controlled primarily by relative slip range δ. However, since frictional forces (which do appear to control fretting fatigue strength) are related to δ, for a particular arrangement of components, and the values of K_{ff} are based on experimental data, the use of such a route might be justified on an empirical basis. Experimental fretting fatigue results for turbogenerator materials (4)(6)(10)(12) show that the values of K_{ff} given by the ESDU route are conservative for steel specimens but highly non-conservative for aluminium ones.

The ESDU route can be applied to predict the likelihood of cracking in plant. However, such application does not prove successful for complex failures such as those occurring in turbogenerator rotors. For example, recent generator rotor failures were the result of fretting fatigue. The failure investigation showed that failure occurred at a fretting contact between mild steel filler blocks, each 150 mm in length, and the rotor shaft. The ESDU route specifies that the relative slip range δ is given by the equation $\delta = \sigma S/E$ where S is the length of the component (150 mm in this case), σ is the fatigue stress amplitude and E is the Young's Modulus of the rotor steel. The applied fatigue stress amplitude was the self-weight bending stress, calculated to be approximately ± 20 MPa. Using the above equation, a slip range of $\delta = 14.5$ μm is obtained. Since the mean stress was low (≈ 0) the ESDU route specifies $K_{ff} = 5$. The fatigue strength of the rotor steel was approximately ± 300 MPa, which when factored by K_{ff} gives a fretting fatigue strength of ± 60 MPa. Since the applied stress amplitude was not greater than ± 20 MPa at any time, the ESDU route would not have predicted the occurrence of fretting fatigue in this case. Similar cases have also been analysed, and on no occasion has the ESDU route been able to predict the occurrence of fretting fatigue.

Fretting fatigue strength reduction factors

A potentially improved method of applying a reduction factor to allow for fretting fatigue is to measure fretting fatigue reduction factors K_{ff} directly on materials relevant to the fretting fatigue cracking situation being assessed and then to apply these to predict cracking. Such measurements have been made on turbogenerator materials in a number of studies (4)(6)(12)(20). However, for rotor steels, the values of K_{ff} obtained were in general lower than those given by the ESDU route. Hence, since the ESDU route did not predict failure, fretting fatigue cracking would similarly not have been predicted in the in-service failures using these reduction factors.

The value of measuring K_{ff} is both in simulating as closely as possible the conditions which occur during fretting fatigue and in comparing the fretting fatigue strengths of different materials. Such comparisons have been made as part of materials selection exercises, for example for the 18%Mn/18%Cr Generator End Ring steel (20). A second use of such testing is in assessing the effectiveness of surface palliative treatments. For example, Nix and Lindley (10) investigated the application of various surface treatments to improve the fretting fatigue resistance of aluminium alloy field lead wedge material. A similar investigation was made of the effects of inserting layers of insulating material between the two fretting surfaces on the fretting fatigue strength of a rotor steel (21). In all such investigations where surface layers or coatings are involved, it is of the greatest importance to simulate as closely as possible the fretting conditions existing in service, in particular the range of relative slip δ. This is because δ controls the wear rate, which is critical in determining the integrity of any coating or surface layer. If such a coating or layer could wear away in service then any improvement in fretting fatigue strength would be lost.

Fracture mechanics based procedures

As discussed above, the majority of experimental and analytical investigations of fretting fatigue have concluded that the fretting fatigue strength of a component is controlled by the stresses existing at the point of fretting contact. Factors such as surface wear are of secondary importance for the fretting problems considered here. The implication from this is that if the stresses at a fretting contact could be calculated, a reliable assessment of fretting fatigue could be made. For some idealized contacts where Hertzian theory applies, such as spheres and cylinders pressed onto a surface, such calculations are possible using analytical methods (8)(9). However, for other contacting geometries which occur much more commonly in service, such as two nominally flat surfaces in contact, such analysis is not possible. Recourse could be made to Finite Element Analysis, but such analysis is highly complex where frictional contacts are involved and to date the authors are not aware of any successful fretting fatigue analysis made using such techniques.

Experimental investigations of fretting fatigue have shown that small cracks (<1 mm deep) initiate early in fatigue life (20)(22). In addition, non-propagating cracks have been observed in specimens tested at stresses below the fretting fatigue limit (4)(12)(15). The implication is that small cracks initiate very readily under fretting conditions and hence it is likely under most circumstances that fretting fatigue strength is determined by the conditions for the continued propagation of such defects. For this reason, and taking account of the difficulties in applying a stress based approach, a number of workers have developed assessment procedures based on fracture mechanics which set out to assess the likelihood of the continued propagation of small defects initiated by fretting fatigue. In most cases these methods have employed the 'nominal' stresses at the fretting location, not incorporating any stress concentrating effect due to fretting. The crack tip stress intensity factor values resulting from the fretting stresses have then been determined using methods such as a Green's function method (Rooke and Jones, 1977), which enable K, due to fretting, to be calculated from the surface contact forces at the fretting interface. The total applied K has then been found by adding the component due to fretting to that resulting from the remotely applied stresses.

There are two possible approaches to the assessment of fretting fatigue using fracture mechanics: the first is to calculate the rate of propagation of a crack over a range of crack depths. By integration, the life of a component can then be evaluated. The second, potentially simpler, method is to evaluate the critical minimum size of crack which could grow under fretting conditions by applying fatigue threshold concepts.

The first case in which fracture mechanics was applied to fretting fatigue was the work of Edwards, Ryman, and Cook (22). Stress intensity factors were evaluated using Green's function method developed by Rooke and Jones (23). Crack growth rates were then calculated using a crack growth law of the form

$$da/dN = C\Delta K^m$$

where da/dN is the crack growth rate per cycle, ΔK is the applied stress intensity factor, C and m are constants. The route then integrated the crack growth rate information to obtain crack propagation lives for an experimental testing assembly. The method was subsequently developed to enable its application to bolted joints in aircraft structures (24). A difficulty experienced in the application of this model was in the accurate prediction of crack growth rates for small cracks. For crack depths less than about 4 mm, it was felt that crack growth rates might be higher than predicted by fracture mechanics due to the 'short crack' effect. This effect results from the breakdown in applicability of fracture mechanics when the scale of crack tip plasticity is of the same order as the crack depth. Unfortunately, the time taken for cracks to grow up to 4 mm depth, which occurred very slowly, tended to be a dominant factor in determining the total life of the component. Hence, any errors in the predictions for this phase gave rise to large errors in predicted life. It was necessary to apply an empirical correction to obtain reasonable results. These errors may be in part due to the 'short crack' effect but also to the extreme sensitivity of fatigue crack growth rate to ΔK in the very slow 'near threshold' crack growth regime (where $m = 7$). Very small errors in calculated ΔK could then give large errors in crack growth rate and hence predicted life.

The route developed by Edwards et al. (22) was developed to enable its application to the structural integrity assessment of turbogenerator rotors and other rotating plant by Nix and Lindley (1). In the case of the aircraft components addressed by Edwards (24), which are subject to fatigue loading of variable amplitude, fatigue threshold concepts are difficult to apply. However, for rotating plant, subject to self-weight bend loading, the magnitude of fatigue load cycles is approximately constant. In such a situation it is relatively straightforward to apply threshold concepts: the applied stress intensity factor range ΔK is calculated over a range of crack depths and compared with the threshold stress intensity factor range ΔK_{th} (a material property), Fig. 7. Fretting has a large effect in increasing ΔK and also results in a slight increase in threshold ΔK_{th}. The increase in threshold results from the fact that the normal contact pressure between two components induces compressive axial stresses

FRETTING FATIGUE

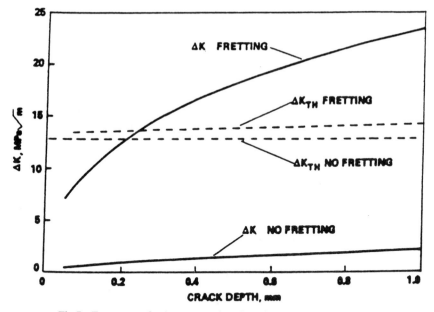

Fig 7   Fracture mechanics computations for a fretting fatigue case history

into the surface of the component which acts to depress the stress ratio (R) and hence elevate ΔK_{th}. The point of intersection of the ΔK and ΔK_{th} curves gives the critical defect size for crack propagation a_{crit}. If this critical size is less than the size of small defects likely to be initiated by fretting fatigue, the component is assessed as at risk from fretting fatigue. Such a technique enables the component to the assessed as 'safe' or 'unsafe' and avoids the complications associated with integration of crack growth rate information at low growth rates, which can lead to substantial errors in life predictions. Fretting fatigue strengths were calculated for an experimental fretting assembly using the model and compared with experimental results. Reasonable agreement was obtained between prediction and measurement.

Other workers have developed similar fracture mechanics based models for fretting fatigue (9)(13)(25)(26). These tend to be based on similar principals to those adopted by Edwards et al. (22) and by Nix and Lindley (1). In all cases, critical defect sizes for crack growth are calculated or total lives estimated on the basis of crack propagation rates. Although refinements are included in some cases, the overall conclusion from all these studies is that fracture mechanics can be used successfully to predict the conditions for the onset of cracking in fretting fatigue.

The fracture mechanics method has been developed into an assessment procedure for fretting fatigue in turbogenerator rotors (27). The procedure includes the fracture mechanics calculation techniques in addition to estimation methods for evaluating the frictional forces at the fretting contact. The model is relatively simple to apply, requiring a knowledge of the nominal applied fatigue stresses (e.g., the self-weight bending stress in a rotor) excluding any stress concentration effects due to fretting, the normal contact pressure between the components in contact, and the fatigue threshold parameter, ΔK_{th} for the material of interest. The procedure has been applied to a number of rotor cracking and integrity assessments (27). The route has been employed both to assess the risk of cracking at a particular location in a rotor and, in the case of failures, to assess the most damaging operating conditions which might have contributed to the failure.

Required improvements in fretting fatigue modelling

The fracture mechanics based assessment procedure discussed above provides a relatively simple method for the quantitative assessment of fretting fatigue. The range of influence of fretting on crack growth is, however, an aspect of fretting fatigue which cannot currently be accounted for. From observations of fretting fatigue cracks both in the laboratory and in plant components it is clear that fretting influences the propagation of a crack only over a limited range of crack depths. Fretting cracks grow initially at a very shallow angle to the surface of a component (4), the growth direction then gradually changes until the crack achieves a crack growth plane perpendicular to the direction of maximum tensile stress in the component. This occurs because fretting imposes high shear stresses on the surface of the component. For small cracks the contribution to the total crack tip stress from fretting is high, resulting in the shallow crack angle. As the crack deepens, the contribution to the total crack tip stress from fretting decreases, resulting in the gradual change in crack growth direction.

This reduction in the influence of fretting-induced stresses means that for cracks greater than a certain depth, crack growth occurs solely under the influence of the stresses within the body of the component, with no significant contribution from fretting. In some situations where the body stresses within a component are low, the body stresses might not be sufficient to give continued crack growth beyond the point at which the influence of fretting is lost. This could lead to the formation of a non-propagating crack, possibly 1–2 mm in depth. Such cracks have been observed in turbogenerator rotors. The current fretting fatigue assessment procedure cannot take account of this since it can take no account of the influence of crack growth on the stresses generated at the fretting contact: as crack depth increases it is likely that crack opening would reduce displacements at the contact, resulting in a reduction in frictional forces and hence in fretting generated stresses. This means that the current fretting fatigue route is likely to be unduly pessimistic under certain circumstances.

An estimation of the depth of influence of fretting and the likelihood of non-propagating crack formation would be of particular value in the integrity

assessment of plant components subject to low fatigue stresses within the body
of the component but potentially high fretting stresses at a contact with a
second component. An extreme example of such a situation would be a button
drive in a turbine disc: if high fretting stresses were generated then a crack
could form by fretting. Such a crack would not continue to grow by fatigue
since the fatigue stresses in a turbine disc are negligibly low. However, the
steady stresses in a disc are high and so the formation of a stress corrosion
crack at such a defect could occur if the crack were large enough for the crack
tip stress intensity factor at the crack tip to exceed K_{1scc}, the critical value for
the initiation of stress corrosion cracking.

These aspects of the depth of influence of fretting and the interaction
between a developing crack and the fretting generated stresses are currently
being addressed using Finite Element Analysis methods. The aim is not to
develop a method which would require the use of finite element analysis in a
plant assessment, but to update the current fretting model to incorporate a
term which addresses the range of influence of fretting.

Conclusions

(1) The fretting fatigue strength of a component is determined principally by
    the stresses generated by fretting at the point of fretting contact. The
    factors which influence fretting fatigue are, therefore, those which control
    these stresses. These are the component geometry and the externally
    applied stresses, including the normal clamping pressure between com-
    ponents.
(2) Other factors, such as the magnitude of relative slip between components
    and the composition and mechanical properties of contacting materials,
    are of secondary importance under most circumstances of relevance to
    turbine and generator rotors.
(3) The only methods which can quantitatively assess the fretting fatigue
    strength of a component are those which either evaluate the stresses at the
    point of fretting or employ fracture mechanics to quantify propagation of
    fatigue cracks from small fretting defects. Since a full stress analysis based
    approach is likely to be highly complex in a plant component assessment,
    a method which employs fracture mechanics is most readily applicable.
(4) Experimental measurements of the reductions in fatigue strength due to
    fretting fatigue cannot be used to quantitatively assess the likelihood of
    fretting fatigue cracking in a component. The value of such measurements
    is in materials selection and in determining the value of palliative treat-
    ments such as surface coatings.
(5) The current fracture mechanics based assessment routes cannot easily
    evaluate the range of influence of fretting on a propagating crack. They
    cannot, therefore, be used to assess the possibility of a crack developing
    from a fretting defect but then becoming non-propagating. This aspect is

currently being addressed with a view to improving the current assessment procedure.

References

(1) NIX, K. J. and LINDLEY, T. C. (1985) The application of fracture mechanics to fretting fatigue, *Fatigue Fracture Engng Mater. Structures*, **8**, 143–160.

(2) FIELD, J. E. and WATERS, D. M. (1967) Fretting fatigue strength of En26 steel, NEL Report, No 275.

(3) NISHIOKA, K. and HIRAKAWA, K. (1969) Fundamental investigations of fretting fatigue, Part 5, *Bull. JMSE*, **12**, 692–697.

(4) NIX, K. J. and LINDLEY, T. C. (1988) The influence of relative slip range and contact material on the fretting fatigue properties of 3.5%NiCrMoV steel, *Wear*, **125**, 147–162.

(5) GAUL, D. J. and DUQUETTE, D. J. (1980) The effect of fretting and environment on fatigue crack initiation and early propagation in quenched and tempered 4130 steel, *Met. Trans.*, **11A**, 1555–1561.

(6) SPINK, G. M. (1990) Fretting fatigue of a 2.5% NiCrMoV low pressure turbine shaft steel: the effect of different contact pad materials and of variable slip amplitude, *Wear*, **136**, 281–297.

(7) JOHNSON, K. L. and O'CONNOR, J. J. (1964) Mechanics of fretting, *Proc. Instn Mech. Engrs*, Part J, **178**, 7–21.

(8) CHIVERS, T.C. and GORDELIER, S. C. (1985) Fretting fatigue and contact conditions: a rational explanation of palliative behaviour, *Proc. Instn mech. Engrs*, **199**, 325–337.

(9) HILLS, D. A., NOWELL, D., and O'CONNOR, J. J. (1988) On the mechanics of fretting, *Wear*, **125**, 129–146.

(10) NIX, K. J. and LINDLEY, T. C. (1986) Palliatives to avoid fretting fatigue in 2014A aluminium alloy, Fatigue prevention and design, Engineering Materials Advisory Service, Warley, pp. 343–352.

(11) WATERHOUSE, R. B. (1972) *Fretting corrosion*, Pergamon Press.

(12) KING, R. N. and LINDLEY, T. C. (1980) Fretting fatigue in a 3.5NiCrMoV rotor steel, Fifth International Conference on Fracture, pp. 631–640.

(13) HATTORI, T., NAKAMURA, M., and WATANABE, T. (1984) Fretting fatigue analysis by using fracture mechanics, ASME Paper 84–WA/DE-10.

(14) RAYAPROLU, D. B. and COOK, R. (1992) A critical review of fretting fatigue investigations at the Royal Aerospace Establishment, *ASTM STP 1159*, ASTM, Philadelphia, pp. 129–152.

(15) NISHIOKA, K. and HIRAKAWA, K. (1972) Some further experiments on the fretting fatigue strength of medium carbon steel, Proceedings International Conference Mechanical Behaviour of Materials, Vol. III, pp. 308–318.

(16) WHARTON, M. H., WATERHOUSE, R. B., HIRAKAWA, K., and NISHIOKA, K. (1973) The effect of different contact materials on the fretting fatigue strength of an aluminium alloy, *Wear* **26**, 253–260.

(17) WRIGHT, G. P. and O'CONNOR, J. J. (1972) The influence of fretting and geometric stress concentrations on the fatigue strength of clamped joints, Proceedings Instn Mech. Engrs, **186**, 827–834.

(18) ESDU, (1967) Engineering sciences data unit item No 67012, notes on the influence of fretting fatigue, ESDU.

(19) ESDU, (1968) Engineering sciences data unit item No 68005, shafts with interference fit collars, ESDU 4.

(20) LINDLEY, T. C. and NIX, K. J. (1982) The role of fretting in the initiation and early growth of fatigue cracks in turbo-generator materials, *ASTM STP 853*, ASTM, Philadelphia, pp. 348–360.

(21) NIX, K. J. and LINDLEY, T. C. (1994) To be published.

(22) EDWARDS, P. R., RYMAN, R. J., and COOK, R. (1977) Fracture mechanics prediction of fretting fatigue under constant amplitude loading, RAE Technical Report No. 77056.

(23) ROOKE, D. P. and JONES, D. A. (1977) Stress intensity factors in fretting fatigue, RAE report No. 77181.

(24) EDWARDS, P. R. (1984) Fracture mechanics application to fretting in joints, Sixth International Conference on Fracture, pp. 3813–3836.

(25) TANAKA, K., MUTOH, S., SADOKA, S., and LEADBEATER, G. (1985) Fretting fatigue in 0.55C spring steel and 0.45 carbon steel, *Fatigue Fracture Engng Mater. Structures* **7**, 129–135.

(26) SATO, K., FUJI, H., and KODAMA, S. (1986) Stress intensity factor for fretting fatigue and crack propagation behaviour, *Conference on Fatigue Prevention and Design*, Engineering Materials Advisory Service Publishers, Warley.

(27) NIX, K. J. and LINDLEY, T. C. (1994) A fracture mechanics based procedure for the assessment of fretting fatigue, to be published.

A. Cardou*, A. Leblond*, S. Goudreau*, and L. Cloutier*

Electrical Conductor Bending Fatigue at Suspension Clamp: a Fretting Fatigue Problem

REFERENCE Cardou, A., Leblond, A., Goudreau, S., and Cloutier, L., **Electrical conductor bending fatigue at suspension clamp: a fretting fatigue problem,** *Fretting Fatigue*, ESIS 18 (Edited by R. B. Waterhouse and T. C. Lindley) 1994, Mechanical Engineering Publications, London, pp. 257–266.

ABSTRACT Overhead electrical conductors are subjected to aeolian vibrations which may induce aluminium wire fatigue breaks at line suspension clamps. Prediction of conductor fatigue endurance is generally based on an alternating stress approach although it is well known that it is a fretting fatigue problem. New tests are performed with a typical conductor–clamp combination. Fatigue diagrams and fracture patterns within the conductor are reported. In particular, fractured wire distribution within conductor cross-section and with respect to clamp tends to show that the usual bending approach could be improved by taking into account conductor–clamp interaction. In future work, they could be useful in helping to predict conductor fatigue directly from individual wire fretting fatigue endurance.

Notation

d_i Wire No i cross-section centre distance from conductor neutral axis
(EI) Conductor equivalent bending stiffness
R_n Radius of layer No n
RTS   Conductor rated tensile strength
Y_b Imposed peak-to-peak bending amplitude at given distance from clamp
σ_a Alternating bending stress

Introduction

Under wind excitation, overhead electrical conductors undergo small amplitude vibrations (aeolian vibrations) which can lead to conductor fatigue failures at suspension points. In practice, such failures are prevented by minimizing bending amplitudes near clamps. Procedures for amplitude characterization have been standardized; for example, in North America the IEEE practice is to measure alternating amplitude at a distance of 88.9 mm (3.5 in) from last point of contact (LPC) between conductor and clamp. This amplitude is then converted to a nominal alternating stress level σ_a by a formula such as the Poffenberger–Swart formula (1). This formula is based on classical beam theory and takes into account some geometrical and material features of the conductor, as well as the static tensile load and vibration amplitude at the selected reference point. Such stress values are useful for assessing the fatigue

---

* Department of Mechanical Engineering, Université Laval, Québec, G1K 7P4, Canada.

strength of widely different conductors. In fact, design curves given by Cigré (Conférence Internationale des Grands Réseaux Électriques) are based on such a stress parameter (2)(3).

Although this approach is of practical value for estimating vibration severity for a given conductor, it neglects an important factor in fatigue performance evaluation, namely clamp–conductor interaction. Such interaction has been known for a long time (4) and it has been clearly shown by Ramey and co-workers (5)(6). In particular, it presents rather a different problem from that of free field, or even, near termination, cable bending fatigue (7)(8). The objectives of the work reported here is thus to present new experimental results on aluminium wire fracture patterns in the clamp–conductor contact region.

Fatigue testing programme

Fatigue testing is conducted on a typical clamp–conductor combination. The conductor is a 35.05 mm diameter conductor consisting of three aluminium layers wrapped around a seven-strand steel core (Fig. 1). Its Canadian designation is 42/7 ACSR Bersimis, where ACSR stands for 'Aluminium Conductor Steel Reinforced'. Its Rated Tensile Strength (RTS) is 154 kN. The clamp is a 267 mm cast aluminium clamp with a bolted keeper (Fig. 2). The fatigue test systen has already been described in previous publications (9)(10). For completeness, its main features will be reviewed (Fig. 3). The bending amplitude is imposed on the conductor specimen by an adjustable eccentric, equivalent to a slider-crank mechanism. The suspension clamp is attached to the slider. Both ends of the specimen are fixed to a tensioning system which allows slight axial displacements in order to maintain a nearly constant tensile load. The up and

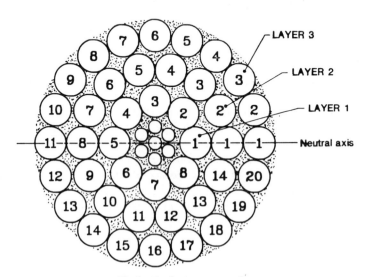

Fig 1   Conductor cross-section

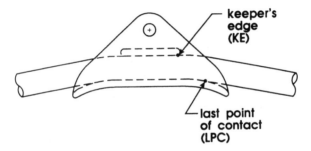

Fig 2   Conductor-clamp system

down clamp motion induces a slight variation of the conductor angle at the clamp mouth. This corresponds, for a given point of the conductor axis, to an alternating amplitude with respect to the clamp, equivalent to the vibration taking place in the field. Two such benches are available.

Specimens 3.6 m long are subjected to an average tensile load T (three levels at 20, 25, and 30 percent RTS). The imposed alternating vertical amplitudes are set to obtain amplitudes of 0.2 to 0.5 mm at the reference point. Based on its minimum theoretical bending strength (all wires acting independently) of 64.8 N m², the relationship between peak-to-peak amplitude Y_b and nominal alternating stress σ_a, for each tensile load is

$\sigma_a = 34.68 \ Y_b$ for 20 percent RTS

$\sigma_a = 36.36 \ Y_b$ for 25 percent RTS

$\sigma_a = 38.38 \ Y_b$ for 30 percent RTS

where Y_b is in mm and σ_a is in MPa.

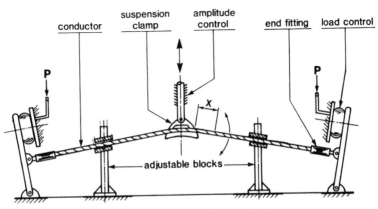

Fig 3   Principle of fatigue test bench

Fatigue test results

Aluminium wire fracture is recorded during each test using the rotation technique described in (9). The first four wire breaks are shown in Figs 4–6 (different data point symbols correspond to separate tests). Bending amplitude is shown in terms of both Y_b and σ_a. Endurance limits are estimated through Weibull analysis (12). Based on the number of cycles to first fracture, they are about 0.27 mm (20 percent RTS) 0.30 mm (25 percent RTS) and 0.22 mm (30 percent RTS). Translated into alternating stress with the Poffenberger–Swart equation (1), they correspond to 9.4 MPa, 10.9 MPa, and 8.4 MPa, respectively. Discrepancy between 20 and 25 percent RTS results is noted but may be considered as normal in this kind of test. Moreover, the results are merely estimates based on short duration tests.

A detailed fracture location analysis is performed on each failed specimen. Thus each fractured wire cross-section is positioned both axially (distance from clamp axis) and transversely (within conductor cross-section). Figures 7–9 show axial locations for each aluminium layer. It is easily seen that most of the fractures occurs in the region between the keeper edge (KE) and last point of contact (LPC) (Fig. 2). Only the external layer shows a majority of fractures occurring at the keeper edge.

Weighted fracture distribution on each layer with respect to non-dimensional distance from conductor neutral axis is shown in Fig. 10. For a given layer (radius R_n), wire distance d_i/R_n varies from 0 on the neutral axis to 1 at the highest and lowest points. A weighting factor is applied to the number of observed fractures to take into account the number of wires in each layer. As already observed in (9)(10), fracture patterns are closer to solid section

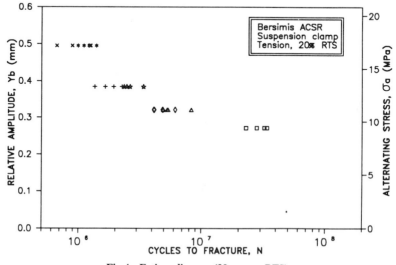

Fig 4   Fatigue diagram (20 percent RTS)

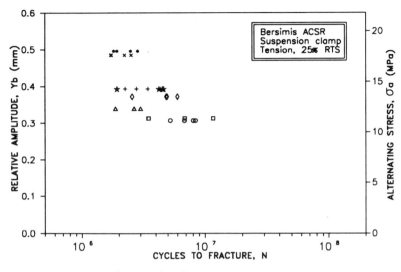

Fig 5   Fatigue diagram (25 percent RTS)

bending behaviour for inner and intermediate layers, where fracture sensitivity increases with distance from the neutral axis. The outer layer fracture pattern, which does not show such a feature, is closer to independent wire behaviour. However, these data mostly represent high amplitude behaviour; low amplitude tests generate very few fractures. A separate analysis is shown in Fig. 11 for three tests where fracture was first obtained above 10 million cycles. These tests gave 16 broken wires with the following distribution per layer: 0 percent

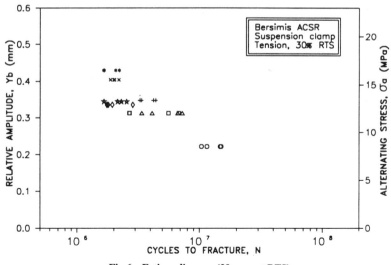

Fig 6   Fatigue diagram (30 percent RTS)

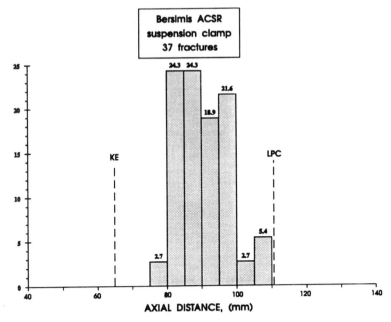

Fig 7   Axial fracture location – internal layer (37 fractures)

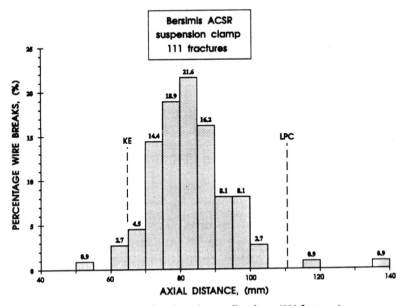

Fig 8   Axial fracture location – intermediate layer (111 fractures)

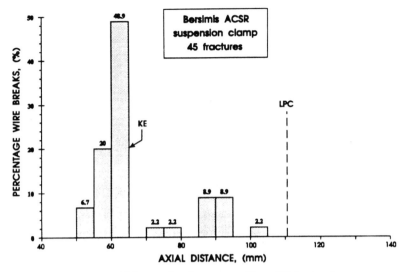

Fig 9   Axial fracture location – external layer (45 fractures)

internal, 41 percent intermediate, 59 percent external. It can be inferred that, at low amplitude, which is the normal case, only the intermediate and external layers move relative to each other. In cross-section, weighted fracture distribution shows a maximum at intermediate wire positions and not at the highest points.

Two other observations are as follows.

(1) All fractures originate at a fretting mark generally located at the outer contact point, that is, the contact point with the next upper layer. Outer layer fractures mostly occur at keeper contact points.

(2) Almost all fractures (95 percent) occur above the conductor 'neutral axis'.

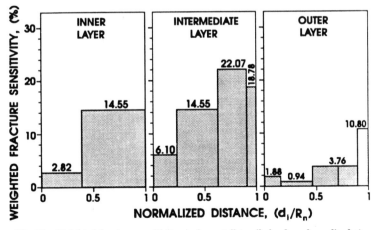

Fig 10   Weighted fracture sensitivity per layer (all tensile loads and amplitudes)

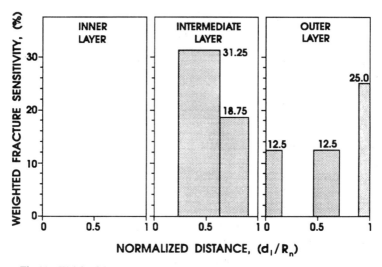

Fig 11   Weighted fracture sensitivity per layer (all tensile loads, low amplitudes)

Clamp–conductor contact

These experimental results show that the bending fatigue concept which is generally used is inadequate if one wants to be able to correlate global conductor strength and individual wire properties. Firstly, the fact that most fractures occur between the keeper edge and LPC, and not in the LPC region, corroborates Ramey's findings (5)(6) that clamp geometry plays a definite role in conductor fatigue strength. Static contact is, therefore, studied on the same clamp–conductor system (11). A combination of strain gauge measurements and FEM modelling is used to evaluate contact pressure distribution between clamp and conductor. It is found that maximum contact pressure occurs near LPC, on the clamp side, pressure being apparently lower between the LPC and the keeper edge. From Figs 7–9, it can be seen that, except for the outer layer, most fractures occur in the region in which relative motion is less constrained than in the LPC region. While the LPC is easy to find in practice and is a good point of reference for vibration amplitude monitoring, it does not seem to correspond to a critical section.

Secondly, stress levels obtained from the Poffenberger–Swart formula are difficult to reconcile with material fatigue properties. Wires are made of EC–H19 aluminium (electrical conductivity aluminium cold-drawn to 90 percent). Its fatigue properties have been studied by Lanteigne et al. (13), under various mean tensile loads and for plain as well as fretting test conditions. For an average axial load of 891 N, which is of the same order as single aluminium wire load when global conductor load is 25 percent RTS, the lowest endurance that was obtained at 100 million cycles was about 60 MPa for plain specimens and 40 MPa with fretting. It is not certain, of course, that test conditions were close enough to those prevailing in a conductor carrying a static tensile load, a static bending moment, a contact pressure on its lower external layer, and undergoing a small cyclic transverse motion on the span side of LPC. However, the estimated conductor fatigue endurance of about 10 MPa is probably not a realistic value.

An improved value can be obtained by analysing the fracture pattern distribution within the cross-section. The Poffenberger–Swart formula was based on strain measurements made on the external layer. Observations made above tend to show that part of the conductor behaves as a solid section, while the rest is constituted with wires acting independently. Thus, equivalent conductor bending stiffness (EI) is much higher than $(EI)_{min} = 64.8$ N m^2 in which all wires are taken independently. For example, considering that the section is solid up to the second aluminium layer, $(EI)_2 = 1054$ N m^2 and the corresponding endurance becomes 43 MPa. This value is in the same range as single wire fretting fatigue endurance reported in (13). However, it is based on stress calculated at the intermediate layers highest and lowest points. Figure 11 shows that a higher number of fractures were observed at mid-distance from the neutral axis, i.e., for an alternating stress of about 20 MPa. Indeed, it is known (8) that maximum relative slip between layers occurs near the neutral axis and decreases with distance from the neutral axis. Thus it appears that conductor bending fatigue is a typical fretting fatigue problem in which both the alternating stress and micro-slip levels have to be considered.

Conclusions

Fatigue testing of large diameter overhead electrical conductors is expensive and time-consuming as fatigue performance is also related to clamp geometry and static tensile load. Thus, it would seem a worthwhile objective to obtain conductor fatigue strength from single wire fretting fatigue behaviour. This implies: (1) experimental data on conductor fatigue behaviour; (2) experimental data on such single wire behaviour; (3) theoretical models describing conductor bending and conductor–clamp mechanical interaction.

The first stage, fatigue testing of conductors, has been carried out for several years in various laboratories. However, to be useful in fretting fatigue analysis, not only the number of cycles to fracture has to be considered but also fracture location patterns. Some new results have been reported in the present work. The analysis was based on completely ruptured wires only. In particular fracture patterns at high and low amplitudes were found to differ. They were used to connect conductor endurance limit to published single wire fretting fatigue values. Even more information on crack initiation process could be obtained from microscopic examination of fretting marks. This kind of work is under way and will be reported separately. However, since inter-layer fretting mechanisms differ at high and low bending amplitudes, more fracture pattern data should be generated based on amplitudes near endurance limit level. The difficulty is, of course, the small number of broken wires generated in these long duration tests. The influence of conductor's number of layers on fatigue strength should also be clarified.

Acknowledgements

Financial support from the Natural Sciences and Engineering Research Council of Canada, Grant No. A8905, as well as collaboration from the Hydro-Québec power utility, are gratefully acknowledged.

References

(1) RAWLINS, C. B. (1979) Fatigue of overhead conductors, Transmission line reference book, wind-induced conductor motion, EPRI Res. Project 792, Electrical Power Research Institute, CA, Chapter 2.
(2) Cigré Working Group 04, Study Committee 22 (1979) Recommendations for the evaluation of the lifetime of transmission line conductors, *Electra*, **63**, 103–145.
(3) Cigré Working Group 04, Study Committee 22 (1985) Guide for endurance tests of conductors inside clamps, *Electra*, **100**, 77–86.
(4) CLOUTIER, L. and HARDY, C. (1988) Effect of suspension clamp design on conductor fatigue life, Report No. ST–178, Canadian Electrical Association.
(5) RAMEY, G. E. and TOWNSEND, J. S. (1981) Effects of clamps on fatigue of ACSR conductors, *J. Energy Engng*, **107**, 103–119.
(6) McGILL, P. B. and RAMEY, G. E. (1986) Effect of suspension clamp geometry on transmission line fatigue, *J. Energy Engng*, **112**, 169–184.
(7) RAOOF, M. (1992). Free bending fatigue of axially preloaded spiral strands, *J. Strain Analysis*, **27**, 127–136.
(8) RAOOF, M. (1992) Free bending fatigue life estimation of cables at points of fixity, *J. Engng Mech.*, **118**, 1747–1764.
(9) CARDOU, A., CLOUTIER, L., LANTEIGNE, J., and M'BOUP, P. (1990) Fatigue strength characterization of ACSR electrical conductors at suspension clamps, *Electric Power Syst. Res.*, **19**, 61–71.
(10) CARDOU, A., CLOUTIER, L., ST-LOUIS, M., and LEBLOND, A. (1992) ACSR electrical conductor fretting fatigue at spacer clamps, *Standardization of fretting fatigue test methods and equipment, ASTM STP 1159*, (Edited by M. Helmi Attia and R. B. Waterhouse), ASTM, Philadelphia, pp. 231–242.
(11) CARDOU, A., LEBLOND, A., and CLOUTIER, L. (1993) Suspension clamp and ACSR Electrical Conductor Contact Conditions, *J. Energy Engng*, **119**, 19–31.
(12) LEBLOND, A. (1992) *Interaction mécanique entre conducteur électrique aérien et pince de suspension: conditions de contact et fatigue*, MSc Thesis, Université Laval, Québec, Canada.
(13) LANTEIGNE, J., CLOUTIER, L., and CARDOU, A. (1986) Fatigue life of aluminium wires in all-aluminium and ACSR conductors, Report No. 131–T–241, Canadian Electricity Association.

M. Raoof *

Free Bending Fatigue Prediction of Steel Cables with the Effect of Interwire Fretting Taken into Account

REFERENCE Raoof, M., **Free bending fatigue prediction of steel cables with the effect of interwire fretting taken into account**, *Fretting Fatigue*, ESIS 18 (Edited by R. B. Waterhouse and T. C. Lindley) 1994, Mechanical Engineering Publications, London, pp. 267–281.

ABSTRACT The importance of design against free bending fatigue at the points of fixity to steel cables used in various applications is emphasized. It is suggested that there is a pressing need for re-examination of the traditional extreme fibre stress (or strain) approaches.

Based on the newly proposed theoretical model, the present paper reports the final formulations from an extensive series of theoretical parameteric studies on a number of realistic strand constructions with widely different cable (and wire) diameters and lay angles.

A straightforward method for design against free bending fatigue, aimed at practising engineers, which takes interwire/interlayer fretting fully into account, is presented. The proposed design method is amenable to hand calculations using a pocket calculator.

The significant shortcomings of certain previously reported twin-wire fretting experiments aimed at predicting and/or improving steel cable fatigue life is discussed. It is suggested that realistic fretting fatigue tests on single or twin wires should be carried out under the modes of deformation representing those operative inside the cables.

Notation

A	Wire cross-section area
c	$= \sqrt{\{(r_1 - r_w)^2 + k^2\}}$
E_s	Steel Young's modulus
k	Pitch of the helix divided by 2π
k_s	$= \dfrac{T}{T_{ult}}$
K_{ln}	Interlayer no-slip shear stiffness
M_0	Maximum bending moment (at the fixed end)
n_2	Number of wires in the second (penultimate) layer
r_1	Helix radius in the outer layer
r_w	Wire radius in the outer layer
s	Standard deviation of the fitted σ_x
s'	Estimated standard deviation of fitted σ_x
S'_1	Cable mean axial strain
T	Cable mean axial tension
T_{ult}	Cable ultimate breaking load

* Civil and Building Engineering Department, Loughborough University of Technology, Loughborough, Leicestershire, LE11 3TU, UK.

U	The computed interlayer relative displacement near the neutral axis at a distance equal to the interlayer contact patch spacing along the wire from the assumed ideally fixed end
x	Trellis contact patch spacing along any individual wire in the outer layer
α_1, α_2	Lay angles in the outer and second (penultimate) layers
θ	Polar angle in the strand
θ_0	Polar angle in the strand at the fixed end
θ_0^{max}	Polar angle in the strand at the fixed end where the direct (axial) stress (or strain) is maximum
κ	Interlayer no-slip shear stiffness $= K_{ln}$
μ	Coefficient of interwire friction
v	Poisson's ratio for the wire material
ρ	Radius of curvature at the fixed end
σ_x	The computed tensile stress at the trailing edge of the interlayer contact patches between wires with opposite lay
σ_x^u	The upper bound estimate of σ_x
σ_x^f	The mean of the data points relating to the value of the tensile stress at the trailing edge of the interlayer contact patch
ψ	$= \alpha_1 - \alpha_2$

Introduction

A detailed study of the available literature strongly suggests the paucity of reliable information on fatigue performance of steel cables in free bending. There has been (since early this century) a considerable body of mainly experimental works reported on the influence of pulley and sheaves as sources of wear and fatigue damage to wire ropes. However, once one addresses the bending effects in axially preloaded spiral strands in the absence of sheaves, fairleads, or other formers, the available literature seems to be very limited. Indeed one is faced with a considerable degree of uncertainty as regards the basic underlying assumptions of various approaches aimed at predicting the free bending fatigue life of a cable whose radii of curvature at points of fixity are not predetermined.

Steel cables are widely used as hangers to suspension bridges and stays in cable-stayed structures. They are also employed as structural elements in guyed masts and offshore floating platforms. Another important area where restrained bending fatigue of helical strands is a source of concern (and not infrequent failures) is that of electromechanical cables where the armour wires are often made from materials different from steel (e.g., aluminium). In all these applications, restrained bending fatigue failures invariably occur very close to terminations and wherever a rapid change in curvature is imposed. As an example of restrained bending fatigue problems under service conditions one may refer to the widely publicized failures of inclined hangers of the Severn Bridge in the UK.

For the free bending of long cables under an approximately steady axial load, it is common to introduce a mathematically convenient constant effective bending rigidity, $(EI)_{\text{eff}}$, for the cable, from which the radii of curvature at the points of restraint are calculated. The maximum bending strains in individual wires are then found on the basis of often semi-empirical formulations of a questionable nature. These strains are further assumed to govern the strand's bending fatigue life. Experimental and field observations, on the other hand, suggest the interwire fretting between often counterlaid wires in various layers as the principal cause of wire fractures (1)–(3).

In a series of fairly recent publications by the present author and his associates, the traditional (i.e., extreme fibre stress or strain) approaches for design against cable restrained bending fatigue have been re-examined and (as discussed next) considerable shortcomings have been identified. Indeed, interwire/interlayer fretting has been identified as the main source of wire fractures. The purpose of the present paper is to present a brief account of these findings. Moreover, a straightforward design method, aimed at practising engineers, which takes the effect of interwire friction and fretting fully into account, is proposed. This alternative, and simple, approach is shown to largely overcome various shortcomings of the previously available maximum stress theories.

This paper is in the form of a review paper relating to recent work conducted by the author on fretting problems: further details may be extracted from papers by the author which are quoted in the reference list of this paper.

Background

The literature on free bending of cables has already been reviewed elsewhere by Raoof and Hobbs (4) and Raoof (5)(6). Perhaps it will suffice to mention some fairly recent attempts by Nowak (7), Knapp (8), and Costello and Butson (9) who ignored the presence of interwire/interlayer friction and attempted to estimate the magnitude of wire direct stresses in spiral strands under uniform bending to a constant radius of curvature away from points of fixity, such as, for example, end terminations. Among those attempting to include interwire friction in their model are Lutchansky (10) and Feld et al. (11) with the latter extending Lutchansky's model to include the bending of individual wires about their own axes. Neither reference, however, provide analytical (or realistic experimental) means of obtaining values of the shear stiffness, κ, which defines the interaction shear force between the outer-layer wires and the underlying core which is assumed to suffer from no shear distortion under bending of the cable to a constant radius of curvature. Consequently, both references only quote tentative values for κ, and their linear models cannot predict the interwire slippage phenomena, which are of prime importance in the analysis of the interwire fretting process near terminations.

Following the experimental observations of Scanlan and Swart (12) and the present author (5)(13), the pre-tensioned large diameter spiral strand is modelled as being composed of a solid core, which undergoes plane-section bending, covered by a single layer of helical wires. For a given strand geometry, the shear interaction between the outer wires and the core has been found (14) to be a function of the mean axial load on the strand and the state of curvature near the clamp. For the severe state of curvature in the very near vicinity of the clamp, results are obtained for outer wire axial stresses and for the associated extent of relative slippage between the wires and the core (14), in addition to the strains in the outer layer wires bending about their own neutral axes (15).

In the newly proposed model, outer wire-core shear interaction is treated as an extended version of the model suggested by Lutchansky (10). In his work, Lutchansky assumed that the interface shear was linearly proportional to the difference between the local movements of the outer wire, u, and the undergoing core particle, u^p, adjacent to the wire surface. Unlike Lutchansky's work, the present model includes a quantitative treatment of the transition between no-slip and full-slip frictional interaction between the wires (14). In particular, it is shown that although axial wire strains (and stresses) at the clamp reach a maximum near the extreme fibre position, interwire/interlayer slippage (and hence fretting) is greatest in the vicinity of what may be considered as the neutral axis. This then throws considerable light on the (perhaps) initially puzzling experimental observations of Hobbs and Ghavami (3), who found that the first wire fractures in their free bending fatigue tests invariably occurred not at the extreme fibre position but near the neutral axis where direct wire strains are very small (4). It is, therefore, concluded that interwire or rather interlayer fretting is more likely to be the cause of wire breakages and is more relevant to a discussion of free bending fatigue life than the maximum wire stresses near the extreme fibres as has traditionally been assumed. The following section examines the viability of this suggestion in the light of experimental findings reported by others.

Critical examination of the previous approaches

Figure 1 shows some data reported by Cunninghame *et al.* (16) for the restrained bending fatigue lives of a number of axially preloaded spiral strands whose fatigue life is defined as the number of cycles to fifth outer wire fracture. The plots of the range of angular deflection, $\Delta\theta$, against fatigue life are found to suffer from a considerable degree of scatter to an extent that it is not even possible to realistically define the general trend of the test data. A very weak correlation is found between $\Delta\theta$ and fatigue life. In other words, the angular deflection is found not to be the underlying controlling parameter.

In a fairly recent publication by Hobbs and Smith (17), large-scale test data were reported for a limited number of 39 mm diameter axially preloaded spiral strands. These steel cables were subjected to steady lateral cyclic loading. Two specimens carried nominally constant axial loads of $T/T_{ult} = 17$ percent, with the other four strands tested with mean axial loads of $T/T_{ult} = 34$ percent,

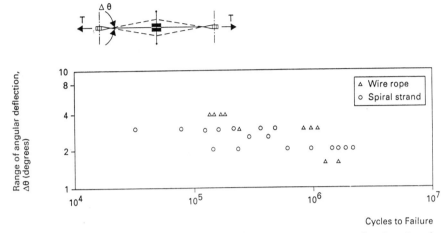

Fig 1   **Bending fatigue performance of socketed spiral strands and wire ropes plotted against the range of angular deflection – Cunninghame *et al.* (16)**

where T_{ult} = ultimate breaking load. For their boundary conditions and pattern of external loading, Hobbs and Smith (17) assumed that

$$\frac{M_0}{(T_{ult}\, EI_s)^{0.5}} \approx 1.1 \tan \theta'(K_s)^{0.5} \tag{1}$$

where EI_s = assumed constant cable bending stiffness, θ' = the nominal angular deflection at the fixed end, and $K_s = T/T_{ult}$. Using the rather limited number of data points, Hobbs and Smith (17) attempted to correlate the so-obtained values of the moment at the fixed end, M_0, with the restrained bending fatigue life (which they assumed to be the number of cycles to first outer layer wire fracture). This approach then enabled them to make reasonable predictions for other values of mean axial load, T, and angular deflection, θ', for this same strand. They did, however, point out that the use of their fatigue curve for other strands with similar construction details and end terminations is rather tentative in the absence of more fatigue data. At a later date, Strzemiecki and Hobbs (18) and the present author reported some free bending fatigue data on 40 and 41 mm diameter spiral strands and a 40 mm diameter multi-strand rope with the tests carried out in the same test rig used by Hobbs and Smith (17). The main difference between these tests and those on the 39 mm strand was the widely differing values of the lay angles with the outer diameters for all the specimens kept nominally constant. It is interesting to note that, similar to the 39 mm strands, Strzemiecki and Hobbs (18) found the initial outer layer wire fractures to invariably occur very close to what one may refer to as the cable neutral axis (in terms of the simple beam bending theory). Plots of the non-dimensionalized moment at the fixed end, $M_0/(T_{ult}\, EI_s)^{0.5}$ versus number of cycles to first outer layer wire fracture are given in Fig. 2 which cover the full set of data on 39, 40 and 41 mm diameter

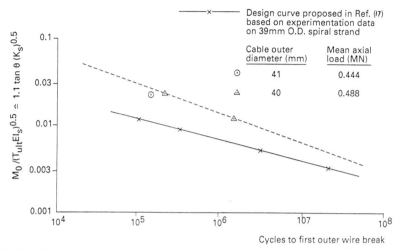

Fig 2   Plot of non-dimensionalized moment at the clamped end against free bending fatigue life for
a number of spiral strand (and one multi-strand rope) constructions – reference (22)

spiral strands. An examination of results in Fig. 2 suggests that the data relating to 40 and 41 mm diameter strands do not lie close to the originally proposed design curve of Hobbs and Smith (17) (based on data relating to the
39 mm strand). Indeed, there appears to be a weak correlation between the
bending moment at the fixed end, M_0, and fatigue life to first outer layer wire
fracture where data points relating to different spiral strands appear not to
follow a single curve.

Raoof (14) recently developed a theoretical model for estimating the direct
(axial) strain of helical wires at points of fixity to spiral strands. It was shown
that the maximum wire axial strain, $(du/ds)_{max}$, for working levels of cable
mean axial load occur fairly close to the extreme fibre position (in terms of
simple plane section bending) with values of du/ds close to the so-called
neutral axis being negligibly small. Indeed, some carefully conducted large-
scale tests on the 39 and 40 mm diameter spiral strands (5)(13) have provided
very encouraging support for the newly developed theoretical model of reference (14). These tests employed nominally identical loading and boundary
conditions to those in references (3)(17) and (18). Perhaps it is worth mentioning that the parameter (du/ds) only relates to the direct wire axial strains
caused by interlayer shear interaction and does not include the bending of the
wires about their own neutral axis. This latter problem has been addressed by
Raoof and Huang (15): it was found that (at least at the fixed end), the values
of (du/ds) are (for large diameter strands) much higher than the corresponding
values of strains due to bending of wires about their own neutral axis which
may safely be ignored. Finally, Fig. 3 presents plots of maximum (extreme
fibre) direct strains, $(du/ds)_{max}$ versus fatigue life to second outer layer wire
fracture for the 39, 40, and 41 mm diameter axially preloaded spiral strands. It

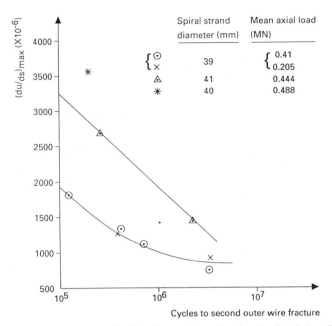

Fig 3 Plot of maximum (extreme fibre) bending strain against free bending fatigue life to second outer wire fracture for a number of spiral strand constructions – reference (22)

is shown that even when one goes through a very careful process of obtained reliable values of maximum direct wire strains, the data for various spiral strand constructions do not follow a single (unified) curve and a considerable degree of scatter is still found. Apparently, this parameter is (at its best) capable only of defining a single curve for fatigue data relating to a *specific* strand construction subjected to a range of mean axial loads with different strand constructions leading to very different (du/ds) against fatigue life curves. It then follows that the widely adopted (traditional) extreme fibre design approach does not yield a generalized fatigue curve which is applicable to any type of spiral strand construction. It is a significant point to bear in mind that very often the experts in the field carry out their tests on one type of strand construction covering a range of other parameters such as mean axial load and/or angular deflection at the point of fixity. Indeed if they are careful in estimating and/or measuring the appropriate values of extreme fibre (i.e., maximum) direct strains, they should be able to propose a fatigue curve with acceptable data scatter for the particular cable used in their tests. The underlying problem, however, is that their proposed fatigue curve is unlikely to be applicable to other types of spiral strand constructions. In other words, such a time-consuming and expensive route has to be followed every time one tries to use a new type of strand construction: with an almost unlimited number of possible spiral strand geometrical make-ups, the practical shortcomings of such an approach are obvious.

As discussed below, in order to develop a generalized fatigue curve, there is a need for an approach which takes the interwire contact forces and slip between wires fully into account with emphasis on the fretting fatigue behaviour over the individual interwire/interlayer contact patches where wire fractures have mostly been found to occur.

Figure 4 presents correlations between a newly developed contact stress–slip parameter, $\sigma_x U/x$, and fatigue life to second wire fracture where a single (generalized) curve can be drawn through data for various helical strands as reported by Hobbs and his associates [3][17][18] and the present author. The observed degree of scatter is quite acceptable with the plot covering a wide range of lay angles, levels of mean axial load, and imposed angular deflection at the fixed end. The contact stress–slip parameter is composed of three components, as listed below.

U = estimated interlayer relative displacement (in the vicinity of the so-called neutral axis) at a distance equal to trellis contact patch spacing

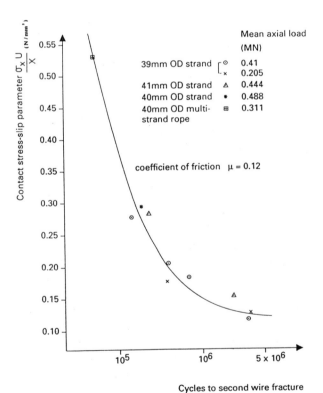

Cycles to second wire fracture

Fig 4   Plot of the newly proposed contact stress–slip parameter against free bending fatigue life to second outer wire fracture for a number of spiral strand (and one multi-strand rope) constructions – reference (22)

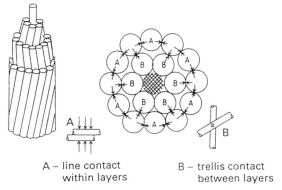

A – line contact
within layers

B – trellis contact
between layers

Fig 5   Pattern of interwire/interlayer contact forces in an axially preloaded spiral strand

along the outer layer helical wire, from the point of fixity to the cable;

x = trellis contact patch spacing along the individual outer layer helical wires, Fig. 5;

σ_x = maximum tensile stress at the trailing edge of the interlayer trellis contact patches between outer and penultimate layers which have opposite lay directions, Fig. 6. The magnitude of this parameter is estimated assuming linear elastic behaviour of wire material.

The coefficient of friction assumed for data in Fig. 4 is $\mu = 0.12$ which has been found to be a reasonable value for galvanized steel wires (19). The plots in Fig. 4 should prove of value as a predictive tool for design against restrained bending fatigue failures of spiral strands. Perhaps it is worth mentioning that the restrained bending fatigue data was obtained by cyclic applications of a point load acting transversely to an axially preloaded cable with both ends fixed against rotation. One end of the cable was nominally pinned

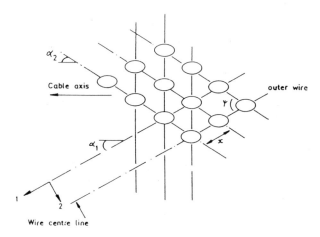

Fig 6   Pattern of interlayer trellis contact patches – reference (22)

with the other end fixed against bending with the point load located nearest to this latter end. All wire fractures were found to occur at the mouth of the fixed sockets where bending moment was a maximum.

Appropriate factors for realistic interwire fretting experiments

According to the above, the mode of interwire fretting for the strand bending fatigue at points of fixity is one of a sawing action at the trellis points of the interlayer contact patches as opposed to the torsional mode over such trellis contact patches (20) for the axial case. Theoretical studies suggest that for the average strand constructions subjected to typical working axial loads, interwire fretting under restrained bending conditions appears to reach a significant level at rather large values of radii of curvature at the restrained socket. For example, for the 39 mm diameter strand under a mean axial load of about 33 percent of the ultimate breaking load, interwire slippage between the outer and penultimate layers can reach the full-slip (gross-sliding) condition (i.e., relative interwire displacement ≈ 0.00085 mm), well within the operational radii of curvature (of the order of 10–20 metres) at the restrained end. Full torsional slippage over the trellis points of contact is, however, very unlikely to occur in this same strand, once one addresses the axial fatigue problem. The use of the *same modes* of interwire contact deformation as those occurring inside spiral strands, in any interwire fretting experiment aimed at predicting and/or improving cable fatigue life is, therefore, strongly recommended. The appropriate levels of input parameters, such as the normal loads and relative deformations, for such experiments may now be realistically determined by the straightforward routines recommended by Raoof (21).

Lubrication is an important factor in cable fatigue performance. In a number of previously reported twin wire experiments which have been carried out to identify the significance of various lubricants as regards cable fatigue performance, the range of input parameters such as magnitude of imposed normal forces and/or relative displacements have not been chosen with due regard given to those that are operative within realistic steel cables. Very often, the modes of interwire contact deformations have been determined by the ease of setting-up the testing equipment with little attention paid to adhering to the most appropriate modes of fretting movements as determined by the actual cable loading conditions. Future tests, however, should take such effects into account.

Finally, there is, at present, an ever increasing demand for larger diameter spiral strands outside the traditional range of cable diameter which tend to employ wire diameters up to, say, 8 mm. There is currently very limited fretting fatigue data available on such large diameter wires. The majority of published axial fatigue S–N curves for bare wires have also been obtained on specimens with often much smaller diameters. Axial fatigue tests on larger diameter wires should, therefore, be carried out in order to identify possible size effects which may also affect the interwire fretting fatigue performance specially under sea-water corrosion conditions.

Calculation of contact stress–slip parameter

Details of an extensive theoretical parametric study which employed a number of realistic spiral strand constructions subjected to a wide range of mean axial loads, is given elsewhere **(22)(23)**. In particular, in view of the lengthy mathematical formulations as presented by Raoof **(4)**, which are not suitable for use by practising engineers, a straightforward method based on the parametric studies for obtaining reasonable estimates of the contact stress–slip parameter has been developed **(22)(23)**. The final formulations based on the proposed straightforward method, which are amenable to hand calculations using a pocket calculator, are presented below.

Taking the simple one first, the value of trellis contact patch spacing, x, is given by **(24)**, Fig. 6.

$$x = \frac{2\pi(r_1 - r_w)\cos \alpha_2}{n_2 \sin (\alpha_1 - \alpha_2)} \tag{2}$$

where, r_1 = helix radius in the outer layer, r_w = wire radius in the outer layer; α_1 and α_2 = lay angles in the first (outer) and second layers, respectively, with right hand lay being +ve; and n_2 denotes the number of wires in the second (penultimate) layer.

The value of the interlayer relative displacement, U, at the ideally fixed end with polar angle θ_0 is

$$U = U'(\sin \theta - \sin \theta_0) \tag{3}$$

with

$$U' = \frac{k^2(r_1 - r_w)}{\rho c} \tag{4}$$

where r_1 = helix radius of the outer layer; r_w = outer layer wire radius; ρ = radius of curvature at the point of fixity; $k = p/2\pi$ with p = outer wire pitch; and $c = \sqrt{(r_1 - r_w)^2 + k^2}$. The magnitude of the radius of curvature, ρ, may be obtained by the closed-form solutions given elsewhere **(22)(23)**. θ = polar angle of the helical outer layer wire measured from the point at which the wire first crosses the extreme fibre position of the toroid, with θ_0 = polar angle at the fixed end **(4)(10)**, Fig. 7. The magnitudes of θ_0 and θ are estimated as follows

$$\theta_0 = \theta_0^{max} - \frac{\pi}{2} \tag{5}$$

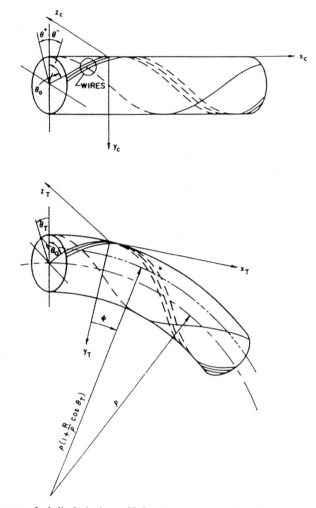

Fig 7 Geometry of a helical wire in a cable bent to a constant radius of curvature – reference (10)

and

$$\theta = -\left(|\theta_0| - \frac{x}{c}\right) \tag{6}$$

with **(10)**

$$\theta_0^{\max} = -\tan^{-1}\left\{\frac{1}{c}\sqrt{\left(\frac{AE_s}{K_{1n}}\right)}\right\} \tag{7}$$

where, θ_0 is always negative and E_s = steel Young's modulus; K_{1n} = no slip interlayer shear stiffness; and A = outer layer wire cross-section area. Extensive numerical results have shown that the magnitude of K_{1n} is, for all practical purposes, given by

$$\frac{K_{1n}}{E_s} = -0.002\,83 + 0.000\,109\psi + 0.0977(S_1')^{0.3179} \tag{8}$$

$$\left(\begin{array}{l} 22 \text{ degrees} < \psi < 42 \text{ degrees} \\ 0 < S_1' < 0.0045 \end{array} \right)$$

where, E_s = Young's modulus for steel, $\psi = \alpha_1 - \alpha_2$ (in degrees), and S_1' = strand mean axial strain.

Finally, the following fitted equation defines variations of σ_x with the cable overall axial strain, S_1' and the total angle between counterlaid wires in the outer and penultimate layers, ψ

$$\sigma_x^f = -124 + 4.41\psi + 2069(S_1')^{0.317} \tag{9}$$

$$\left(\begin{array}{l} 22 \text{ degrees} < \psi < 42 \text{ degrees} \\ 0 < S_1' < 0.0045 \end{array} \right)$$

where, σ_x^f = the mean of the data points relating to the value of the tensile stress in (N/mm^2) at the trailing edge of the interlayer contact patch between wires in opposite lay – i.e., at the edge of the so-called 'trellis' contact, $\psi = \alpha_1 - \alpha_2$ is in degrees, and S_1' = strand mean axial strain.

An upper bound estimate of σ_x may simply be obtained from the following equation (23)

$$\sigma_x^u = \sigma_x^f + t\sqrt{(s'^2 + s^2)} \tag{10}$$

where, for a 95 percent confidence limit to the present number of 95 samples, $t = 1.99$. In equation (10), σ_x^u = upper bound to σ_x (in N/mm^2) whose fitted mean as calculated from equation (9) is σ_x^f, s = standard deviation of the fit, and s' = estimated standard deviation of fitted σ_x. In equation (10), $s = 40.96$, and $s' \approx 7.12$, with both parameters having units of N/mm^2. Note that the theoretical numerical results reported in this paper relate to wire materials made of high tensile steel with Young's modulus $E_s = 200$ KN/mm^2 and Poisson's ratio $v = 0.28$.

Once one has calculated values of σ_x^u, U, and x, the upper bound to contact stress–slip parameter may simply be calculated as $\sigma_x^u U/x$. Using this value of the parameter, the plot in Fig. 4 gives a conservative estimate of the free bending fatigue life to second outer layer wire fracture in the absence of corrosion. It must be borne in mind that corrosion is often a problem in practice and its treatment is outside the scope of the present paper.

A numerical example for using the above simplified approach is given elsewhere (23).

Conclusions

The importance of design against free bending fatigue at the points of fixity to steel cables used in various applications has been emphasized.

A brief review of the available methods of design against free bending fatigue has highlighted the significant shortcomings of the previously available maximum stress theories (and their variations) in the light of some recently reported experimental observations on steel cables. It can be concluded that there is a pressing need for developing a more realistic and appropriate design method which takes the effects of interwire fretting into account.

The final formulations based on a rather extensive theoretical parametric study which has employed a number of realistic spiral strand constructions subjected to a wide range of mean axial strains, have been presented. These provided a simple method for obtaining reasonable estimates of the contact stress–slip parameter for predicting free bending fatigue life.

Finally, possible shortcomings of some previously reported twin-wire fretting experiments aimed at predicting and/or improving cable fatigue life were discussed. In particular, guidelines were provided for carrying out appropriate twin-wire fretting experiments under realistic modes of movements which are believed to depend on the type of fatigue loading on cables.

Acknowledgements

Thanks are due to Bridon Ropes personnel, Doncaster, UK, for their patient help with information over a number of years.

References

(1) CAPADONA, E. A. and COLLETI, W. (1967) Establishing test parameters for evaluation and design of cable and fittings for VDS towed systems, Transactions of the New Trust Seaward, Third Marine Technology Society Conference, pp. 641–675.
(2) HONDULUS, B. (1964) Comparative vibration fatigue tests 84/19 ACSR 'Chukar' versus 61-strand 5005, IEEE Winter Power Meeting.
(3) HOBBS, R. E. and GHAVAMI, K. (1982) The fatigue of structural wire strands, *Int. J. Fatigue*, **4**, 69–72.
(4) RAOOF, M. and HOBBS, R. E. (1984) The bending of spiral strand and armoured cables close to terminations, *J. Energy Resources Technol.*, **106**, 349–355.
(5) RAOOF, M. (1989) Free bending tests on large spiral strands, *Proc. Inst. Civ. Engrs*, Part II, **87**, 605–626.
(6) RAOOF, M. (1992) Free bending fatigue life estimation for cables close to points of fixity, *J. Engng Mech.*, **118**, 1747–1764.
(7) NOWAK, G. (1974) Computer design of electromechanical cables in ocean applications, Proceedings of the Tenth Annual MTS Conference, pp. 293–305.
(8) KNAPP, R. H. (1988) Helical wire stresses in bent cables, *J. Offshore Mech. Arctic Engng*, **110**, 55–61.
(9) COSTELLO, G. A. and BUTSON, G. J. (1982) Simplified bending theory for wire rope, *J. Engng Mech.*, **108**, 219–227.
(10) LUTCHANSKY, M. (1969) Axial stresses in armour wires of bent submarine cables, *J. Engng Ind.*, **91**, 684–693.
(11) FELD, G. *et al.* (1991) Power cables and umbilicals conductor strain under pure bending, First International Offshore and Polar Engineering Conference, Vol. II, pp. 228–235.

(12) SCANLAN, R. H. and SWART, R. L. (1968) Bending stiffness and strain in stranded cables, IEEE Winter Power Meeting, Vol. 68, CP43–PWR.

(13) RAOOF, M. (1992) Free bending fatigue of axially preloaded spiral strands, *J. Strain Analysis*, **27**, 127–136.

(14) RAOOF, M. (1990) Free bending of spiral strands, *J. Engng Mech.*, **116**, 512–530.

(15) RAOOF, M. and HUANG, Y. P. (1992) Wire stress calculations in helical strands undergoing bending, *J. Offshore Mech. Arctic Engng*, **114**, 212–219.

(16) CUNNINGHAME *et al.* (1992) Strengthening and refurbishment of Severn Crossing – Part 4: TRRL research on Severn Crossing, *J. Structures Buildings, Proc. Instn. Civ. Engrs*, **94**, 37–49.

(17) HOBBS, R. E. and SMITH, B. W. (1983) The fatigue performance of socketed terminations to structural strands, *Proc. Inst. Civ. Engrs*, Part II, **25**, 35–48.

(18) STRZEMIECKI, J. and HOBBS, R. E. (1988) Properties of wire ropes under various fatigue loadings CESLIC Report SC6, Department of Civil Engineering, Imperial College, London.

(19) RAOOF, M. (1990) A multi-contact problem, Proceedings of the Nineth International Conference on Experimental Mechanics, Vol. 1, pp. 103–111.

(20) RAOOF, M. (1990) Axial fatigue of multi-layered strands, *J. Engng Mech.*, **116**, 2083–2099.

(21) RAOOF, M. (1993) Influence of interwire contact forces on the overall behaviour of spiral strands, *Proceedings of Workshop on 'Length Effect on Fatigue of Wires and Strands'*, International Association for Bridge and Structural Engineering (IABSE), pp. 279–291.

(22) RAOOF, M. (1993) Design of sheated spiral strands against free bending fatigue at terminations, *J. Strain Analysis*, **28**, 163–174.

(23) RAOOF, M. (1993) Design of steel cables against free bending fatigue at terminations, *The structural engineer*, Institution of Structural Engineers, Vol. 71, pp. 171–178.

(24) PHILLIPS, J. W., MILLER, R. E., and COSTELLO, G. A. (1980) Contact stresses in a straight cross-lay wire rope. First Annual Wire Rope Conference, pp. 177–200.

G. A. Kopanakis*

Effects of Fretting Fatigue on Flexed Locked Coil Ropes

REFERENCE Kopanakis, G. A. **Effects of fretting fatigue on flexed locked coil ropes,** *Fretting Fatigue,* ESIS 18 (Edited by R. B. Waterhouse and T. C. Lindley) 1994, Mechanical Engineering Publications, London, pp. 283–294.

ABSTRACT Over recent years, during regular non-destructive testing, a number of track ropes have shown an unexpected degree of damage in the areas subject to stretching and alternate bending at the roller chain. As a result of the findings the Institute of Lightweight Structures and Ropeways of the Swiss Federal Institute of Technology, Zurich began investigations of locked coil track ropes both in service and in a specially constructed full-scale fatigue testing machine.

Introduction

A counterweight can be used to compensate for potential load variations of the track rope of a reversible aerial tramway caused by fluctuating ambience temperature, cabin load, and cabin position. In most larger aerial tramways the track rope is directly deflected to the counterweight over a roller chain. The compensating movement of the counterweight causes a motion of the rope and the roller chain over the deflection saddle **(1)**.

During regular non-destructive testing, a number of track ropes have shown an unexpected degree of damage within the areas subject to stretching and alternate bending at the roller chain. The calculation of the multiaxial state of stresses of a single wire in a bent steel wire rope is still not satisfactory, which makes a theoretical approach almost impossible **(2)**. Therefore, the Institute of Lightweight Structures and Ropeways of the Swiss Federal Institute of Technology, Zurich started investigations of locked coil track ropes both in service and in a specially constructed full-scale fatigue testing machine in order to answer the following questions **(3)–(5)**:

(1) how does the damage nucleate and develop within the flexed area of the rope (see **(6)**);
(2) is there a visible or acoustic indication, giving a reliable warning of a critical amount of damage;
(3) to what extent are the results of gamma ray examination applied in this area reliable;
(4) how is the fatigue behaviour of the track rope over roller chain influenced by:
  – the rope itself (design and construction, lubrication);

* Institute of Lightweight Structures and Ropeways, Swiss Federal Institute of Technology, Zurich, Switzerland.

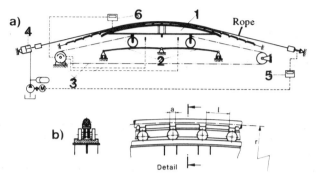

Fig 1   **Layout of the roller chain testing machine**
   1   Saddle and roller chain (fabricated by the Swiss Ropeways Industry) fixed on a carriage
   2   Track (of the carriage)
   3   Drive system
   4   Tensioning cylinder
   5   Automatic control of the tensile force
   6   Automatic control of the carriage velocity, adjusted to the rope temperature

   – the ropeway installation (geometrical parameters of the deflection saddle
     and roller chain, number, and length of the spans);
   – the ropeway operation (trip frequency, cabin load).

Experimental procedures

Test equipment

The testing machine (Figs 1 and 2) was designed to test ropes up to a diameter
of 55 mm, reproducing the actual stress condition of a locked coil track rope
flexed at a roller chain of a reversible tramway. The rope specimen was bent

Fig 2   **View of the roller chain testing machine**

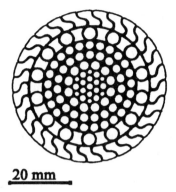

20 mm

Fig 3   Rope R_1. Diameter: 50.7 mm. Construction: $1\frac{1}{2}$ × locked coil $1 + 6 + (6 + 6) + 12 +$
19 + 25 + 14 × I + 14 × 0 + 30 × Z. Minimum breaking load: 2640 kN

over the roller chain and tensioned by the hydraulic cylinder. The shape of the
carriage tracks was calculated to invert the movement which occurs in practice. Thus it is not the roller chain and the rope which rolled over the saddle
but the saddle which moved under the roller chain and rope. For the application of service life relubricant, drip oil lubrication was used (7). In order to
perform the necessary regular NDT it was ensured that the rope was totally
clear of the roller chain along each flexed length under the regular test load
(8).

Tests

The fatigue tests were conducted using specimens taken from two different
ropes (Figs 3 and 4) which had been manufactured with three different initial
lubricants (L_1, L_2, and L_3).

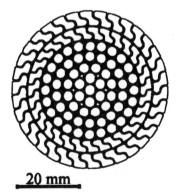

20 mm

Fig 4   Rope R_2. Diameter: 50 mm. Construction: 2 × locked coil $1 + (6 + 6) + 12 + 18 +$
24 + 31 × Z + 34 × Z. Minimum breaking load: 2590 kN

Two different roller chain/saddle sets were used in the test.

RK_1: radius = 5500 mm (r_1/d = 108.5), link length = 185 mm, link distance
      = 40 mm, link form (1) = curved ($r_{link} = r_{saddle}$), link form (2) =
      straight, link lining (1) = Polydur, link lining (2) = steel;
RK_2: radius = 5000 mm (r_2/d = 98.5), link length = 360 mm, link distance
      = 45 mm, link form = curved ($r_{link} = r_{saddle}$), link lining = Polyacetate.

The flexed length was about 1200 mm ($\sim 3 \times$ the lay length of the outer layer)
and the length of the constantly bent area to 1600 mm ($\sim 4 \times$ the lay length of
the outer layer) (Fig. 5).

A completed test covered 600 000 bending cycles and lasted about three
months: one bending cycle (bc) is completed after a stretched cross-section has
been bent and stretched again once. The frequency of bending cycles chosen
ensured that the temperature of the rope surface never exceeded 40°C during
the tests.

The development of the number of wire breaks was followed by means of
gamma ray examination which was applied regularly during test life. In order
to keep close to practice, the chosen NDT methods, as well as the equipment
required, were those used on tramways for regular examinations. Thus an
Ir^{192} source was used throughout all the experiments (8). At the end of each
test the rope specimen was carefully dismantled and the damage found was
registered.

Although the reproducibility of wire rope fatigue tests is usually not high,
even when using the same rope, the tests mentioned here show that in this
particular case of locally high resultant stress a high degree of consistency is
possible. Figure 6 shows the results of two identical tests. The plots in Figs 6,
10–16 show the development of the number of wire breaks as a function of
bending cycles. The number of wire breaks is plotted on the Y axis, and the
number of cycles is plotted on X axis. The lines drawn are an exponential
approximation.

During the dismantling of the rope the relative position of the wires in the
rope was registered in order to enable reconstruction of the local configuration
in the surroundings of the breaks.

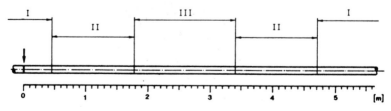

**Fig 5   Areas of the rope specimen. I Constantly stretched area (0 bc); II Flexed area (n bc); III
Constantly bent area (1 bc)**

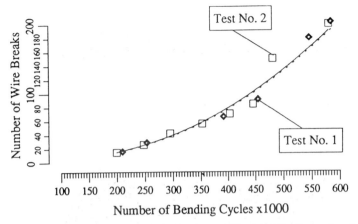

Fig 6 **The development of the number of wire breaks as a function of bending cycles. An example of good reproducibility**

Results of the tests

The first observation to be made is that the damage first starts and propagates in the second layer. The first wire breaks start in the 'lower part' (i.e., at the rope side contacting the roller chain links) or/and in the 'upper part' of the rope (i.e., the opposite rope side) (Fig. 1(b)).

Audible crackle emissions starting during the early stage of a test were always soon followed by the first wire breaks occurring, as well as by a high amount of total damage at the end of a test. However, the fact that during other tests no crackle emissions could be heard did not mean that no wire breaks occurred. In those cases the damage started at a later stage and resulted in fewer wire breaks. The lubricant condition found after terminating these tests was always better.

After the first breaks had occurred, more wire breaks followed – nearly in the same cross-section. They first remained in the same layer and then propagated ringwise **(8)(9)**. Visible helical distortion generated during service life, always indicated a concentration of wire breaks within the same area in the second layer. Wire beaks in the first (outer) layer always followed the previous advanced damage (all the wires of the second layer multiply broken) in the vicinity of the damage zone in the second layer. After terminating the test, the rope specimens were dismantled. The condition found at the second layer described below, was constantly repeated throughout the tests.

Almost all the wires of the second layer were found to have multiple breaks (Fig. 7), as well as several cracks next to the existing breaks (Fig. 8). An extreme local lack of lubrication within the contact area of two wires of different layers (Fig. 9) and traces of fretting or frictional wear within these areas characterized the damage.

Fig 7   A 60 mm long piece of 'O' wire ($d = 4.5$ mm) taken from the second layer, broken five times

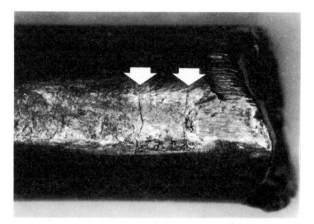

Fig 8   A piece of 'O' wire taken from the second layer with a fatigue break on the left. The arrows indicate the cracks nucleated next to the fatigue break

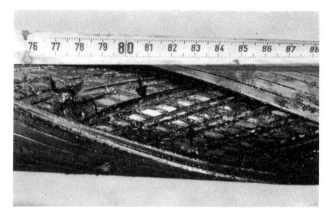

Fig 9   Picture of the second layer after partly removing the first layer. The arrows indicate the contact areas with the wires of the first layer with visible local lack of lubrication

The correspondence between the diagnosed (by NDT) sum of wire breaks and the actual figure was found to be satisfactory. Even in cases of considerable damage, 80 percent of the wire breaks were identified. However, the existence of numerous multiple wire breaks made the correlation between the diagnosed number of wire breaks and the actual cross-section loss practically impossible. The fatigue behaviour was influenced by the various parameters (mentioned at the beginning), as follows: the life expectancy of the rope construction R_2 (2 × locked coil) was, for this particular application and the ropes tested, almost 80 percent longer than that of R_1 ($1\frac{1}{2}$ × locked coil) (Fig. 10).

The amount of fretting corrosion found, depended on the quality and quantity of the lubricant used, i.e., in cases of a better lubricant, for this particular application there was hardly any fretting corrosion to be found, this also resulted in a life improvement. For the lubricants used, rope construction, and deflection radius the service life using initial lubricant L[2] was about 40 percent longer than the one using initial lubricant L[1]. Figure 11 shows the comparison between the two initial lubricants.

The influence of service relubrication depends on its compatibility to the initial lubricant, as well as on its ability to penetrate to the second layer of the locked coil wire rope. An example is shown in Fig. 12(a); in this case a life increase of about 40 percent was achieved.

Figure 12(b) shows that no significant improvement can be expected if there is no compatibility between initial and service relubrication. The above results show that the choice of the initial lubricant, as well as the application of a proper service relubricant, has a significant influence on the fatigue behaviour of the rope under these resultant stress conditions. These findings initiated a detailed investigation of the influence of this particular parameter on the fatigue behaviour of the locked coil rope. As expected the reduction of service load significantly influenced the fatigue behaviour of the rope (Fig. 13). In this

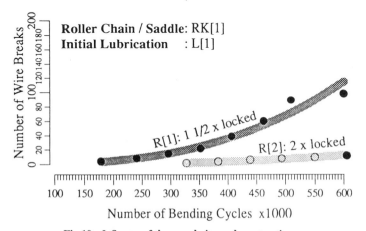

Fig 10   Influence of the rope design and construction

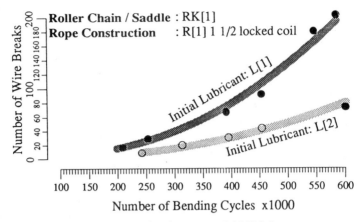

Fig 11   Comparison between two initial lubricants

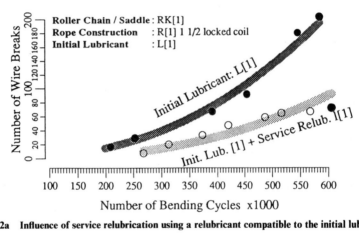

Fig 12a   Influence of service relubrication using a relubricant compatible to the initial lubricant

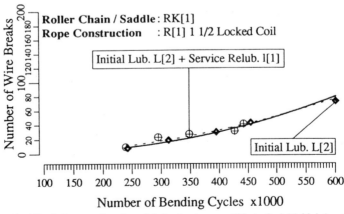

Fig 12b   Influence of service relubrication incompatible to the initial lubricant

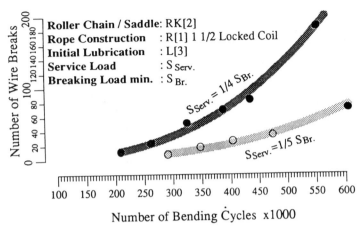

Fig 13 Influence of service load

case a service load reduction of 20 percent resulted in a life increase of approximately 40 percent.

The two roller chain/saddle constructions were compared (Fig. 14). Although the number of the wire breaks in the case of **RK(2)** was higher than in the case of **RK(1)**, the effective damage was the same in both cases (the second layer was totally broken). The difference was a result of the varying number of the existing multiple wire breaks.

For the tested rope construction, lubrication, roller chain construction and deflection radius no significant service life differences were found by comparing the curved and the straight chain link form (Fig. 15). Figure 16 shows the

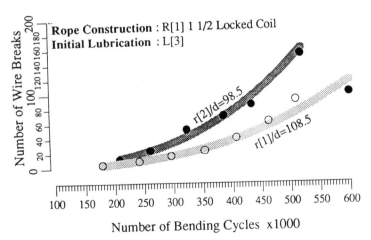

Fig 14 Influence of the deflection radius

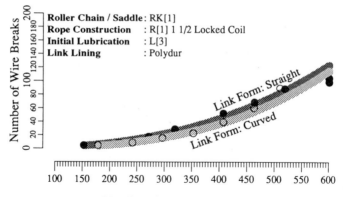

Fig 15   Influence of the link form

comparison between two different link linings. The higher local pressure at the edges of the steel link lining results in a shorter service life.

All the above findings correspond well to the results of the metallographic examinations carried out by the Swiss Federal Laboratories for Materials Testing and Research, according to which, with the exception of some single wires, all the origins of the first cracks were situated within the intensive wear zones, mainly caused by high local pressure (Fig. 17). On the other hand, as indicated by the fractographical findings, the crack propagation is caused by combined bending and torsion and secondarily by tension **(10)**.

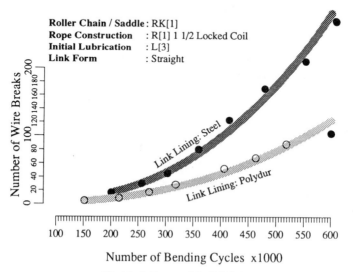

Fig 16   Influence of the link lining

Fig 17 All the origins of the first cracks were situated within the areas subjected to fretting or intensive wear

Conclusions

The following conclusions can be drawn.

(1) The damage starts and initially remains in the second layer. When this layer reaches a significant amount of damage, visible damage in the first (outer) layer occurs.

(2) Audible crackle emission indicates crack propagation, and visible helical distortion of the rope surface indicates existing wire breaks.

(3) If the application of gamma ray examination procedure is possible, satisfactory results can be expected concerning the diagnosed number and location of the wire breaks which occur.

(4) All the improvements resulting in a reduction of high local pressure and fretting between the single wires like lower service load, suitable rope construction, and/or lubrication, larger deflection radius, as well as avoiding any local reduction of the deflection radius, cause a significant life improvement.

Acknowledgements

The author wishes to thank BAV (Swiss Federal Department of Traffic), the Swiss Rope and Ropeways Industry, and SVS (Swiss Ropeways Association) for supporting this project. Particular thanks are owed to EMPA (Swiss Federal Laboratories for Materials Testing and Research) for the very useful cooperation, namely for all the NDT and especially the detailed metallographic examinations, as well as for all the useful information given.

References

(1) CZITARY, E. (1962) *Seilschwebebahnen*, Springer Verlag.
(2) LEIDER, M. (1975) *Die Bestimmung der Zusatzspannungen bei der Biegung von Drahtseilen und ihr Einfluss auf die Seillebensdauer*. Dissertation TH-. Karlsruhe.
(3) KOPANAKIS, G. (1991) On the behaviour of ropes on roller chains, *International Aerial Lift Review*, 1, 6–10.
(4) KOPANAKIS, G. (1991) Fatigue behaviour of locked coil ropes over roller chains, OIPEEC Technical Meeting 'Safety, Acceptance and Discard of Ropes'.
(5) KOPANAKIS, G. (1992) The procedure of fatigue and failure of a flexed locked coil wire rope, IABSE Workshop on 'Length effect on fatigue of wires and strands'.
(6) WOODTLI, J. (1994) Microscopical identification of growth and wear due to fatigue in locked coil wire ropes, *Fretting Fatigue*, Mechanical Engineering Publications, London, pp. 297–306. *This volume.*
(7) OPLATKA, G. (1985) Nachschmierung von Drahtseilen, *Draht*, 36, 375–378.
(8) LUETHY, T. and BLASER, E. (1989) Radiographic Examination of Steel Wire Ropes, OIPEEC Round Table Conference.
(9) CHAPLIN, C. R. and TANTRUM, N. R. H. (1985) The Influence of Wire Break Distribution on Strength, OIPEEC Round Table Conference.
(10) WOODTLI, J. and KOPANAKIS, G. (1993) Fractographical Evaluation of Fatigue Damage in Locked Coil Ropes. *Fourth International Conference on Structural Failure, Product Liability and Technical Insurance*, Elsevier Science Publishers, Amsterdam, pp. 521–528.

IDENTIFICATION AND EFFECTS

*J. Woodtli**

Microscopical Identification of Fretting and Wear Damage due to Fatigue in Locked Coil Wire Ropes

REFERENCE Woodtli, J., **Microscopical identification of fretting and wear damage due to fatigue in locked coil wire ropes,** *Fretting Fatigue*, ESIS 18 (Edited by R. B. Waterhouse and T. C. Lindley) 1994, Mechanical Engineering Publications, London, pp. 297–306.

ABSTRACT In order to examine fatigue behaviour of locked coil ropes bent over a roller chain, full-scale fatigue tests are run using two different ropes (1). For better understanding of how the fatigue damage develops, microscopical investigations of single wires, which are broken first, are carried out. The fatigue cracks of all locked coil wire ropes tested start in the second layers. The morphology of the damage clearly reveals multiple mechanisms of crack nucleation and propagation. By using SEM and metallographical investigation a differentiation of the predominant damage mechanism involved is established. The fatigue crack nucleation is a result of friction wear and fretting under high compressive stress. Typical features of each damage mechanism are described. It is shown that the local fretting and abrasive wear, which lead to crack nucleation, are mainly the result of misfit in the construction of the ropes.

Introduction

In aerial tramways the track ropes used are deflected over the roller chain and their life time is determined by the breaks within the rope area flexed at the roller chain in the terminal. These breaks start in the second layer and show typical features of fatigue loading. Over recent years, during regular non-destructive testing (2) it has been found that a number of track ropes have an unexpected degree of damage within this critical area. Therefore, the Institute of Lightweight Structures and Ropeways at the ETH Zürich, together with the Swiss Federal Laboratories of Materials Testing and Research EMPA, have examined fatigue behaviour of a locked coil rope in simulated tests. The results of these tests are described in (3).

It is obvious, that the fatigue loading spectrum within this area is very complex and consists of tensional, bending, torsional, and compressive components. The analytical solutions for the determination of the damage caused by multi-componental loading in a single wire are still not satisfactory, and hence a theoretical approach is almost impossible. Since the fracture surface is a detailed record of the failure history, an attempt has been undertaken to determine the significant loading components leading to the cable damage by means of fractographical analysis (4)(5). Therefore, EMPA started microscopical investigations in order to estimate the mechanism of the damage. The exact location of the primary break was determined using the γ-ray recordings. The results obtained are discussed in relation to both the crack initiation and the propagation in single wires.

* Swiss Federal Laboratories for Materials Testing and Research, CH 8600 Dübendorf, Switzerland.

Material and tests

The test equipment and parameters used are extensively described in the paper
by Kopanakis (3). The ropes were stretched by a tensile force, which corresponds to 1/4 of the breaking load of the rope and cyclically bent over the
roller chain (bend radius $R = 98.5 \times d$) over about three months with 600 000
load cycles. This testing simulates approximately twenty years of service life.
To be able to investigate the primary wire fractures, the ropes were periodically checked by γ-ray during the fatigue loading and the position of the
indicated cracks recorded. Using this method, the primary wire breaks could
be traced and the development of the crack propagation in single wires within
the clusters was established (1)(2). The γ recordings of the damage development in the ropes avoided the confusion of primary and secondary fractures
within the crack clusters. In the following the nucleation of primary wire
breaks and spreading of wire breaks in clusture arrangements are discussed.

Two different ropes were investigated (Fig 1). The specifications of the wires
used are shown in Tables 1(a) and (b).

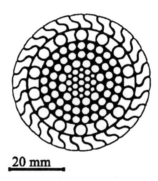

20 mm

Fig 1(a)   Rope R_1. Diameter: 50.7mm. Construction:1 1/2 × locked coil 1 + 6 + (6 + 6) +
12 + 19 + 25 + 14 × I + 14 × 0 + 30 × Z. Minimum breaking load: 2640 kN

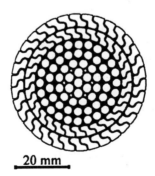

20 mm

Fig 1(b)   Rope R_2. Diameter: 50mm. Construction: 2 × locked coil 1 + (6 + 6) + 12 + 18 +
24 + 31 × Z + 34 × Z. Minimum breaking load: 2590 kN (Fig 4)

Table 1(a)

R_1: ϕ 50.7 m	R_{eff} (N/mm^2)	C (percent)	Mn (percent)	Si (percent)	S (percent)
'I' profile wire	1770	0.8	0.6	0.23	0.02
'O' profile wire	1860	0.78	0.68	0.23	0.02
'Z' profile wire	1670	0.73	0.54	0.23	0.02

Table 1(b)

R_2: ϕ 50 mm	R_{eff}(N/mm^2)	C (percent)	Mn (percent)	Si (percent)	S (percent)
'Z' profile wire 1st L.	1730	0.8	0.73	0.23	0.015
'Z' profile wire 2nd L.	1850	0.8	0.6	0.28	0.009

Morphological investigations

By means of morphological study the following points were clarified:

– which mechanisms contribute to nucleation of the primary cracks?
– which forces are responsible for the crack propagation in single wires?
– why are cracks locally concentrated in nest-like zones?
– how could the nucleation of cracks be retarded?

Macroscopical findings

Before the ropes were dismantled, the relative positions of the wires with regard to the neighbouring wires in the same layer as well as to the neighbouring wires in the superimposed layer were recorded. After the first layer was dismantled it was established that the sites with the breaks showed a local lack of lubrication. The intensive friction left locally red coloured oxide debris. Further investigation showed that the friction oxidation debris was partly due to consequential damage phenomena and, therefore, of limited importance for the indication of the prime crack initiation. In all ropes investigated the cracks and fractures were always in nest-like clusters. Two configurations dominated; either in a circumferential direction (around the cross-section, within the second layer), or in a helical direction (along the contact line of the superimposed wire with the second layer) (Fig. 2).

With a few exceptions, the breaks appeared within the intensive wear zones. Typical of the 1 1/2 × locked coil wire ropes were crack origins in the 'I' wires on the surface facing the first layer, whereas the 'O' wires showed crack origins mostly on the interface with the adjacent 'I' wire (Fig. 3). The first layer of these ropes suffered fatigue cracks with origins on the rope surface and not, as presupposed, on the interface with the second layer. The fatigue breaks in the 2 × locked coil wire ropes always originated on the surface of the second layer. During the tests conducted no cracks were found in the first layer of these ropes.

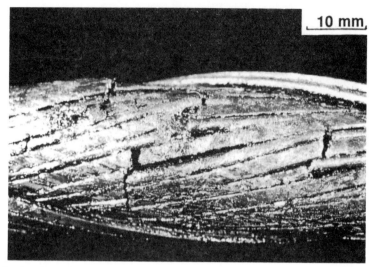

Fig 2   **Partly dismantled rope** R_1**. View of the second layer with nest-like configurated crack clusters**

It was observed in both types of ropes that single wires in the first and/or the second layers were locally shifted from their ideal position which resulted in a geometrically reduced contact area where primary cracks were nucleated. The shifts were more prominent in the 2 × locked coil ropes. Figure 4 illustrates the type of reduced contact area between the first and the second layer as found in the 2 × locked coil rope.

Fig 3   **Arrows indicate crack origins in 'I' and 'O' wires in rope** R_1

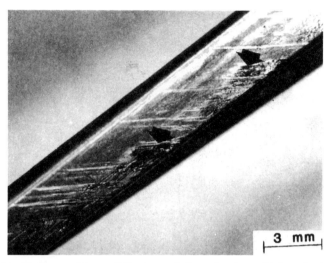

Fig 4 **'Z' wire of the second layer of R_2. Arrows indicate locations with intensive contact areas. The reduction of the contact area is a result of the shift of super-imposed wires**

A further aim of the macroscopical investigation was the evaluation of the fracture features, such as beach marks, steps, and shearlips. In the case of 'O' profile wires, the results indicated the direction and the mode of loading (4). It was found, that bending and torsion were definitely the driving forces behind crack propagation.

Microscopical findings

The microscopical investigations involved SEM observations of fractures and surfaces of single wires as well as evaluation of metallographical sections. All fatigue fractures showed irregular oriented striations, which allowed a comparison with the dimples in overload ruptures, this is a clear difference between the fatigue crack and overload residual rupture.

The investigation of the crack origin zones on the wire surfaces revealed the dominating role of the compressive loading component on the crack initiation in the second layers in both ropes. The contact zones between 'I' and 'O' wires showed the typical structure of fluctuating compressive loading as illustrated in Fig. 5. The hammered-looking surface topography can be contributed to contact fretting loading with small lateral amplitude. On the other hand the contact areas of the 'I' wires and 'Z' wires from the second layer of the 2 × locked ropes showed distinct wear traces as typical features of abrasive wear, as revealed in Fig. 6. These abrasive wear scars were accompanied by surface break-up of the material.

A detailed metallographical investigation was carried out on numerous longitudinal sections which were taken both from the prime fracture regions and

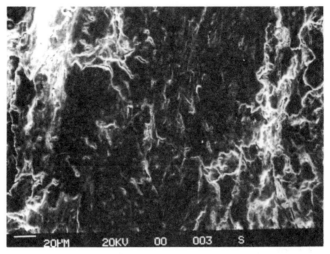

Fig 5   Fretting scar within the contact area of an 'O' wire with an 'I' wire; hammered surface

neighbouring contact areas. An examination of these sections revealed formation of white etching areas in the subsurface regions prior to fatigue cracks.
Within the contact zones of the 'O' wires the formation of white etching areas
predominantly appeared along the shear planes as shown in Fig. 7. The crack
origin sites in the 'I' and 'Z' wires from the second layer showed a severely
worn surface as a result of frictional wear which was also accompanied by the
creation of white etching areas. The typically rounded shape of the layer of
white etching areas in an 'I' wire is shown in Fig. 8. In the 'Z' wires of the
2 × locked coil ropes the transformed zones of white etching areas showed

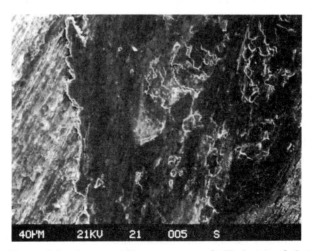

Fig 6   Traces of abrasive wear within conact area on the 'I' wire towards first layer

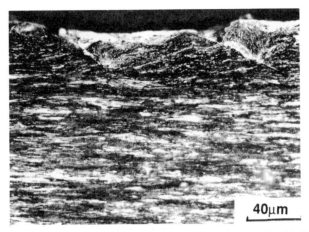

Fig 7 Cross-section through the contact area with fretting scar in 'O' wire; frictional martensite along slip bands

an irregular shape, Fig. 9. The fractographical investigation showed that the orientation of initial fatigue crack growth correlates with the morphology of the white etching areas, whereas in the 'I' and 'Z' wires the fatigue cracks are cone-like, and run in a normal plane in 'I' wires (arrow in Fig. 3).

Discussion

Based on the morphological findings, the crack nucleation in the second layers can be divided into two groups: (a) crack origins due to friction wear under high compressive loading; and (b) crack origins due to fretting.

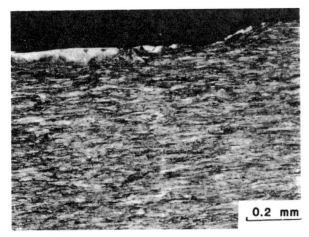

Fig 8 Cross-section through the contact area with traces of abrasive wear in 'I' wire; irregularly shaped frictional martensite

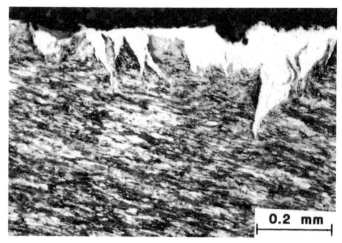

Fig 9 Cross-section of 'Z' wire from the second layer of R_2. Multiple transformation to frictional martensite in the same area

Severely worn surface-areas with traces of sliding are typical features of *friction wear*. The microstructure of the metallographical sections shows cold worked areas with local, irregularly distributed, white etching areas. Such microstructural changes strongly influence the local mechanical properties and are responsible for the nucleation of fatigue cracks. This mode of damage was primarily found at the interface of the second and first layers of 'I' (1 1/2 × locked coil) and 'Z' wires (2 × locked coil). This frictional wear is a result of mainly axial relative motion of the superimposed layer during the bending of the rope under tension. The intensity of this wear damage was locally enhanced due to the helical misfit of single wires, which caused reduction of the contact area.

On the other hand, the fatigue nucleation in the 'O' wires shows typical features of *fretting*. The hammered-looking surface and transformation of the microstructure along slip planes into white etching areas in sub-surface zones are typical findings for tight-fitting surfaces subjected to oscillation at very small lateral amplitude under high vertical compressive stresses. The Hertzian stresses, as revealed by the metallographical findings, are the result of both radial forces and circumferential dislocation of the second layer during the bending of the rope under tension. This damage mechanism was also locally intensified by a diminished contact area with regard to the ideal geometry.

As shown in (3), the 2 × locked coil wire ropes attain an 80 percent higher service life than the 1 1/2 × locked coil ropes. All damage is restricted to the second layer, which has a greater contact surface with the first layer than that of the 1 1/2 locked coil ropes, and is consequently subjected to lower compressive stresses. It has been metallographically shown, that the transformations in the microstructure due to intensive friction and pressure multiply repeatedly,

Fig 10   View of the second layer of the rope R_2. Crack origins (arrows) in a helical configuration caused by the reduced connecting area due to shifting of the superimposed wires

so that the white etching areas are tempered, broken out, and repressed on the surface. The non-transformed matrix also suffered repeated cold working, as revealed by the metallographical sections.

The nest-like concentration of the cracks in the second layer of all ropes tested can be explained in one of two ways.

(1) The circumferentially configured clusters of cracks were clearly created by the notch effect and stress raisers, respectively, caused by fractures in the neighbouring wires.
(2) The helical concentrations of crack clusters beneath one wire (Fig. 10) clearly show the influence of changes in relative position due to the shifted or helically displaced wires. On the one hand, these displacements are caused by a local deviation of the tolerances with respect to the cross-section or lay length which originated during production, and on the other hand they are a consequence of the wire breaks and the relaxation of the elastic internal stresses in the rope caused by displacements.

Based on these fractographical and metallographical results, the following ways of prolonging the service life of the investigated rope constructions can be suggested.

(1) The crack initiation period can be retarded by reduction of Hertzian stresses either by improved construction or adjustments when in service.
(2) The speed of the fatigue crack propagation can be decelerated by reducing the local bending stresses.

The important palliative role of lubrication is the subject of further research.

Conclusions

(1) In both rope constructions tested the fatigue damage starts in the second layer.

(2) The 'O' wires in R_1 suffer fatigue within the fretting scars. The fretting is a result of Hertzian stresses caused by radial forces and circumferential dislocation of the first layer during bending of the rope under tension.

(3) The fatigue cracks in 'I' wires in R_1 originate within the traces of abrasive wear caused by lateral displacement of the superimposed layer.

(4) The higher fatigue service life of R_2 is a result of less fretting and wear damage due to the larger contact area between the first and the second layer.

(5) Based on the fractographical results the dominating forces behind the crack propagation are bending and torsion.

(6) Improvements in service life can be achieved by using proper tolerance in the construction with particular regard to the first and the second layers.

References

(1) KOPANAKIS G. (1991) On the behaviour of ropes on roller chains, *Int. Aerial Lift Revue*, 1.
(2) LUTHY, Th. and BLASER, E. (1989) Radiographic examination of steel wire ropes, OIPEC Round Table Conference.
(3) KOPANAKIS G. (1994) Effects of fretting fatigue on flexed locked coil ropes, *Fretting Fatigue*, Mechanical Engineering Publications, London, pp. 283–294.
(4) Metals handbook (1987) Vol. 12, Fractography, ASM.
(5) WOODTLI–FOLPRECHT J. and FICHTER R., (1975) Fraktographie von Brüchen an Seildrähten aus Stahl, *Draht* 5.
(6) WOODTLI J. and KOPANAKIS G. (1991) *Structural failure, product liability, and technical insurance.* vol. 4 Elsevier pp. 521–528.

M. H. Attia*

Friction-Induced Thermo-Elastic Effects in the Contact Zone due to Fretting Action

REFERENCE Attia, M. H., Friction-induced thermo-elastic effects in the contact zone due to fretting action, *Fretting Fatigue* ESIS 18 (Edited by R. B. Waterhouse and T. C. Lindley) 1994, Mechanical Engineering Publications, London, pp. 307–319.

ABSTRACT Due to the thermal constriction resistance at the micro-contact spot, frictional heat results in a steep temperature gradient and high thermal stresses in the subsurface layer. This study provides a theoretical analysis of the characteristics of the temperature field in the contact zone and the resulting thermal stresses. Analysis of the results shows that the von Mises thermal stresses are quite significant and may lead to plastic deformation and the development of tensile residual stresses. Under certain operating conditions, the friction-induced thermal stresses are found to be comparable to the isothermal stresses produced by mechanical loading (normal contact stresses and shear traction).

Introduction

Due to the nature of engineering surfaces, the physical and tribological reactions between solids are limited to the highest asperities. Therefore, frictional heat results in a highly localized temperature rise and steep temperature gradient at the micro-contacts, which in turn may lead to high thermal stresses. Following the early work on the flash temperature concept, e.g., **(1)–(4)**, special attention has been directed in recent years towards the thermo-mechanical phenomena observed in uni-directional sliding systems, e.g., thermal asperities (localized thermal bulging), and thermo-elastic transition and instability **(5)–(8)**. In fretting, the relative motion is oscillatory and its amplitude is of the same order as the size of the micro-contact. Based on the analysis carried out by Hirano *et al.* for predicting the temperature rise in reciprocating motion **(9)**, Attia *et al.* **(10)** have developed a model to predict the rise in the contact temperature due to fretting. This model was further refined in **(11)** to include the effect of thermal interaction between adjacent asperities and to investigate the nature of the thermal constriction phenomenon in fretting **(12)**. No data are available, however, regarding the thermal stresses and surface deformation in fretting. The objective of the present study is to bridge this gap and to provide an theoretical analysis of the three-dimensional stress field in the vicinity of a micro-contact spot. The paper also provides an assessment of the relationship of the thermal stresses to the isothermal stresses produced by mechanical contact pressure and shear traction.

Three-dimensional temperature field in the micro-contact zone

To determine the three-dimensional temperature field and temperature gradient in the contact zone, the mutual thermal interaction of adjacent micro-

* Materials Technology Unit, Ontario Hydro Technologies, Toronto, Ontario, Canada.

contacts must be taken into consideration. Therefore, the first step in the analysis is to define the contact configuration, i.e., the average radius of the micro-contact area and the average spacing between two neighbouring contact asperities.

Contact model

Under fretting fatigue conditions, the ratio ε^2 between the real and apparent contact areas, A_r and A_a, respectively, could be as large a 0.05 to 0.10. Due to surface roughness and waviness, discrete real contact spots, known as 'micro-contacts', are usually clustered in a smaller number of bounded zones, known as 'contour areas' A_{cn}. The contact model of solids 1 and 2 is based on the following assumptions.

(1) The distributions of the heights of the asperities of the two rough surfaces are Gaussian with standard deviation s_1 and s_2. The mean slopes of the asperities are m_1 and m_2.
(2) The contact between the two rough surfaces is equivalent to the contact between a perfectly flat rigid plane and an equivalent rough surface, whose parameters are s_e and m_e.
$$s_e = \sqrt{(s_1^2 + s_2^2)} \quad \text{and} \quad m_e = \sqrt{(m_1^2 + m_2^2)}$$
(3) The contact pressure and shear traction, σ_f and $\mu\sigma_f$ respectively, are assumed to be uniformly distributed over the micro-contact area. The flow stress of the softer material σ_f and the coefficient of friction μ are assumed to be constant.
(4) The micro-contacts are squares (sides, $2L$) and are uniformly distributed over the contour area A_{cn}, as shown in Fig. 1(a).

The average characteristic length L the micro-contact and the average spacing S between adjacent micro-contacts are obtained from the following relation **(13)(14)**

$$L = \sqrt{\left[\frac{1}{2\sqrt{2}}\frac{s_e}{m_e} e^{\zeta^2} \text{erfc}(\zeta)\right]} \quad S = \frac{2L}{\varepsilon} \tag{1}$$

The parameter ζ is related to the applied p_a and the flow stress σ_f by **(13)(14)**

$$\text{erfc}(\zeta) = 2\varepsilon^2 = 2\frac{A_r}{A_a} = 2\frac{p_a}{\sigma_f} \tag{2}$$

From the contact mechanics viewpoint, one can treat the solids in contact as periodic structures, which are composed of typical volumetric unit cells. By applying the proper boundary conditions and satisfying the compatibility requirements, the stresses can be computed using a reduced model of a single cell (Fig. 1(b)). From the heat transfer point of view, the volume that encompasses a single micro-contact and extends some distance into the solid is known as the elemental 'heat flow channel' **(15)**.

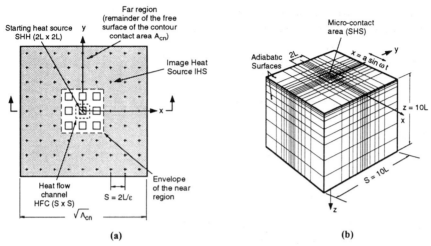

Fig 1 Idealized contact configuration within a single contour area A_{cn}, (a) view of the array of micro-contact areas (frictional heat sources), and (b) typical unit cell surrounding a single micro-contact area (HFC)

Heat transfer model

For the purpose of clarity and completeness, the heat flow channel HFC model developed in **(12)** to predict the contact temperature in fretting is summarized in this section. This model is based on the following assumptions:

(1) except for the micro-contact areas, the asperities and the control surface of the heat flow channel are adiabatic;
(2) the semi-infinite body oscillates in a simple harmonic motion;
(3) the thermal and mechanical properties of the material are temperature-independent;
(4) the contact configuration does not change with the temperature rise.

Following the method proposed by Barber **(16)** for sliding contact, and by Beck **(17)** and Negus *et al.* **(18)** for static contact, the temperature rise $\theta(x, y, z, t)$ at any point within the HFC is a linear superposition of the contribution of its own frictional heat source θ_{shs} (labelled as the starting heat source in Fig. 1(a)) and that of the neighbouring image sources θ_{ihs}. The contribution of the image sources is grouped into two regions to simplify the analysis; the near and far regions θ_{nr} and θ_{fr}, respectively. Therefore

$$\theta(x, y, z, t) = [\theta_{shs}] + \sum_{j=1}^{N_t - 1} [\theta_{ihs}]_j = [\theta_{shs}] + \sum_{j=1}^{N_{nr} - 1} [\theta_{ihs}]_j + [\theta_{fr}] \qquad (3)$$

where N_t and N_{nr} are the total number of micro-contacts within the contour area and the near region, respectively. Each of the terms given in equation (3); namely, θ_{shs}, θ_{nr} and θ_{fr}, can be obtained from the fundamental solution for the temperature rise $\theta(x, y, z, \tau)$ at point $P(x, y, z)$ and time τ due to a single square heat source ($2\delta \times 2\delta$), located at point $Q(x_s, y_s, 0)$ and oscillates in a simple harmonic motion over a semi-infinite body **(11)(12)**

$$\Theta(\bar{x}, \bar{y}, \bar{z}, \tilde{t}) = \frac{\sqrt{\pi}}{8} \sqrt{Fo} f_p \int_{\Delta\tilde{t}=0}^{\Delta\tilde{t}=\tilde{t}} [\{F(\bar{z})\}\{F(\bar{x})\}\{F(\bar{y})\}]$$

$$\times \frac{|\cos 2\pi(\tilde{t} - \Delta\tilde{t})|}{\sqrt{\Delta t}} \, \mathrm{d}\Delta\tilde{t} \qquad\qquad (4)$$

where

$$F(\bar{x}) = \{\mathrm{erf}\,(A_{\bar{x}}) + \mathrm{erf}\,(B_{\bar{x}})\}, \quad F(\bar{y}) = \{\mathrm{erf}\,(C_{\bar{y}}) + \mathrm{erf}\,(D_{\bar{y}})\},$$

$$F(\bar{z}) = \exp\left(-\frac{\bar{z}^2}{4\Delta\tilde{t}Fo}\right)$$

$$A_{\bar{x}} = \frac{\bar{x} + \bar{A}\sin\,(2\pi\tilde{t}) + \bar{A}\sin\,2\pi(\tilde{t} - \Delta\tilde{t}) + \bar{\delta}}{2\sqrt{[\Delta\tilde{t}Fo]}},$$

$$B_{\bar{x}} = \frac{-\bar{x} + \bar{A}\sin\,(2\pi\tilde{t}) - \bar{A}\sin\,2\pi(\tilde{t} - \Delta\tilde{t}) + \bar{\delta}}{2\sqrt{(\Delta\tilde{t}Fo)}}$$

$$C_{\bar{y}} = \frac{\bar{\delta} + \bar{y}}{2\sqrt{(\Delta\tilde{t}Fo)}}, \quad D_{\bar{y}} = \frac{\bar{\delta} - \bar{y}}{2\sqrt{(\Delta\tilde{t}Fo)}}$$

In equation (4), the dimensionless temperature parameter Θ and other dimensionless variables are defined as:

the temperature rise parameter:
$\quad \Theta = \theta[k/L\bar{q}] = \theta[k/\{4L(\mu\sigma_f)(af)\}]$
the amplitude parameter
$\quad \bar{A} = a/L$
the time parameters
$\quad t = \alpha t/L^2, \tilde{t} = t/\tau$
Fourier modulus
$\quad Fo = \alpha\tau/L^2 = \alpha/f L^2$
position parameters
$\quad \bar{x}, \bar{y}, \bar{z} = x/L, y/L, z/L$
the characteristic length of the heat source
$\quad \bar{\delta} = \delta/L$

The variables a and f are the amplitude and frequency of oscillation, respectively, while k and α are the materials thermal conductivity and diffusivity, respectively. The quantity $\bar{q}$ represents the average frictional heat flux, $\bar{q} = 4af\mu\sigma_f$. The symbols τ and τ_c stand for time and cycle duration, respectively. The dummy time variable $\Delta\tilde{t} = (\tau - \tau')/\tau$. At the beginning of the quasi-steady state cycle, i.e., at the centre of the stroke, the dimensionless time parameter $\tilde{t} = 0$. The Peclet number, $P\acute{e}$, is related to Fourier modulus by the following relation: $P\acute{e} = 2af L/\alpha = 2\bar{A}/Fo$.

Equation (4) can only be solved numerically. Details of the formulation and the numerical treatment of the problem are given in (12).

Thermal stress field and deformation in the micro-contact zone

Thermal stress analysis

Thermal stresses are produced in the subsurface layer by two distinctive effects (19): (a) the deviation of the temperature field from that which is a linear function of the space coordinates; and (b) the mechanical constraints imposed at boundaries of the domain being investigated. Due to the nature of the thermal and mechanical loading on the micro-contact area, the thermal and stress fields are only symmetrical about the x axis. Although the adiabatic surfaces of the HFC, which are perpendicular to the x axis, remain parallel all the time, they will be curved and deviate from their original rectangular shape, particularly near the contact interface (as shown in Fig. 8(b)). The curvature is expected to vary within the quasi-steady state cycle, depending on the instantaneous location of the heat source. The solution of the stress field $\sigma(x, y, z)$ is obtained numerically due to this complex boundary condition system and because no analytical solution exists for the temperature field (equation (4)).

In the present analysis the following conditions are assumed:

(1) the material behaves elastically and its properties are temperature-independent;
(2) the heat produced through small deformations of the HFC is negligible and, therefore, the thermo-elastic problem is decoupled;
(3) the cyclic time variation of the temperature field within the HFC is relatively slow to justify ignoring the inertia effect.

Similar analysis of thermal stresses σ_{th} in solids shows that σ_{th} is related to the temperature rise by the following general relation (19)(20)

$$\frac{\sigma_{th1}}{\sigma_{th2}} = \frac{\Delta\Theta_1 \cdot E_1 \cdot \alpha_1 \cdot (1 - v_2)}{\Delta\Theta_2 \cdot E_2 \cdot \alpha_2 \cdot (1 - v_1)} \tag{5}$$

Therefore, the thermal stresses results obtained in this analysis are normalized with respect to a reference stress level $\sigma_{ref, th}$; $\bar{\sigma}_{th} = \sigma_{th}/\sigma_{ref, th}$, where $\sigma_{ref, th} = \Theta_{ref}\{E\alpha/(1 - v)\}$. The reference temperature rise Θ_{ref} is defined as: $\Theta_{ref} = \bar{q}L/k$. Therefore, $\bar{\sigma}_{th} = (\sigma_{th}) \cdot (\bar{q}L/k) \cdot \{E\alpha/(1 - v)\}$. Using the distortion energy flow criterion, the dimensionless von Mises thermal stress $\sigma_{th, vm}$ is defined as

$$\sigma_{th, vm} = \frac{[\sqrt{\{(\sigma_1 - \sigma_2)^2 + (\sigma_1 - \sigma_2)^2 + (\sigma_1 - \sigma_2)^2\}}]}{\sqrt{2} \cdot \sigma_{ref, th}} \tag{6}$$

The thermal and mechanical stresses σ_{th} and σ_m, respectively, are also presented in relation to the yield point of the material σ_y; $\hat{\sigma}_{th, vm} = \sigma_{th, vm}/\sigma_y$ and $\hat{\sigma}_{m, vm} = \sigma_{m, vm}/\sigma_y$. Although the present analysis is limited to the elastic domain, the distortion energy flow criterion can be used as an indication of the extent of the plastically deformed zone.

Finite element model

The finite element model of the HFC and the coordinates system are shown in Fig. 1(b). The domain is discretized using a mesh of three-dimensional solid elasticity elements. At the contact region, very fine elements (as small as 5 μm $\times$ 5 μm $\times$ 5 μm) are used to accommodate the steep stress gradients. Larger elements are used away from the contact interface. The stress analysis part of the solution was carried out using a general purpose finite element program ALGOR (21). Since the FEM is used to obtain the thermal stress field within the basic HFC cell without discretizing the whole domain, special considerations have been exercised in expressing the boundary conditions along the control surfaces of the HFC. Because of the symmetry about the x axis, only a half-model ($-1 \leqslant \bar{x} \leqslant +1$, $0 \leqslant \bar{y} \leqslant +1$, $\bar{z} \geqslant 0$) is required to fully determine the stress field. The boundary conditions are:

(1) $u_y = 0$ at $\bar{y} = 0$, $\bar{y} = +1$, and $\bar{z} \geqslant 0$
(2) $u_x, u_y, u_z = 0$ at $\bar{z} = 10$
(3) $[u_x, u_y, u_z]_{\mathrm{X}=-1} = [u_x, u_y, u_z]_{\mathrm{X}=+1}$ for $\bar{y} \geqslant 0$, and $\bar{z} \geqslant 0$

where, u_x, u_y, and u_z are the displacements in the x, y, and z directions, respectively. To prevent a rigid body motion, the model is constrained in the z direction through the boundary condition (2). To satisfy boundary condition (3), nodes of the same $\bar{y}$ and $\bar{z}$ coordinates on surfaces $\bar{x} = -1$ and $\bar{x} = +1$ are connected with 'linear rigid elements' (shown in Fig. 5(b) and (c)). This idealization provides a physically meaningful representation of the system. If the surfaces $\bar{x} = \pm 1$ were assumed to be *free*, the structure would be more flexible, which in turn under-estimates the stress levels. On the other hand, if these surfaces were *clamped*, i.e., symmetric boundary condition, the stresses would be over-estimated. The approach proposed by Cifuentes *et al.* to deal with a finite array of a periodic structure (22), in which the following condition is imposed: $[u_x]_{\mathrm{X}=-1} = [u_x]_{\mathrm{X}=+1}$ = constant, is not suitable for the present case since the HFC volume is surrounded by elastic material, which hinders thermo-elastic expansion of the outermost boundaries of the volume associated with the contour area A_{cn}.

Results and discussion

Thermal field

Using equations (3) and (4), the three-dimensional temperature field in the heat flow channel is obtained for the following conditions:

- the motion variables: $a = 100$ μm, and $f = 20$ Hz;
- the ratio between the real and apparent contact areas, $\varepsilon^2 = 0.04$;

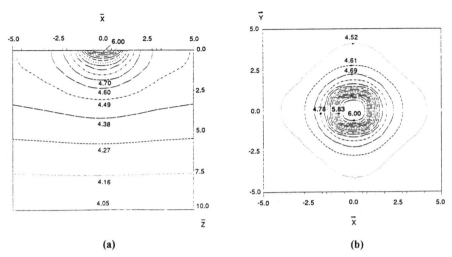

(a)   (b)

Fig 2   Friction-induced temperature field; (a) $\theta(\bar{x}, \bar{z})$ and (b) $\theta(\bar{x}, \bar{y})$ at $\tilde{t} = 0$, and $\varepsilon = 0.2$, $A = 5$, and $P\acute{e} = 0.01$

 - Peclet number, $P\acute{e} = 0.01$ (based on $L = 20 \ \mu m$);
 - the relative size of the contour area A_{cn}, $\gamma = [\sqrt{A_{cn}}]/L = 100$.

At the beginning of the quasi-steady state cycle, $\tilde{t} = 0$, Fig, 2(a) shows the temperature field in the $\bar{x}$–$\bar{z}$ plane, at $\bar{y} = 0$, while Fig. 2(b) shows the field in the $\bar{x}$–$\bar{y}$ plane for $\bar{z} = 0$. The figure shows clearly that the lateral diffusion of thermal energy $\partial\Theta/\partial\bar{x}$, and $\partial\Theta/\partial\bar{y}$ are very steep near the centre of the micro-contact. The cyclic variation in the temperature field in the $\bar{x}$–$\bar{z}$ plane is also given in Fig. 3. The figure shows that the temperature gradient $\partial\Theta/\partial\bar{z}$ is damped out rapidly at a depth of only fivefold the radius of the micro-contact, i.e., of the order of 100 μm. Therefore, the thermal stresses associated with this temperature field is expected to be quite significant. After one-quarter of the quasi-steady state cycle, $\tilde{t} = 0.25$, the temperature gradient $\partial\Theta/\partial\bar{z}$ becomes nearly zero, releasing all the thermal stresses instantaneously. Since the material experiences thermal cycling at twice the frequency of oscillation, the

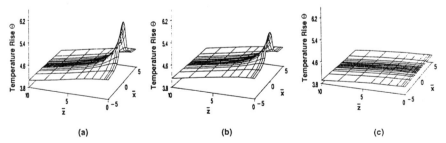

(a)   (b)   (c)

Fig 3   Cyclic time variation of the temperature field $\theta(x,z)$ for the conditions indicated in Fig. 2, and for (a) $\tilde{t} = 0$, (b) $\tilde{t} = 0.125$, and (c) $\tilde{t} = 0.25$

contribution of thermal loading to the fatigue response behaviour of the material cannot be ignored.

To demonstrate the nature of the temperature gradient in the subsurface layer and the significance of its effect on thermal stresses, the results shown in Figs 2 and 3 are used to predict the temperature field when two bodies, made of AISI I030(b) carbon steel in quenched and tempered condition, are fretted under similar conditions. The following mechanical and thermal properties of the material (23) are assumed: the yield stress $\sigma_y = 600$ MPa, the flow stress σ_f at the surface $= 5.58 \times 10^3$ MPa, the modulus of elasticity $E = 96.5 \times 10^2$ MPa, Poisson's ratio $v = 0.334$, the coefficient of thermal expansion $\beta = 5.85 \times 10^{-6}$ K^{-1}, and the thermal conductivity $k = 0.015$ W/mm°C. The coefficient of friction μ is assumed to be 1.25. It is also assumed that the frictional heat is partitioned equally between the two bodies.

The temperature distribution in the subsurface layer $\theta(z)$ at $\bar{t} = 0$, and at $\bar{x} = \bar{y} = 0$, is shown in Fig. 4(a). At a depth $z > 100$ μm ($\bar{z} > 5$), the temperature profile takes a linear form, i.e., constant gradient g_0. Within the region $0 \leqslant \bar{z} \leqslant 5$, which is defined as the 'thermally disturbed zone', the temperature profile and temperature gradient g are highly non-linear. The temperature deviation δ_θ between the local temperature rise $\theta(z)$ and the extrapolated linear temperature profile outside the disturbed zone is the thermal driving force required to overcome the thermal constriction resistance at the micro-contact and to generate thermal stresses. Therefore, one can argue that the functional relationship $\theta(z)$, $0 \leqslant \bar{z} \leqslant 5$, should be described in terms of the temperature deviation δ_θ. As Fig. 4(b) indicates, the distribution of δ_θ in the sub-surface layer is exponential

$$\delta_\theta = \delta_{\theta,0} \exp{(cz)} \tag{7}$$

where, $\delta_{\theta,0}$ is the temperature deviation at $z = 0$, and c is a constant. The change in the temperature gradient $\partial\theta/\partial z$ with depth z can, therefore, be expressed as

$$\frac{\partial\theta}{\partial z} = \delta_{\theta,0}\{c \exp{(cz)}\} + g_0 \tag{8}$$

Figure 4(c) shows that the temperature gradient $\partial\theta/\partial z$ at $z = 0$ can be > 1000°C/mm, even when the maximum temperature rise at the centre of the micro-contact is only in the order of 100°C.

Thermal stresses and deformation

Figure 5(a), shows the dimensionless von Mises thermal stresses $\bar{\sigma}_{th,vm}(\bar{x}, \bar{y}, \bar{z})$ resulting from the temperature field shown in Figs 2–4. The maximum von Mises thermal stress is produced over the micro-contact area and reaches a value equivalent to 0.54 σ_y of the material. If the operating conditions are changed to double the temperature field, e.g., by increasing the frequency from

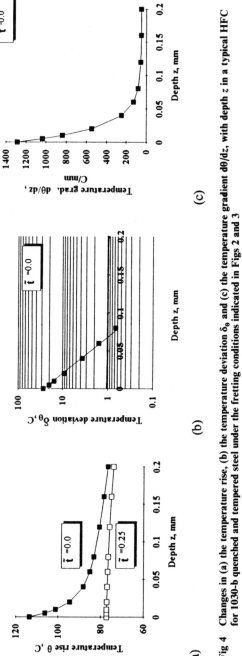

Fig 4   Changes in (a) the temperature rise, (b) the temperature deviation δ_θ and (c) the temperature gradient $d\theta/dz$, with depth z in a typical HFC for 1030-b quenched and tempered steel under the fretting conditions indicated in Figs 2 and 3

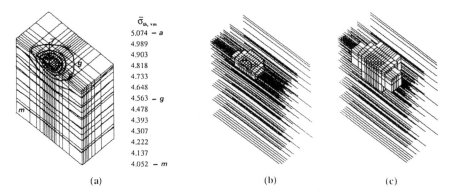

Fig 5   Von Mises thermal stresses $\sigma_{th, vm}$ due to (a) the temperature field $\theta(x, y, z)$ described in Figs 2–4, and the extent of the zone in which $\sigma_{th, vm} > \sigma_y$ as a result of (b) doubling $\theta(x, y, z)$ and (c) imposing a rigid mechanical constraint over the micro-contact area

20 to 100 Hz, the maximum von Mises thermal stress exceeds the distortion energy flow criterion, i.e., the yield point σ_y of the material. The extent of this zone is shown in Fig. 5(b). As indicated before, thermal stresses are strongly dependent on the mechanical constraints imposed on the system. One of the issues which is of a concern to researches dealing with fretting fatigue testing is the rigidity of the normal loading system used to apply the normal contact pressure. To demonstrate the significance of this aspect on fretting fatigue test results, Fig. 5(c) presents the extent of the contact zone where $\sigma_{th, vm} > \sigma_y$ when the micro-contact area is fully restricted from expansion in the z direction by a perfectly rigid member.

The thermal displacement and localized bulging of the contact interface is shown in Fig. 6. The upward expansion of the centre of the micro-contact is in the order of 1 μm. The compressive stresses developed in the subsurface layer,

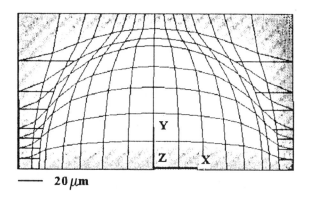

deformation scale: 2.5E03

Fig 6   Thermal displacement (bulging) over the contact interface due to the friction-induced temperature rise $\theta(x, y, z)$

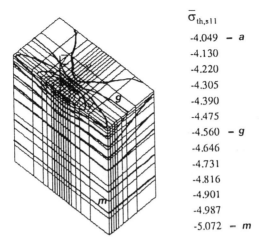

$\bar{\sigma}_{th,s11}$

-4.049 – a
-4.130
-4.220
-4.305
-4.390
-4.475
-4.560 – g
-4.646
-4.731
-4.816
-4.901
-4.987
-5.072 – m

Fig 7   Distribution of the principal thermal stress S_{11} in the HFC due to $\theta(x, y, z)$

as a result of restricting this thermal expansion, are quite significant as shown in Fig. 7, which maps the principal thermal stress $\bar{\sigma}_{th, s11}$ in the $\bar{x}$ direction, parallel to the fretting motion. It is well established that if a localized region is deformed plastically, but is surrounded by a large mass of material that remains elastic, then residual tensile stresses are set up in the deformed region when the condition causing the plastic deformation is released **(24)(25)**. If the contribution of the thermal stresses is sufficiently high to cause plastic deformation, then this mechanism will take place twice during the fretting cycle, at $\bar{t} = 0$ and 0.25, as the temperature gradient and the resulting stresses approach zero.

To appreciate the significance of the thermal stresses in relation to the isothermal stress field due to mechanical loading, the same finite element model was used to calculate $\hat{\sigma}_{m, vm}$. The normal and shear traction over the microcontact, are assumed to be σ_f and $\mu\sigma_f$, respectively. The resulting stress field in the $\bar{x}$–$\bar{z}$ and $\bar{x}$–$\bar{y}$ planes are shown in Fig. 8, in which the darkened area represents the region where $\sigma_{m, vm} > \sigma_y$. The $\bar{x}$–$\bar{y}$ projection of the deformed HFC (Fig. 8(b)) shows the curvature of the deformed surfaces $\bar{x} = \pm 1$, and hence the importance of using the linear rigid elements in constructing the FE model. Comparison of Figs 5 and 7 with Fig. 8 indicates that, under certain operating conditions, the friction-induced thermal stresses could be as high as the isothermal stresses produced by mechanical loading.

Conclusions

(1) The distribution of the temperature gradient $\partial\theta/\partial z$ in the subsurface layer is exponential, resulting in an extremely high temperature gradient near the centre of the micro-contact area. For the conditions assumed in this study $\partial\theta/\partial z > 1000°C/mm$.

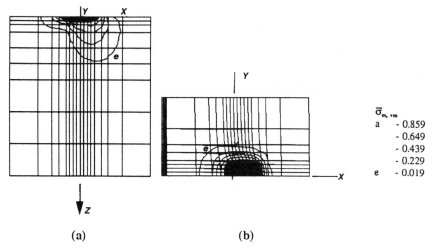

(a)   (b)

Fig 8 Von Mises stresses $\sigma_{m, vm}$ in the HFC due to mechanical loading (normal contact pressure and friction traction) over the micro-contact area

(2) Under certain operating conditions, the von Mises thermal stresses in the contact region could be as high as the isothermal stresses produced by mechanical loading, resulting in a plastically deformed zone near the contact interface.

(3) The combination of the cyclic variation of the temperature field and the plastically deformed zone results in releasing the thermal stresses at double the frequency of oscillation. This mechanism introduces tensile residual stresses in the contact region, which affect the fatigue response behaviour of the material.

(4) The mechanical constraint introduced by the rigidity of the mechanism used to apply normal contact pressure in fretting fatigue testing has a significant effect on the induced thermal stresses and consequently the variability of test results.

Acknowledgement

The author wishes to acknowledge the financial support of the National Sciences and Engineering Research Council of Canada.

References

(1) BLOK, H. (1937) Theoretical study of temperature rise at surfaces of actual contact under oiliness lubricating conditions, *Proc. General Discuss. on Lubrication and Lubricants*, Vol. 2, Inst. Mech. Engrs, London, pp. 222–235.

(2) JAEGER, J. C. (1942) Moving sources of heat and the temperature at sliding contact, *Proc. R. Soc. NSW*, **56**, 203–224.

(3) HOLM, R. (1948) Calculation of the temperature development in a contact heated in the contact surfaces, and the application to the problem of the temperature rise in a sliding contact, *J. appl. Phys.*, **19**, 361–366.

(4) FUREY, M. J. (1964) Surface temperature in sliding contact, *ASLE Trans.*, 133–146.

(5) BURTON, R. A. (1980) Thermal deformation in frictionally heated contact, *Wear*, **59**, 1–20.

(6) BARBER, J. R. (1969) Thermoelastic instabilities in the sliding of conforming solids, *Proc. R. Soc. Lond, Ser. A*, **312**, 381–391.

(7) BARBER, J. R. (1980) The transient thermoelastic contact of a sphere sliding on a plane, *Wear*, **59**, 21–29.

(8) DOW, T. A. (1980) Thermoelastic effects in a thin sliding seal-*a* review, *Wear*, **59**, 31–52.

(9) HIRANO, F. and YOSHIDA, S. (1966) Theoretical study of temperature rise at contact surface for reciprocating motion, Proceedings of the Third International Heat Transfer Conference, Vol. IV, pp. 127–132.

(10) ATTIA, M. H. and D'SILVA, N. S. (1985) Effect of mode of motion and process parameters on the prediction of temperature rise in fretting wear, *Wear*, **106**, 203–224.

(11) ATTIA, M. H. (1989) Friction-induced temperature rise in fretting–elemental heat flow channel model, Proceedings of the International Congress on Tribology, (Edited by K. Holmberg and I. Nieminen), Vol. 3, pp. 22–29.

(12) ATTIA, M. H. and CAMACHO, F. (1993) Temperature field in the vicinity of a contact asperity during fretting, ASME Symposium on Contact Problems and Surface Interaction in Manufacturing and Tribological Systems, PED, **67**/Trib, **4**, 51–61.

(13) MIKIC, B. B. (1971) Analytical studies of contact of nominally flat surfaces; effect of previous loading, *J. Lubric. Technol.*, **93**, 451–519.

(14) MIKIC, B. B. (1973) Thermal contact conductance; theoretical considerations, *Int. J. Heat Mass Transfer*, **17**, 205–214.

(15) YOVANOVICH, M. M. (1986) Thermal contact resistance: theory and applications, Lecture Notes, Mechanical Engineering Department, University of Waterloo, Canada.

(16) BARBER J. R. (1970) The conduction of heat from sliding solids, *Int. J. Heat Mass Transfer*, **13**, 857–869.

(17) BECK, J. V. (1979) Effect of multiple sources in the contact conductance theory, *J. Heat Mass Transfer*, **101**, 132–136.

(18) NEGUS K. J., YOVANOVICH, M. M., and DEVAAL, J. W. (1985) Development of thermal constriction resistance for anisotropic rough surfaces by the method of infinite images, *National Heat Transfer Conference,*, ASME, NY.

(19) BOLEY, B. A. and WEINER, J. H. (1985) *Theory of thermal stresses*, Krieger Publishing Company, USA.

(20) HOVANESIAN, J. D. and KOWALSKI, H. C. (1967) Similarity in thermoelasticity, *Exp. Mech.*, **7**, 82–84.

(21) Algor process reference manual, Algor Interactive Systems, PA.

(22) CIFUENTES, A. O. and STIFFLER, S. (1992) Modelling thermal stresses in periodic structures: some observations regarding the boundary conditions, ASME Winter Annual meeting, Paper No. 92–WA–EEP–3.

(23) *Metals handbook*, Tenth edition, Properties and selection: irons, steels and high-performance alloys Vol. 1, ASM International.

(24) KENNEDY, Jr., F. E. and LING, F. F. (1974) Elasto-plastic indentation of a layered medium, *J. Engng. Mater. Technol.*, **96**, 97–103.

(25) KENNEDY, Jr., F. E. (1982) Thermocracking of a mechanical face seal, *Wear*, **79**, 21–36.

RESPONSE OF MATERIALS

L. Vincent*

Materials and Fretting

REFERENCE Vincent, L. **Materials and fretting,** *Fretting Fatigue* ESIS 18 (Editing by R. B. Waterhouse and T. C. Lindley) 1994, Mechanical Engineering Publications, London, pp. 323–337.

ABSTRACT Fretting wear and fretting fatigue are commonly associated with damage of quasi static loaded assemblies and decrease in lifetime. Any search for palliatives is made difficult as the contact parameters are not generally identified and the material responses cannot be predicted from well-established wear mechanisms. In fact two main types of damage can appear under both fretting wear and fretting fatigue conditions: (a) wear induced by fretting (WIF); and (b) cracking induced by fretting (CIF)

Too often, fretting is only related to the observation of debris in or around the contact point while the most detrimental reductions in durability properties can occur without any formation of debris. These two kinds of damage depend on whether the material fails due to overstraining (WIF) or to overstressing (CIF). Considering these two possibilities, one can clearly understand that the required material properties are not the same for WIF or CIF and that contradictory assessments can be made about the effects of parameters such as hardness, surface roughness, residual stresses, etc.

This paper proposes to define the two kinds of damage and to locate them on fretting maps. Of the several map domains, the material properties will be discussed.

Introduction

According to the ASM glossary of terms, fretting is a small amplitude oscillatory motion, usually tangential, between two solid surfaces in contact **(1)**. Due to the nature of the degradation, often with the presence of oxide debris or a predominant chemical reaction, the term fretting corrosion is used. When the movement is the result of external vibrations applied to surfaces not submitted to imposed displacement, the term 'fretting wear' is used. If the relative movement is a consequence of a cyclic loading of one of the components, the loading is called 'fretting fatigue'. Several syntheses have been proposed to justify the fretting type degradations **(2)–(6)** and to propose palliatives **(7)(8)**.

When considering the material behaviour under these different types of fretting loading, the following should be considered:

- for general wear, fretting strength is not an intrinsic property of materials;
- contact parameters, such as movement amplitude and frequency, can strongly modify the material behaviour **(9)–(12)**;
- corrosion and oxidation of debris are generally consequences of the initial damage and are not at the origin of the degradation **(6)(13)**;

* Département 'Matériaux-Mécanique Physique', URA CNRS 447, Ecole Centrale de Lyon, BP 163, 69131 Ecully Cédex, France.

- the type of contact is one of the most important parameters as it governs the trapping of the debris (14);
- surface roughness and residual stresses have contradictory effects (6);
- fretting is a third-body contact problem as, once formed, debris must stay in contact before possible ejection;
- in most cases, good material choice can only delay or modify the fretting damage formation but cannot do away with any fretting damage.

More generally, good or bad anti-fretting materials do not exist. As is the case for any wear analysis, the triplet 'mechanisms–first bodies–third body' is to be considered (15). Very often material effects are explained by the way the velocity is accommodated in the third-body layer. This requires a good knowledge of the insitu formed debris bed (e.g., oxide debris), rheology, and trapping (16)(17).

This paper deals with damage analysis of the contacting first bodies. It applies to both fretting wear and fretting fatigue, and relies on numerous experimental results obtained in the cases of homogeneous and heterogeneous fretting contacts which will not be detailed here (18)–(22)

Damage and materials

The brief overview of the fretting parameters given above shows the apparent complexity of establishing a guideline in the choice of materials. However, the designer must have a good knowledge of the materials which will be used. The main properties that can be obtained are:

- constitutive laws (for example elastic, elasto viscoplastic) and physical values (moduli, Poisson's coefficient, conductivity, thermal expansion, and so on) from which the material loading (instantaneous strain and stress), is estimated;
- limit criteria (yield or failure stresses, ductility, sometimes fatigue limits or crack propagation rates) from which damage or failure risks are detected.

Generally when attempting to predict contact damage, one considers the final stage only: that is, failure or material loss. This final process can include several material histories and, for instance, competition between several failure mechanisms.

To rationalize the material approach in a contact problem, it is useful to go back to the definition of what damage is and then to begin by describing contact behaviour from the first detected damage. Damaging and damage laws have been analysed by both mechanical engineers and physicists. From a mechanical point of view, Kachanov and Rabotnov's approaches are often referred to as the starting point for modelling. These authors proposed a continuous variable related to the density of formed defects which can predict time to failure, initially in the case of the creep behaviour of metals (23).

For physicists, research has described the damage of metals through: (a) the formation of cavities around non-metallic inclusions or carbides, or by vacancy diffusion and concentration at high temperature (creep); and (b) the decohesion of interfaces and their growth up to failure. In the early 1950s, slip bands were shown on the polished surfaces of plastically deformed FCC metals.

To follow the degradation processes, the material life can be separated into three stages.

(1) The first stage is composed of microstructural alterations or the accommodation process in which creation, movement, interaction and array of dislocations are the main physical stages (localized or generalized phenomena).

(2) This is followed by the creation of new surfaces within elementary volumes; this is the damage defined as in-volume discontinuities. Considering the atomic scale, these new surfaces are created by cleavage, slip or cavitation.

(3) The final stage involves the growth or propagation of these nanodefects. This depends on the density and the spacing of the created discontinuities. The failure can be instantaneous (e.g., brittle fracture) or delayed (fatigue, creep, embrittleness, environmental effects).

Among the several manifestations of the local destruction of the continuity of the material, i.e., the damage creation, the local cleavage of the brittle failure or the local high deformation at the origin of the ductile failure is considered to be due to instability around point, linear, or surface defects. In the case of fatigue, the applied stress is generally far below the yield stress, but local stress concentrations caused by defects (notch, non-metallic inclusion) induce local plastic microdeformation. The crack nucleation is mainly a microscopic phenomenon located in one or a few grains (micrometer scale). At this scale, the material can no longer be considered to be homogeneous and isotropic. Therefore, the local constitutive laws are generally unknown. Localized plastic deformation, vacancies, and persistent slip bands were shown to nucleate the volumic discontinuities. From the several discontinuities, one or a few cracks develop up to failure.

Sometimes the applied strains are higher and the plastic deformation effects the entire bulk (it is no longer a localized phenomenon). In this latter loading the prediction of the lifetime depends on both the damage induced by the plastic strain cumulation and on the metal characteristics which can be strongly modified by the global plastic loading. A large number of cracks are created very early and the main crack inducing failure is generally noted in less than 5 percent of the total life time. Manson–Coffin's law is the basic tool for life prediction; it is related to the cumulative plastic strain. The low cycle fatigue theories are very helpful in explaining the local damage, even in the case of the high cycle fatigue. They can describe the initiation process in the persistent slip bands and the material behaviour at the crack tip during the propagation stage (24)(25).

It is not easy to separate the stage of the nucleation from the stage of the damage propagation. According to K.J. Miller's proposal (26) the nucleation of the damage will be defined as:

– the nucleation itself;
– the cracks whose lengths are in the order of the microstructure size (one or two grains);
– the so-called short cracks for lengths below which the elastic fracture mechanics cannot be applied.

High cycle fatigue is often described by Wohler's diagrams and it would be useful to separate the Wohler limiting life diagram into two domains to better identify the nucleation period. Using French's curve (27)(28), the left domain is defined as a time region in which nucleation is not yet produced. Unfortunately very few experimental results are presently available.

It is clear that dislocation structures produced by cyclic loading display specific features when compared to other kinds of loading (29). However, we will first consider whether the damage is localized or global, and whether the loading is time dependent or not. Hence local damage criteria are mainly related to the accommodation of local overstresses (brittle or high cycle failure), while global damage criteria are related to the accommodation of the plastic extended deformation. For fatigue, the same physical damage can appear at the initiation point for high cycle fatigue loading or in the whole specimen for low cycle fatigue. For the designer, the former fatigue damage is described using stress criteria (fatigue limit); the latter fatigue damage is described using strain criteria (e.g., the cumulative amplitude of the plastic deformation). This dual approach based upon local overstressing or extended overstraining will be kept in mind in the next wear study.

Fretting processes

The fretting damage can be described as:

(1) wear induced by fretting (WIF): this corresponds to classical material loss;
(2) cracking induced by fretting (CIF): cracks initiated on the surface can propagate up to the final failure of the specimen.

Usually, the WIF damage is related to fretting wear and the CIF to fretting fatigue. It has been shown recently that the two kinds of damage can appear whether an external loading is applied (fretting fatigue) or not (fretting wear) and thus superposed, or not, to the contact loading. Therefore, we prefer to keep the terms fretting wear, fretting fatigue, and fretting corrosion, to describe the loading, and the terms wear induced by fretting (WIF) or cracking induced by fretting (CIF) to describe the degradation. These two processes are briefly described below.

Wear induced by fretting: (WIF)

Classically WIF is related to debris powder (red powder for steels, black powder for aluminium or titanium alloys). For a wide range of metals, alloys, and fretting conditions, WIF begins with the removal of superficial screens and the increasing of the metal-to-metal contact. Then a new microstructure is formed called the Tribologically Transformed Structure (TTS) in the case of non-initially brittle materials (30). This TTS is a nanocrystalline structure which becomes too brittle to accommodate the imposed displacement other than by breaking. The metallic debris is then trapped, crushed, and oxidized in the contact which induces the formation of a powder bed (the red powder for example for steel). The establishment of the third body depends on the possibility of ejection of the debris (i.e., on the nature and the shape of the debris and the vibratory environment). If the powder bed is maintained in the contact, the bulk degradation can be stopped as the velocity is now entirely accommodated in the powder bed (Table 1) A cross-section of the contact shows three zones (Fig. 1): zone I is a plastically deformed layer, while zone II (the TTS) is covered by the debris bed (zone III). An example of the depth of the TTS versus the number of fretting cycles for titanium alloys is given in Fig. 2.

Cracking induced by fretting: (CIF)

Several features have been noted for the cracks formed on the contact surface.

(1) In some cases, few cracks nucleate at each side of the contact limit (as related to the friction movement). Then at each side, a main crack develops which can lead to spalling. The location of these cracks is justified by

Material Accommodation	
	formation of the TTS and first cracking.
Debris Formation	
	debris formation ⇩ fragmentation
Debris Evolution	
	slightly oxidised metallic powder
	oxide formation
	formation of platelets from very small oxide debris

Table 1   Example of third body formation in the case of steel contacts

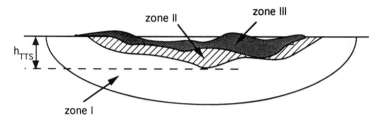

Fig 1   Schematic view of a contact cross-section in the case of WIF

contact mechanics theories but several inputs (local strain and stress following the fretting regimes) remain under discussion.
(2) Cracks can also initiate and be distributed everywhere on the contact
surface. Very often this induces the formation of coarse debris or spalls but
usually only the most external cracks propagate because of the compressive field beneath the surface.

WIF/CIF competition

Material loss and cracking sometimes appear as competitive processes: for
instance material loss can eliminate small superficial cracks or the opening of a
deep crack can accommodate the main part of the imposed displacement and
thus strongly reduce the slip amplitude and consequently the debris formation.
Obviously the fretting processes appear to depend on very complex phenomena. Therefore, it appears very difficult to propose guidelines for the understanding of the material behaviour and for the choice of material. Some ideas
to rationalize this material effect are given below.

Materials and Fretting

It is quite trivial to consider that the material damage depends on the applied
stress and strain fields! However, the damaging phenomena are described

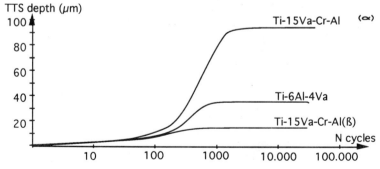

**Fig 2   Depth of the tribologically transformed structure versus the number of cycles for two tita-
nium alloys and two microstructures of the Ti–15V–Cr–Al**

from the analysis of the final contact degradation and with a more or less clear assumption that the stress–strain fields do not change throughout the contact life. To rationalize the material effect, two steps are proposed: local loading and material behaviour.

The local loading

The stress and strain fields must be determined locally under the entire contact. This requires a good knowledge of the fretting regime which cannot be estimated only from the global mechanical conditions (normal load, displacement, frequency, geometry, and contact shape). The material (surface and bulk) contributes to the nature of the regime and physical chemistry appears to be a main approach to justify its establishment. The environment (for example humidity and/or elevated imposed temperatures, developing superficial screens), roughness, hard phase indentation of one of the two first bodies (e.g., normal glass fibres of a composite material contacting an aluminium alloy **(22)**) must be taken into account in order to determine the fretting regime.

Fretting regimes were first mapped by Vingsbo *et al.* **(10)** considering stick, mixed stick and slip, and gross slip regimes. These regimes were defined using a critical amplitude depending on normal load and frequency. Of course, these critical values depended on the testing device and especially on the contact stiffness. In a similar way, from our own experiments run on a specific fretting device described in previous papers **(18)** and using the so-called friction logs (i.e., a three-dimensional plotting of the tangential load F required at each value of the imposed displacement D versus the number of cycles N on a logarithmic scale), we proposed to plot 'running conditions fretting maps' or RCFMs. These RCFMs indicate three fretting regimes in a 'normal load–displacement' diagram.

(1) The stick regime is defined from F–D cycles remaining closed up to a coarse damage initiation characterized by a decrease in the contact stiffness or in the F–D initial slope. Surfaces in the contact are 'stuck' together. Elastic deformation of the device and the sample accounts for the accommodation of the imposed displacement. Small amounts of partial slips can appear but no plastic deformation of the first bodies is noted, even on metallographic sections of the specimens (Fig. 3).
(2) The slip regime is characterized by cycles remaining quasi-rectangular but with possible strong variations of the maximum tangential load obtained at the end of each cycle (friction coefficient). Of course the real displacement in the contact is smaller than the imposed displacement because of the non-infinite rigidity of the devices and of the elastic deformation of the samples (Fig. 4).
(3) A fretting regime characterized by a complex shape for the friction log is called a 'mixed regime'. Closed, quasi-rectangular and very often elliptic cycles are noted during one and the same testing. The elliptic cycle gener-

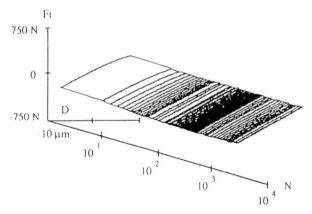

Fig 3   Example of a friction log for a stick regime

ally appears after some time and is described as partial slip associated with plastic deformation in the upper layers of the stuck central contact region. This complex regime is intermediate between the stick and gross slip regimes (Fig. 5).

The above RCFMs are indispensable for any analysis of material behaviour as one can only calculate the stress–strain fields if postulating on clear hypotheses. A comparison of material behaviour when the materials are not submitted to the same fretting regime is of no interest and is considered wrong from a scientific point of view. Furthermore, due to known difficulties in calculating stress and strain fields in contact, except in the case of specific configurations, any comparison of the damage mechanism and the damage level requires that materials be submitted to the same fretting regimes with similar values of the friction coefficient.

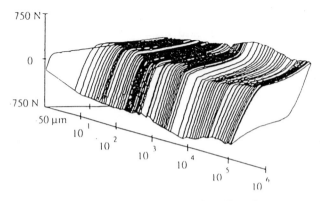

Fig 4   Example of a friction log for a slip regime

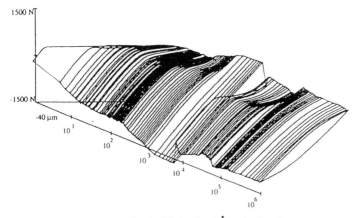

Fig 5   Example of a friction log for a mixed regime

The material behaviour

The second approach is to plot a new set of fretting maps in which the material damage is considered. Three main responses have been identified using optical microscope analyses of contact cross-sections. These are either the absence of damage (noted ND) or the two types of damage described above, that is WIF, considered in this damage diagram as particle detachment (PD), and CIF. (C).

It is very important to consider the first main damage to plot the material response fretting maps or MRFMs, because of the competition suggested above which can cancel the initial damage. The MRFMs provide a way to relate the wear behaviours of materials to some of their properties. Of course the MRFMs depend on the number of cycles as the damage nucleation is generally time-dependent. (Fig. 6). The comparison between the two sets of fretting maps justifies the effect or the absence of effect of specific mechanical properties.

The limit between the non damage and the cracking domains must be estimated through the calculation of stresses at the contact limits and using classical fatigue strength criteria. For the highest number of cycles, it seems possible to compare the tensile skin stress as calculated from Mindlin's theory (31), with fatigue limits because the cracking domain spreads over the stick (mainly elastic) regime of the corresponding RCFM. However, cracks were shown to nucleate first under the mixed regime. In this case the amount of plastic deformation in the stick zone makes the identification and the calculation of the stress responsible for crack nucleation rather difficult. Despite this plastic deformation, it clearly appears that damage criteria have to be related to local stresses, at least in the case of metals.

Several aluminium alloys have been compared under the two 'stick' and 'mixed' regimes and far different behaviour was obtained. They could be

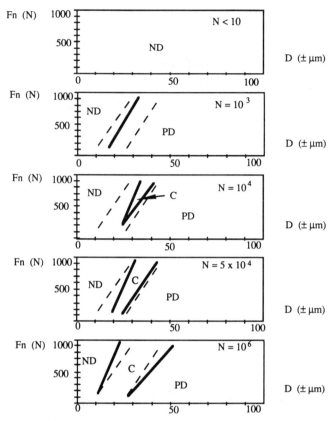

Fig 6 Sets of material response fretting maps for several fretting cycles (aluminium alloy). The heavy black lines are the boundaries of the damage area. The dotted lines represent the boundaries of the fretting regime defined on a RCFM

explained from their resistance to crack nucleation and propagation (20). In a similar way, for a Ti–6Al–4V alloy, the appearance of the cracking domain was clearly related to its fatigue limit. This is easily modified by the use of thermal treatments (19).

Generally, particles detach well before the fatigue crack initiation and it seems reasonable to superpose the particle detachment domain to the gross slip regime. Except for brittle materials with low elongation ratio and low fracture toughness, the particles form from the TTS. Thus for the same alloy, and even for several alloys of a same group, the particle detachment does not depend on initial strength properties resulting from thermo-mechanical history. The cyclic work hardening which induces the mechanical-like re-crystallization, and then the TTS, cancels the initial properties (yield stress, elongation, residual stresses). Here the damage depends on a global plastic deformation effect and wear criteria appear to be related to cumulative strain hardening and thus to be deformation criteria.

Consequently, it is obvious that it is very difficult to rank the wear properties of several alloys of a same group: the nature of the TTS is quite similar for all of them and their wear property mainly depends on the detached particle life, i.e., of their trapping in the contact. The main parameters for WIF resistance are essentially mechanical ones (nature of the contact, vibratory environment, frequency). The material itself is responsible for the depth of the TTS and thus for the amount of debris which can form in the contact (19). It can also govern a part of the trapping through the effect of the chemical environment which can favour the formation of platelets (platelets are harder to eject than free particles).

Discussion

Our proposal is to study material effects in fretting by describing the formation of the initial damage. Overstressing and overstraining concepts are used to localize the region of appearance and to describe the extent to which the damage will develop (32). Overstressing at the limit of the contact zone (stick regime) or at the limit between the stuck, plastically deformed area and the outer slip zone (mixed regime) justifies the crack initiation. The global overstrained superficial zone is the origin of the debris formation (slip regime). Therefore, the explanation of the material behaviour or the choice of palliatives requires good identification of the fretting regime, that is the local loading. For the designer, the first step is to decide what damage is to be avoided and thus guidelines can be obtained in order to look for adequate mechanical properties and to choose good anti-fretting materials fitted to the specific contact problem. The two sets of fretting maps are useful tools, especially when it is difficult to estimate the local loadings of the industrial contacts (this is a very common circumstance).

Table 2 indicates guidelines to help choose surface treatments or coatings as palliatives against failure using the two fretting maps. Coatings can favour gross slip and thus diminish the mixed regime, which is the most detrimental regime for cracking. The problem is to control the maintainence of the coating as long as possible in the contact. Some coatings also induce low friction coefficient values which give lower values for contact stresses available to nucleate cracks (for example, varnishes).

If the RCFM is not modified, higher resistance to crack nucleation (superficial heat treatment) or compressive stresses to arrest cracks (shot peening) can be sought to push the cracking domain up towards the highest values of the normal load.

Sometimes the hard coatings quickly give rise to particle detachment, which prevents the mixed regime by accommodating the displacement in the powder bed. The quick degradation of brittle materials modifies the RCFM by increasing the slip regime.

This approach to fretting, using the fretting maps and the velocity accommodation mechanism concept, can be extended successfully to fretting fatigue

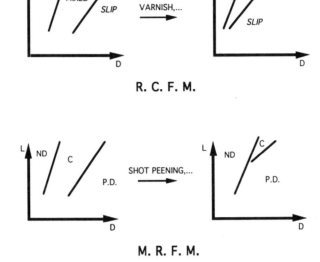

R. C. F. M.

M. R. F. M.

Table 2 Examples of palliative effects justified by fretting maps

(33). In a similar way surface treatments, or claddings, can be chosen as palliatives for their effects on the RCFMs (varnish, solid lubricant, thin film) or on the MRFMs (increase in fatigue limit, introducing of compressive residual stresses). The analysis of the shot peening effect is very interesting:

 – the induced roughness can favour the stick domain of the RCFM and debris formation in the gross slip domain,
 – the compressive residual stresses can push up and reduce the cracking domain of the MRFM,
 – the compressive stresses induced in the sublayers can arrest small cracks formed on the surface (33),
 – the hardened, more brittle peened layers can favour the debris formation in the slip domain, but in this case the residual stresses are cancelled in the very first strokes.

This example shows how important it is to identify the fretting regime before attempting to explain the material behaviour and before making any material ranking.

The above approach to fretting can also be used for other wear analyses. Up to now, this paper has considered the formation of the first damage. In a contact problem, this stage should not be related to dispersive results at all. On the other hand, we can assert that the nucleation period is very reproducible when compared to other fatigue problems. This is explained by the fact that we are not looking for the weakest point(s) in the material: the cracks nucleate at the point where the detrimental stresses are maximum and this point does not depend on material heterogeneities. The same is true for the formation of debris from the TTS. This forms very quickly in the contact and covers a large part of the contact track.

However, in literary surveys, tribological properties are generally considered to be scattered. This results from a second wear stage which we can relate to the effect of the first created damage on the contact loading (wear regime) and on the velocity accommodation mechanisms. For instance, the wear rate depends on the trapping possibilities as justified by the third-body theory. On another hand, crack growth depends on loading at the tip. This is related to the interaction between the various cracks and the local stresses (34).

Conclusion

The complexity of describing material effects under fretting loading is generally associated with many parameters. However, this often results from approaches which only attempt to analyse a phenomenon related to moving local loadings applied to a moving target (i.e., the two materials in contact have properties which vary during the loading time) using the final observation of the degradations in the contact and using the measurement of global tracers (material loss, friction coefficient).

The fretting process is first to be considered as a two-stage phenomenon: (a) an incubation period leading to the formation of the first damage; and (b) a complicated period in which the damage modifies the local loadings and participate in stopping, stabilizing, or increasing the degradation process. In the first stage, the material participates in the establishment of the fretting regime, through the chemistry of its surface, its ability to form a thin layer according to the chemical environment, and its roughness (35). Once the fretting regime is set, the material will suffer damage by overstraining or overstressing. The properties required to delay the damaging will depend on this loading. Contact mechanics theories can provide tools to calculate these local loading conditions (36)(37). Fatigue criteria already exist (38) and can be used to predict crack initiation in the case of overstressing. After this stage of crack initiation, fracture mechanics are used to predict propagation paths, crack growth rates, and lifetimes – even in the case of short cracks, complicated crack networks and fretting fatigue experiments (33)(39)–(43). It will be some time before we can do modelling in the case of a particle detached from the TTS because this requires improvement of our knowledge about it. In a similar way, it remains doubtful that we can describe the competition existing between the two main damage modes.

In this paper, we have suggested the use of two fretting maps, i.e., the running condition fretting maps and the material response fretting maps to rationalize the study of material behaviour in close connection with the mechanical approach.

Acknowledgments

The author wishes to thank all his colleagues from the Ecole Centrale de Lyon and from INSA Lyon who are quoted in the following references. Thanks are also due to Mr A. Barrier for technical support in the fretting tests.

The author would especially like to thank Professor M. Godet for friendly and fruitful discussions at almost all hours of the day and night. This paper is dedicated to Professor Maurice Godet who died in October 1993.

References

(1) ASM Handbook, vol. 18, *Friction, Lubrication and Wear Technology*, P. J. BLAU volume chairman, ASM Int'l. Edit., 1992, p. 9.
(2) McDOWELL, J. R. (1953) Fretting corrosion tendencies of several combinations of materials, *ASTM STP 144*, ASTM, Philadelphia, p. 2439.
(3) WATERHOUSE, R. B. (1974) *Fretting corrosion*, Pergamon Press.
(4) WATERHOUSE, R. B. (1981) *Fretting fatigue*, Applied Science Publishers.
(5) BERTHIER, Y., VINCENT, L., and GODET, M. (1989) Fretting fatigue and fretting wear, *Tribology Int.*, 22, 235–242.
(6) WATERHOUSE, R. B. (1992) Fretting wear, ASM handbook, Vol. 18, *Friction, lubrication and wear technology*, (Edited by P. J. Blau), ASM, pp. 242–256.
(7) CHIVERS, T. C. and GORDELIER, S. C. (1985) Fretting fatigue and contact conditions, a rational explanation of palliative behaviour, *Proc. Instn. Mech. Engrs*, Part C, 199, pp. 325–337.
(8) BILL, R. C. (1985) Selected fretting-wear resistant coatings for Ti — 6A1 — 4Va alloy, *Wear*, 106, 283–301.
(9) KENNEDY, P. J., PETERSON, M. B., and STALLINGS, L. (1982) An evaluation of fretting at small amplitudes, *ASTM STP 780*, ASTM Philadelphia, pp. 30–48.
(10) VINGSBO, O. and SÖDERBERG, S. (1988) On fretting maps, *Wear*, 126, 131–147.
(11) SÖDERBERG, S., BRYGGMAN, U., and Mc CULLOUGH, T. (1986) Frequency effects in fretting wear, *Wear*, 110 19–34.
(12) BERTHIER, Y., COLOMBIE, Ch., VINCENT, L., and GODET, M. (1988) Fretting wear mechanisms and their effects on fretting fatigue, *J. Tribol.*, 110, 517–524.
(13) COLOMBIE, Ch., BERTHIER, Y., FLOQUET, A., VINCENT, L., and GODET, M. (1984) Fretting: load carrying capacity of wear debris, *J. Tribol.*, 106, 194–201.
(14) BERTHIER, Y., COLOMBIE, Ch., LOFFICIAL, G., VINCENT, L., and GODET, M. (1986) First and third-body effects in fretting. A source and sink problem, *Mechanisms and surface distress*, (Edited by D. D. Dowson *et al.*), Butterworths, London.
(15) GODET, M., BERTHIER, Y., LANCASTER, J., and VINCENT, L. (1991) Wear modeling: using fundamental understanding or practical experience?, *Wear*, 149, 325–340.
(16) BERTHIER, Y. (1988) *Mécanismes et Tribologie*, thesis, INSA Lyon.
(17) BERTHIER, Y., VINCENT, L., and GODET, M. (1992) Velocity accommodation sites and modes in tribology, *Eur. J. Mech. A*/Solids, 11, 35–47.
(18) BLANCHARD, P., COLOMBIE, Ch., PELLERIN, V., FAYEULLE, S., and VINCENT, L. (1991) Material effects in fretting wear: application to iron, titanium and aluminium alloys, *Met. Trans*, 22, 1535–1544.
(19) BLANCHARD, P., FAYEULLE, S., and VINCENT, L. (1993) Fretting behaviour of various titanium alloys, *Tribology transaction*.
(20) ZHOU, Z. R., FAYEULLE, S., and VINCENT, L. (1992) Cracking behaviour of various aluminium alloys during fretting wear, *Wear*, 155, 317–330.
(21) DAHMANI, N., VINCENT, L., VANNES, A. B., BERTHIER, Y., and GODET, M. Velocity accommodation in polymer fretting, *Wear*, 158, 15–28.
(22) TURKI, C., SALVIA, M., and VINCENT, L. (1993) Fretting maps of composite-metal contacts, *Advanced composites '93* (edited by T. Chandra and A. K. Dhingra) TMS, pp. 1419–1424.
(23) MONTHEILLET, F. and MOUSSY, F. (1986) *Physique et mécanique de l'endommagement*, Les Editions de Physique, Paris.

(24) COFFIN, L. F. (1962) Low cycle fatigue: a review, *Appl. Mater. Res.*, **1**, 129–141.
(25) PINEAU, A. and PETREQUIN, P. (1980) La fatigue plastique oligocyclique *La Fatigue des matériaux et des structures* (Edited by C. Bathias and J. P. Baïlon), Maloine S. A. Edit. Paris, 1980, pp. 107–161.
(26) MILLER K. J. (1987) The behaviour of short fatigue cracks and their initiation, *Fatigue Fracture Engng Structures*, **10**, 75–91.
(27) FRENCH, H. J. (1993) Fatigue and steel hardening, *Steel Treating*, **21**, 899.
(28) LE BOITEUX, H. (1973) La fatigue des Matériaux, aspects physiques et mécaniques, Edisciences, Paris.
(29) KOCANDA, S. (1978) *Fatigue failure of metals*, Sijthoff and Noordhoff, Poland.
(30) FAYEULLE, S., VANNES, A.B., and VINCENT, L. (1991) Tribological first-body behaviour before debris formation *Debris: from the cradle to the grave*, (Edited by D. Dowson *et al.*) Elsevier pp. 229–235.
(31) MINDLIN, R. D. (1949) Compliance of elastic bodies in contact, *J. Appl. Mech.*, **16**, 259–268.
(32) BERTHIER, Y., DUBOURG, M. C., GODET, M. and VINCENT, L. (1991) Wear data: what can be made of it? simulation tuning, *Debris: from the cradle to the grave*, (Edited by D. Dowson *et al.*) Elsevier.
(33) PETIOT, C., FOULQUIER, J., JOURNET, B., and VINCENT, L. (1994) Fretting fatigue behaviour of a chromium, vanadium, molybdenum steel: influence of surface protection, *Fretting fatigue*, Mechanical Engineering Publications, London, pp. 497–512. *This volume.*
(34) DUBOURG, M. C., GODET, M., and VILLECHAISE, B. Analysis of multiple fatigue cracks. Part II results, *J. Tribology*, **114**, 462–468.
(35) SINGER, I. L. (1992) Solid lubrication processes, *Fundamentals of friction: macroscopic and microscopic processes* (Edited by I. L. Singer and H. M. Pollock) Vol 220, Kluwer Academic, pp. 237–261.
(36) JOHNSON, K. L. (1985) *Contact mechanics*, Cambridge University Press.
(37) CHILDS, T. H. C. (1992) Deformation and flow of metals in sliding friction, *Fundamental of friction: macroscopic and microscopic processes*, (Edited by I. L. Singer and H. M. Pollock) Vol 220, Kluwer Academic, pp. 209–226.
(38) DANG VAN, K. (1973) Sur la résistance à la fatigue des métaux, Sc. et Tech. de l'armement, Mém. de l'Art. Fran., 3ème fascicule.
(39) DUBOURG, M. C. and VILLECHAISE, B. (1992) Stress intensity factor in a bent crack: a model, *Eur. J. Mech. A/Solids*, **11**, 169–179.
(40) KING, R. N. and LINDLEY, T. C. (1981) Fretting fatigue in a $3^{1/2}$ Ni-Cr-Mo-V rotor steel, Proc. ICF5, Vol. 4, (Edited by D. Francois) pp. 631–640.
(41) HATTORI, T., NAKAMURA, M., and WATANABE, M. (1984) Fretting fatigue analysis using fracture mechanics, ASME paper 84.
(42) SATO, K., FUJI, H., and KODAMA, S. (1986) Crack propagation behaviour in fretting fatigue, *Wear*, **107**, 245–262.
(43) NOWELL, D. and HILLS, D. A. (1987) Open cracks at or near free edges *J. Strain Analysis*, **22**, 177–185.

R. B. Waterhouse

Effect of Material and Surface Condition on Fretting Fatigue

REFERENCE Waterhouse, R. B., **Effect of material and surface condition on fretting fatigue,** *Fretting Fatigue*, ESIS 18 (Edited by R. B. Waterhouse and T. C. Lindley) 1994, Mechanical Engineering Publications, London, pp. 339–349.

Introduction

When two metal surfaces are in contact undergoing fretting, the sequence of events may be summarized as follows: (1) disruption of air-formed oxide or other surface films, (2) intimate intermetallic contact with the possible formation of cold welds, local plastic deformation, surface roughening, and the initiation of fatigue cracks, (3) development of wear debris, usually the relevant oxide, which evidence strongly suggests arises from the comminution of plate-like metallic wear particles formed by a delamination process **(1)**. The debris becomes compacted and the surfaces are smoother. However, the wear process continues by sporadic metal-to-metal contact **(2)**.

The behaviour of a particular metal or alloy will, therefore, be dictated by its response to the various stages of the process. Some of the effects can also be seen in the fretting of a ceramic or polymer material in contact with a metal, due usually to the transfer of metal, but more commonly the transfer of oxide to the counter surface **(3)**. In general, however, the cold welding stage does not occur and the possibility of initiating fatigue cracks is much less likely.

Environmentally sensitive materials

Most corrosion resistant materials rely for their corrosion properties on the presence of a passive film, usually some form of oxide which gives the material a much more positive (more noble) position in the electro-chemical series than that of the exposed virgin metal. Materials which fall into this category are the aluminium and titanium alloys and the stainless steels. In aqueous electrolytes, particularly those containing chloride ions, e.g., seawater and body fluids, they are extremely resistant to corrosive attack, but in fretting where the protective film is continuously disrupted, they are particularly vulnerable. Figure 1 shows the change in potential of a titanium alloy in saline solution when fretting is switched on amounting to about 0.5 V **(4)**. Normally when such a surface is scratched or locally damaged the passive film reforms within microseconds, but after fretting the surface is so mashed up that recovery is much slower.

---

* Department of Materials Engineering and Materials Design, University of Nottingham, Nottingham, NG7 2RD, UK.

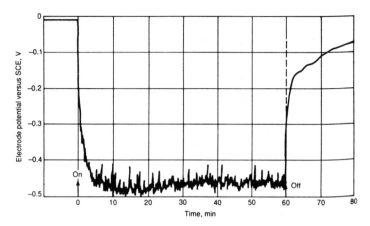

Fig 1   **Potential drop on fretting Ti–6Al–4V in saline solution**

Although the coefficient of friction is usually lower in aqueous solution and, therefore, the alternating shear stress in the contact area is lower than in air, fatigue cracks can still be initiated and their early propagation is of an order of magnitude higher than in air, Fig. 2 **(5)**.

Although an environment such as seawater is particularly virulent, the normal constituents of air, particularly water vapour and oxygen, enhance the fretting damage. In a vacuum, the coefficient of friction is high but the wear is low. As the air pressure is increased, the coefficient of friction falls and the wear increases **(6)**. The initiation of a fretting fatigue crack in vacuo is very much slower than in air **(7)**. These materials, particularly titanium alloys, are therefore extremely sensitive to fretting damage.

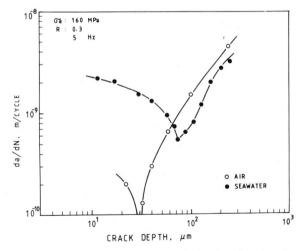

Fig 2   **Crack propagation curves for 0.64C steel wires fretted in air and seawater**

Bulk mechanical properties

In the second stage of intimate inter-metallic contact, the bulk properties of the material are a major factor. Although fatigue strength is usually linearly related to the UTS of the material, this is not so in fretting fatigue where the fretting site can be compared to a machined notch which is characterized by the rapid formation of a propagating crack. Strength reduction factors (SRFs) of between 2 and 5 are quite common.

One method of investigating the ease with which a crack can be initiated by fretting is to use a two-stage test, where fretting is applied for a certain number of cycles in a fatigue test, the fretting device is removed and the test is continued in plain fatigue under the same stressing level. If a propagating fatigue crack has developed in the first stage it will continue to failure in the second stage. By varying the length of the first stage, a curve such as that in Fig. 3 can be constructed indicating N_i, the minimum number of fretting cycles to initiate such a crack. This will, of course, depend on the clamping pressure, and the applied alternating stress and the amplitude of slip. If the clamping pressure is kept constant and the alternating stress varied for a constant size of fretting bridge, a series of curves can be constructed for like-on-like fretting contacts for a number of materials. As the materials vary widely in strength the ratio of the alternating stress ($R = -1.0$) to the yield strength of the material has been plotted. The results are shown in Figs 4 and 5.

The materials in Fig. 4 fall into three categories. The work-hardened materials show the greatest susceptibility to crack initiation. The annealed stainless steel, (En58A), the normalized 0.2C steel, and the annealed 70/30 brass are much more resistant, but the ferrite layer on the decarburized steel appears to give even greater protection. This, however, would not be an advisable method of avoiding fretting fatigue failures since the fatigue strength of the material is

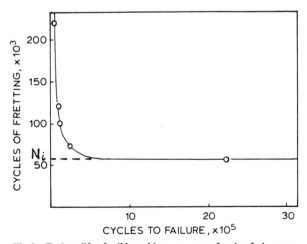

Fig 3 **Fatigue life of mild steel in a two-stage fretting fatigue test**

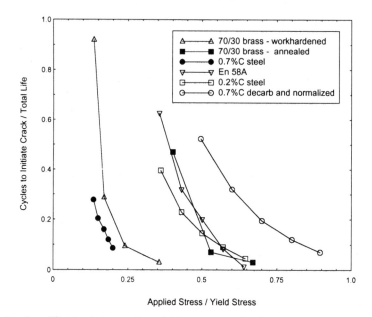

Fig 4   Number of fretting fatigue cycles to initiate a propagating fatigue crack in various materials as a function of normalized applied stress

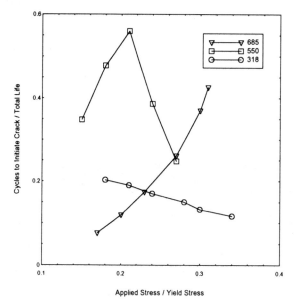

Fig 5   As in Fig. 4 for three Ti alloys

very low. It does, however, point to the possible beneficial effect of a low shear strength material on the surface, e.g., lead or cadmium plating (8) or a bonded polymer (9), or the cladding of aluminium alloys with a layer of pure aluminium (10). The three titanium alloys in Fig. 5 behave somewhat differently. Ti–6Al–4V (318) if plotted on the same scale as Fig. 4 falls closely with the work-hardened alloys. An earlier study showed that the alloy IMI685, which has a basket weave structure, was much more resistant to the initiation of fatigue cracks than the other two alloys (11). IMI550 appeared to be resistant to crack initiation in the early stages of the process, similar to IMI685, but with prolonged fretting its behaviour more closely resembled that of IMI318. As the applied alternating stress is increased, so does the amplitude of slip and with it the wear rate, which, in its turn, makes initiation of propagating fatigue cracks more difficult (12). All the alloys were of approximately the same hardness (365 VHN) but IMI685 had much the lowest impact strength (13). Unfortunately, there is no information available to permit a comparison of their fretting wear behaviour.

It is obvious that microstructure is a major factor in the fretting behaviour of a material and, in particular, the microstructure of the surface. The fretting process itself can modify that structure since both plastic deformation with attendant work hardening, Fig. 6 (14), as well as the generation of heat, with the possible overageing of age-hardened materials, Fig. 7 (15), and even the formation of metastable phases such as the 'white layer', Fig. 8 (16), can occur. In an investigation to produce a surface structure which was insensitive to fretting and these types of change, the author found that by shot-peening an

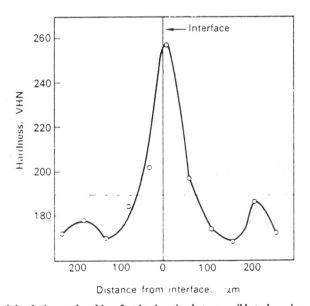

Fig 6   Work hardening produced in a fretting junction between mild steel specimen and bridge

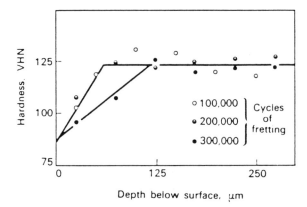

Fig 7   Softening produced by fretting on the surface of a Al–Cu–Mg age-hardened alloy

Al–Cu–Mg alloy in the SHT condition and then ageing, a structure resulted
which was immune to fretting fatigue damage, Fig. 9 **(17)**. This was due to the
increase in dislocation density resulting from the surface plastic deformation
by the shot-peening, which gave a vast increase in the nucleation sites for the
θ' on ageing, and much finer precipitates. This improved surface structure
proved very resistant to fretting damage.

Surface conditions

The important surface conditions which influence fretting behaviour are, initially, the presence of an oxide film (which has been dealt with above), surface
roughness, and residual stress.

Residual stress can arise in a number of ways. It can be the result of the
fabrication method, e.g. deep drawing of sheet materials, wire drawing, hot
and cold forging; thermal treatment, e.g. quenching; machining, e.g. milling or
grinding; or a deliberate process to induce such stresses, e.g., shot-peening or
surface rolling.

In wire drawing the residual stress distribution is tensile in three quadrants
of the circumference and compressive in the fourth, Fig. 10 **(18)**. The behav-

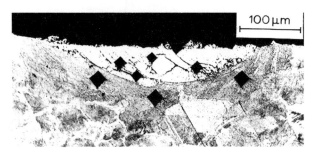

Fig 8   White layer produced by fretting on carbon steel

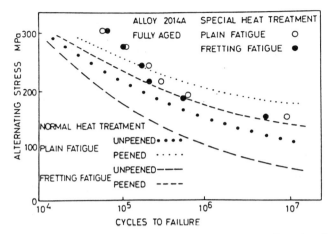

Fig 9 S–N curves in plain fatigue and fretting fatigue of Al–Cu–Mg alloy after shot-peening in SHT condition and re-ageing

iour in fretting fatigue is very dependent on the position of the fretting contact relative to the stress distribution. If fretting bridges are applied at points B, the diametral points where the tensile stress is a maximum, cracks develop at each contact, Fig. 11(a), whereas if the fretting bridges are applied at points A and C, only one crack is initiated at the point where the residual stress is tensile, Fig. 11(b).

The deliberate introduction of residual compressive surface stresses by shot-peening is beneficial in fretting fatigue as in plain fatigue. However, in fretting there are two further advantages, i.e., shot-peening causes a reduction in the coefficient of friction **(19)**, and also that the rough surface gives a further increase in fatigue strength, in contrast to its effect on plain fatigue **(20)**.

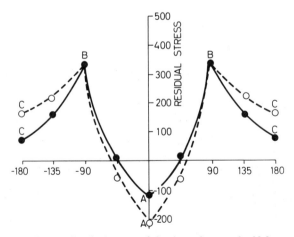

Fig 10 Residual stress distribution around the circumference of cold drawn steel wire

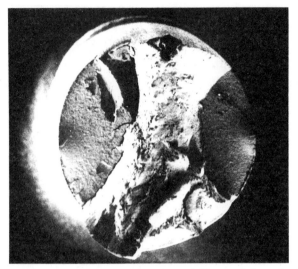

Fig 11 (a)   Fracture surfaces of fretted wire: bridges applied at points B

The author has investigated the effect of surface roughness by shot-peening an age-hardened aluminium alloy to three levels of intensity on the Almen scale. By re-solution heat treating and then re-ageing, the compressive stress and work-hardening effects were removed but the different surface roughness were preserved on specimens with the same mechanical properties. The results of fatigue tests and fretting fatigue tests on the un-peened and the treated

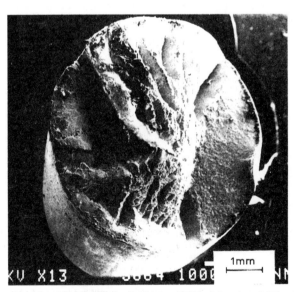

Fig 11 (b)   Fracture surfaces of fretted wire: bridges applied at points A and C (see Fig. 10)

Table 1   Fatigue strengths (in MPa) after removal of residual compressive stress and work hardening

No of cycles	Condition	Plain fatigue		Fretting fatigue		Percent reduction due to fretting
		Fatigue strength	Percent reduction due to roughness	Fatigue fretting	Percent reduction due to roughness	
10^6	Unpeened	150		100		33
	12–16A	130	13	75	25	42
	16–20A	125	17	77	23	38
	8–10C	95	37	82	18	14
10^7	Unpeened	100		50		50
	12–16A	93	7	27	46	71
	16–20A	83	17	33	34	60
	8–10C	50	50	40	20	20

specimens are shown in Table 1(21). It is apparent that the plain fatigue strength decreases with increasing surface roughness, but that the fretting fatigue strength increases. The explanation is that by breaking up the real surface contact areas into small discrete areas, the volume of material affected by the fretting action in any one contact was much reduced and with it the possibility of initiating a crack. This was first put forward by Bramhall who showed that the fretting fatigue strength of an aluminium alloy was doubled when a pad with machined grooves was used as the fretting contact rather than a plain pad (22).

Table 2   Fatigue strengths (fs) and fretting fatigue strengths (ffs) of materials fretted against themselves

Material	UTS MPa	fs at 10^7 cycles MPa	ffs at 10^7 cycles MPa
0.2C steel	509	255	150
0.4C steel	772	266	87
0.7C steel (normalized)	940	216	154
0.7C steel (cold drawn)	1360	525	139
18/9 austenitic stainless steel	598	295	160
70/30 brass (annealed)	328	89	80
70/30 brass (work hardened)	508	139	90
Al–4Cu–1Mg–1Mn	430	108	60
Al–4Mg–0.7Mn	314	100	73
Nickel-based alloy Inconel 718	1316	550 ± 275	550 ± 120*
cp titanium (IMI 130)	675	201	123
Ti–2.5Cu	578	278	123
Ti–6Al–4V	1095	540	154
Ti–4Al–4Mo–2Sn–0.5Si	1125	634	139
Ti–6Al–5Zr–0.5Mo–0.25Si	1054	432	154

* Tested in fluctuating tension

Conclusion

The effects described in this paper relate entirely to the material itself. No account has been taken of applied coatings or chemical changes to the surface. Emphasis is placed on the much greater sensitivity of materials to the environment in fretting fatigue compared with plain fatigue. Generally alloys which derive their enhanced mechanical properties by work-hardening or age-hardening are likely to suffer greater reduction in fatigue strength in fretting than cast or annealed materials. Table 2 gives come comparisons of fatigue and fretting fatigue data. Finally, surface roughness can be an advantage and structural modification of the surface could be a possibility for improving fretting resistance in heat-treatable materials.

References

(1) SUH, Nam P. (1973) The delamination theory of wear, *Wear*, **25**, 111–124.
(2) PENDLEBURY R. E. (1987) Unlubricated fretting of mild steel in air at room temperature. Part II – electrical contact resistance measurements and the effect on wear of intermittent loading, *Wear*, **118**, 341–364.
(3) HIGHAM, P. A., STOTT, F. H., and BETHUNE, B. (1978) Mechanisms of wear of the metal surface during fretting corrosion of steel on polymers, *Corros. Sci.* **18**, 3–13.
(4) WATERHOUSE, R. B. (1970) Tribology and electrochemistry, *Tribology Int*, **3**, 158–162.
(5) TAKEUCHI, M. and WATERHOUSE, R. B. (1990) The initiation and propagation of fatigue cracks under the influence of fretting in 0.64C roping steel wires in air and seawater, *Environment Assisted Fatigue*, EGF7, (Edited by P. Scott and R. A. Cottis), Mechanical Engineering Publications, London, pp. 367–379.
(6) IWABUCHI, A., KAYABA T., and KATO, K. (1983) Effect of atmospheric pressure on friction and wear of 0.45C steel in fretting, *Wear*, **91**, 289–305.
(7) FENNER, A. J. and FIELD, J. E. (1960) A study of the onset of fatigue damage due to fretting, *Trans. N.E. Coast Instn. Engrs. Shipbldrs.* **76**, 184–228.
(8) WATERHOUSE, R. B., BROOK, P. A. and LEE, G. M. C. (1962) The effect of electro-deposited metals on the fatigue behaviour of mild steel under conditions of fretting corrosion, *Wear*, **5**, 235–244.
(9) JOHNSON, K. L. and O'CONNOR, J. J. (1963–64), Mechanics of fretting, *Proc. Instn. Mech. Engrs.* Part J, 7–21.
(10) WATERHOUSE, R. B. (1964–65), An assessment of the fretting fatigue damage produced on mild steel by certain non-ferrous metals and alloys, *Proc. Instn. Mech. Engrs.* Part J, **179** (35), 258–261.
(11) GOTO, S. and WATERHOUSE, R. B. (1980) Fretting and fretting fatigue of titanium alloys under conditions of high normal load, Proceedings of the Fourth Internation Conference on Titanium, (Edited by H. Kimura and O Izumi), *Met. Soc.* **3**, 1837–1844.
(12) VINGSBO, O. and SÖDERBERG, D. (1988) On fretting maps. *Wear*, **126**, 131–147.
(13) WATERHOUSE, R. B. (1981) *Fretting fatigue*, Applied Science Publishers, London, 213.
(14) BETHUNE, B. and WATERHOUSE, R. B. (1965) Adhesion between fretting steel surfaces, *Wear*, **8**, 22–29.
(15) WATERHOUSE, R. B. (1972) *Fretting corrosion*, Pergamon, Oxford, 104.
(16) DOBROMIRSKI, J. and SMITH, I. O. (1987), Metallographic aspects of surface damage, surface temperature and crack initiation in fretting fatigue, *Wear*, **117**, 347–357.
(17) FAIR, G. H., NOBLE, B., and WATERHOUSE, R. B. (1984) The stability of compressive stresses induced by shot-peening under conditions of fatigue and fretting fatigue, *Advances in Surface Treatment*, (Edited by A. Niku-Lari), Pergamon, Oxford, 1, pp. 3–8.
(18) SMALLWOOD, R. and WATERHOUSE, R. B. (1990), Residual stress patterns in cold drawn steel wires and their effect on fretting-corrosion-fatigue behaviour in seawater, *Applied stress analysis*, (Edited by T. H. Hyde and E. Ollerton), Elsevier Applied. Science, pp. 82–90.

(19) WATERHOUSE, R. B. (1986) Residual stresses and fretting: crack initiation and propagation, *Advances in surface treatment*, Vol. 4, Residual stresses, Pergamon, pp. 511–525.

(20) LEADBEATER, G., NOBLE, B., and WATERHOUSE, R. B. (1984) The fatigue of aluminium alloys produced by fretting on a shot-peened surface, Proceedings of the Sixth International Conference on Fracture, *Advances in fracture research*, (Edited by S. R. Valluri), Vol 3 Pergamon, Oxford, pp. 2125–2132.

(21) WATERHOUSE, R. B. and TROWSDALE, A. J. (1992) Residual stress and surface roughness in fretting fatigue, *J. Phys. D: appl. Phys.* **25**, A236–239.

(22) BRAMHALL, R. (1973) *Studies in fretting fatigue*, PhD Thesis, Oxford University, UK.

M. Sumita*, N. Maruyama*, and K. Nakazawa*

Fatigue and Fretting Fatigue Behaviour of SiC Whisker-Reinforced Aluminium Metal Matrix Composite

REFERENCE Sumita, M., Maruyama, N., and Nakazawa, K., **Fatigue and fretting fatigue behaviour of SiC whisker-reinforced aluminium metal matrix composite**, *Fretting Fatigue*, ESIS 18 (Edited by R. B. Waterhouse and T. C. Lindley) 1994, Mechanical Engineering Publications, London, pp. 351–361.

ABSTRACT A study is made of the role of silicon carbide (SiC) whiskers in fatigue and fretting fatigue strengths at high cycles using a 7075–T6 metal matrix composite reinforced with 20%vol. whiskers.

The fretting fatigue strength at 10^7 cycles in the composite is about 60 percent higher than that estimated using a method proposed on the basis of the change of the elastic modulus and the ultimate tensile strength due to the reinforcement. However, the practical value of the plain fatigue strength at 10^7 cycles in the composite is nearly equal to the estimated value.

The effect of SiC whisker reinforcement on the friction coefficient, the main crack initiation site at the fretted area, crack initiation, and crack propagation in the composite material is examined in order to explain fretting fatigue strength of the composite.

Introduction

Aluminium matrix composites reinforced with a discontinuous phase in the form of chopped fibres, whiskers, or particles are considered to be strong candidates for many structural applications by virtue of high specific strength and stiffness, and high workability (1). SiC whisker-reinforced aluminium (SiC_w/Al) composites are typical of this class of material in which significant increases in the modulus of elasticity, yield strength, and ultimate tensile strength are achieved over unreinforced alloys at the sacrifice of ductility and fracture toughness. As the addition of SiC_w to aluminium alloys improves wear resistance performance (2), many potential applications of these materials involve cyclic loading with fretting, and, therefore, fretting fatigue properties are of critical interest.

The purpose of this research was to study the role of the reinforcement phase in a $SiC_w/7075$–T6 aluminium alloy composite under fretting fatigue loading.

Experimental procedures

$SiC_w/7075$ Al alloy composite containing 20%vol. whiskers was fabricated by the power metallurgy method and was subjected to extrusion and T6 heat treatment. Before extrusion the SiC whiskers had a diameter and a length distribution ranging from 0.1 to 1 μm and 5 to 50 μm, respectively. After the

---

* National Research Institute for Metals, 1-2-1 Sengen, Tsukuba, Japan.

Table 1   Mechanical properties of materials used

	SiC_w volume (percent)	$\sigma_{0.2}$ (MPa)	σ_B (MPa)	δ (percent)	Hv	E (MPa)
7075–T6 FRM	20	660	758	1.3	230	120 000
7075–T6	–	689	702	6.7	200	70 600

extrusion, their length distribution was 1–5 μm. The unreinforced 7075–T6 Al was used as a control material. Tensile properties are listed in Table 1. The microstructures of the composite material are shown in Figs, 1 (a) and (b). The SiC whiskers were uniformly distributed throughout the matrix, tending to be closely aligned with the extrusion direction.

The dimensions of fretting fatigue specimens and pads are shown in Fig. 2. The fretting fatigue testing is schematically shown in Fig. 2(c)

A constant normal load was applied to the pads on both flat surfaces of the fretting fatigue specimen. The relative slip between the specimen and the outer edges of the pads was measured using an extension meter, and it was found that the relative slip increased linearly with the increase in stress amplitude. Frictional force between the fatigue specimen and the pad was measured using gauges bonded to the side of the central part of the pad (3). The pads were machined from the same materials as the specimens.

Fretting fatigue tests were carried out under load control using a sine wave frequency of 20 Hz, a stress ratio $R = 0.1$ under tension-to-tension mode, and at a contract pressure of 50 MPa. The plain fatigue tests were carried out under the same testing conditions as those of the fretting fatigue tests. The

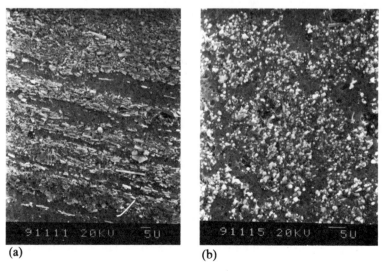

(a)   (b)

Fig 1   SEM photographs of 20%vol. SiC$_w$/7075–T6 Al composite. (a) L direction; (b) T direction

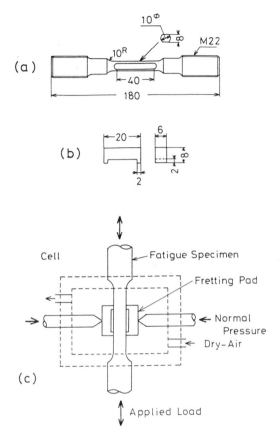

Fig 2 Dimension of (a) a fretting fatigue specimen, (b) a pad, and (c) schematic representation of fretting fatigue test

tests were carried out using an environment cell in which the relative humidity of air was controlled to be less than 0.5 percent at room temperature.

Results

Plain fatigue strength and fretting fatigue strength

The S–N curves are shown in Fig. 3 for the SiC$_w$/7075–T6 Al composite and the 7075–T6 Al alloy under fatigue loading and fretting fatigue loading. Little difference in the fatigue strength at 10^7 cycles is found between the composite material and the control material, the strength being about 155 MPa for both materials. The fretting fatigue strength at 10^7 cycles of the composite material (106 MPa) is about 70 percent higher than that of the control material (62 MPa).

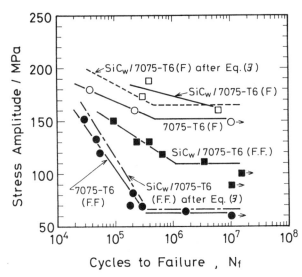

Fig 3   Fatigue and fretting fatigue behaviour of a SiC$_w$/7075–T6 Al alloy composite and a 7075–T6
Al alloy. (F: fatigue, F F: fretting fatigue)

Interrupted fretting fatigue tests

In order to examine the accumulation of fretting damage with the increase in
the number of cycles and its effect on fatigue life, a two-stage test method was
used. The fretting pads were removed at some pre-determined number of fretting fatigue cycles, and then the test was continued in plain fatigue at the same
testing conditions. The results for the 7075–T6 Al alloy at stress amplitudes of
69 and 132 MPa, and those for the SiC$_w$/7075-T6 Al composite at 132 MPa
are shown in Figs 4(a) and (b), respectively. The abscissa and the ordinate in
Fig. 4 represent the fretting period cycles, N_f, and the total life cycles including
the fretting period cycles, N_t, respectively. The data on a straight line of
$N_t = N_f$ represent the pure fretting fatigue life, and those at $N_f = 0$ represent
the pure plain fatigue life without fretting. When $N_f = 0$, N_t exceeds 10^7 cycles
under the testing conditions for both the materials. The fretting fatigue lives
for the 7075–T6 Al alloy are about 3×10^5 cycles and about 5×10^4 cycles at
stress amplitudes of 69 and 132 MPa, respectively. The fretting fatigue life for
the SiC$_w$/7075–T6 Al composite is about 3×10^5 cycles at a stress amplitude of
132 MPa.

The smallest number of fretting cycles to cause the saturation of damage,
N_{fs}, corresponds to the number required to initiate a propagating fatigue
crack without fretting action. Beyond N_{fs}, fretting has no effect on crack propagation. The N_{fs} values are larger than 90 percent of the fretting fatigue lives
at a stress amplitude of 69 MPa for the 7075–T6 Al alloy and at a stress
amplitude of 132 MPa for the SiC$_w$/7075–T6 Al composite. On the other

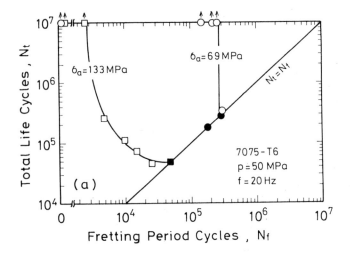

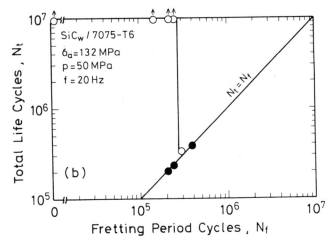

Fig 4 Effect of fretting period cycles on total life cycles. ● : failure in continuous fretting fatigue ○□: two stage tests – fretting fatigue followed by fatigue. (a) 7075–T6 Al alloy; (b) SiC$_w$/7075–T6 composite

hand, N_{fs} is about 50 percent of the fretting fatigue life at a stress amplitude of 133 MPa for the 7075–T6 Al alloy.

Friction coefficient

The stress amplitudes that control fretting fatigue failure are the plain fatigue stress amplitude, σ_a and the friction stress amplitude, $2f_a$.

$$f_a = \mu P \tag{1}$$

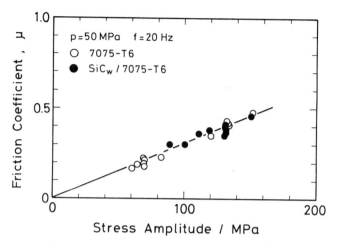

Fig 5   Friction coefficient as a function of stress amplitude

where μ is the friction coefficient and P is the contact pressure. As the friction stress decreases with the decrease in friction coefficient, the smaller the friction coefficient the higher the fretting fatigue strength. The values of μ as a function of σ_a are shown in Fig. 5 for the composite material and the matrix material. The values of μ are not influenced by the addition of SiC_w reinforcement to the 7075–T6 Al matrix in the present study although it has been shown that the presence of SiC particulate in the matrix could reduce values of μ (7).

Main crack initiation site at the fretted area

Under certain testing conditions, a stick region is found at the middle portion of the fretted area and slip regions on either side of it. The main cracks responsible for the failures are often initiated near boundaries between the stick and slip regions. A schematic representation for the fretting damage is shown in Fig. 6 (8). At high contact pressures, the stick region is wide and the main crack is initiated near the outer edge of the pad. However, at low contact pressures, the stick region is narrow, the main crack is initiated at the middle portion of the fretted area, and the life is lower than that expected from the apparent contact pressure. The contact pressure is probably concentrated at the narrow stick region when the stick region is narrow (8).

In the 7075–T6 Al alloy, the main cracks were initiated at the middle portion of the fretted area at stress amplitudes ranging from 60 to 130 MPa. In the SiC_w/7075–T6 Al composite material, the main cracks were initiated near the outer edge of the pad, except at a stress amplitude of 150 MPa. Therefore, the effect of stress concentration of the fretted area on fretting fatigue strength is larger in the unreinforced material than in the composite material. The difference in the main crack initiation site at the fretted area

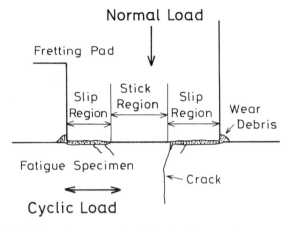

Fig 6   Schematic representation of fretting fatigue on the contact area

corresponds to that in the fretting fatigue strength of 20 to 30 MPa at high cycles range in a high strength steel and a Ti–6Al–4V alloy **(8)**.

Fracture morphology

The fracture surfaces of fatigue and fretting fatigue were mainly facetted fracture in the 7075–T6 Al alloy, and were dimple fracture in the SiC_w/7075–T6 Al composite, as shown in Figs 7 (a) and (b). Striations which are characteristics of fatigue fracture in metals were not observed in either of the materials.

(a)   (b)

Fig 7   Typical crack growth in fretting fatigue. (a) 7075–T6 Al alloy $\sigma_a = 64$MPa, $N_f = 1.65 \times 10^5$; (b) SiC_w/7075–T6 Al composite, $\sigma_a = 111$MPa, $N_f = 3.75 \times 10^6$

Discussion

Fatigue strength and fretting fatigue strength of the $SiC_w/7075-T6$ Al composite

In order to evaluate the role of reinforcement phase in fatigue and fretting fatigue strengths in a metal matrix composite, the following equation **(10)** is used.

$$\sigma_{w \cdot c} = \sigma_{w \cdot m}(\sigma_{b \cdot c}/\sigma_{b \cdot m}) \tag{2}$$

where $\sigma_{w \cdot c}$ is the expected fatigue or fretting fatigue strengths of a composite, $\sigma_{w \cdot m}$ is the fatigue or fretting fatigue strength of the matrix material, $\sigma_{b \cdot m}$ is the practical ultimate tensile strength in a composite, $\sigma_{b \cdot m}$ is the practical ultimate tensile strength in a matrix. The changes of the elastic modulus and the ultimate tensile strength in metal matrix composite due to the reinforcement are included in equation (2).

If the estimated value of the fatigue strength in a composite according to equation (2) is higher than the practical value, it means that the reinforcement phase contributes effectively to the increase in fatigue strength.

In the present study, $\sigma_{b \cdot c}$ is 758 MPa, and $\sigma_{b \cdot m}$ is 702 MPa, as shown in Table 1. Therefore,

$$\sigma_{w \cdot c} = \sigma_{w \cdot m} \times (758/702) = 1.08\ \sigma_{w \cdot m} \tag{3}$$

The estimated values of the fatigue strength and the fretting fatigue strength produced by equation (3) are drawn in Fig. 3 by a broken line and a single dot and dash line, respectively.

Fatigue strength and fretting fatigue strength of the $SiC_w/7075-T6$ Al composite

The practical fatigue strength at 10^7 cycles is about 160 MPa and the estimated value after equation (3) is about 167 MPa. The typical fatigue crack initiation site in the $SiC_w/7075-T6$ Al composite is shown in Fig. 8. The crack was initiated at one of the sub-surface defects whose size was 150–200 μm. This indicates that the fatigue strength of the composite material is controlled by the stress concentration at the defect **(6)**. If these defects in the composite material were removed, a fatigue strength higher than the practical value would be expected.

The practical value of the 10^7 cycles fretting fatigue strength of the composite material is 105 MPa and the estimated value after equation (3) is 67 MPa. The former is 38 MPa (57 percent) higher than the latter. This means that the reinforcement by SiC_w makes a great contribution to increasing the fretting fatigue strength at high cycles range.

Influence of SiC_w reinforcement on crack initiation

In fretting fatigue, crack initiation life and early stage of crack propagation life are influenced by contact pressures and friction stresses. The number of fret-

(a)   (b)

Fig 8   Typical fatigue crack initiation from a defect in the SiC$_w$/7075–T6 Al composite

ting cycles to cause the saturation of damage to fretting fatigue life is higher than 90 percent as shown in Figs 4(a) and (b) in both the materials. So, the fretting fatigue life is dominated by crack initiation life.

The addition of SiC$_w$ reinforcement to a metallic matrix is effective in the prevention of crack initiation under cyclic loading. The mechanisms concerned are schematically shown in Figs 9 (a) and (b). Slip occurring prior to crack initiation is prevented by SiC$_w$ whose effect is similar in principal to that of

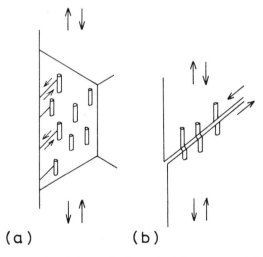

(a)   (b)

Fig 9   Schematic representation of the mechanisms of reinforcement on fatigue crack initiation. (a) barrier against slip; (b) barrier against crack initiation

grain refinement (see Fig. 9(a)). The extension of small cracks is prevented by SiC_w which causes contact shielding through bridging (see Fig. 9(b)).

Influence of SiC_w reinforcement on crack growth

The crack growth rate in the SiC_w/7075–T6 Al composite material is higher than that in the matrix material at low stress intensity range (9). However, as fretting fatigue lives consist mainly of fretting fatigue crack initiation in the high cycles range in both the materials as described before, the crack growth behaviour is not important in terms of the total life.

Conclusions

(1) The fretting fatigue strength at 10^7 cycles of the SiC_w/7075–T6 Al composite material is 106 MPa and that of the 7075–T6 alloy is 62 MPa.
(2) The plain fatigue strengths at 10^7 cycles are 155 MPa in both the materials.
(3) The fretting fatigue life is mostly occupied by crack initiation life at high cycles ranges.
(4) The friction coefficient of the matrix is not influenced by the addition of SiC whisker.
(5) The fretting fatigue strength at 10^7 cycles in the composite was significantly higher than that estimated on the basis given in (4). This suggests that the increase in fretting fatigue strength is probably due to low stress concentration at the fretted area and restriction of crack initiation caused by SiC whiskers.
(6) The practical value of the plain fatigue strength at 10^7 cycles in the composite is nearly equal to the estimated value. The crack is initiated from the sub-surface defect whose size is 150–200 μm.

References

(1) SHANG, JIAN KU, YU, WEIKANG, and RITCHIE, R. O., (1988) Role of silicon carbide particles in fatigue crack growth in SiC-particulate-reinforced aluminium alloy composites, *Mater. Sci. Engng.* **102**, 181–192.
(2) GIBSON, P. R., CLEGG, A. J., and DAS, A. A. (1984) Wear of cast Al–Si alloys containing graphite, *Wear*, **95**, 193–198.
(3) NAKAZAWA, K., SUMITA, M., and MARUYAMA, N. (1989) Fretting fatigue of high strength steels in seawater, EVALMAT' 89 B2–4, pp. 151–158.
(4) NAKAZAWA, K., SUMITA, M., and MARUYAMA, N. (1989) Fretting fatigue of high strength steels for chain cables in seawater, *ISIJ International*, **29**, 781–787.
(5) NAKAZAWA, K., SUMITA, M., and MARUYAMA, N. (1990) Effect of cathodic protection on fretting fatigue of high strength steels in seawater, *Tetsu-to-Hagane*, **76**, 1552–1558 (in Japanese).
(6) SUMITA, M., UCHIYAMA, I., and ARAKI, T. (1974) Relationship between the effect of inclusions on the endurance limits and the work hardening behaviours of carbon steels, *Trans. ISIJ*, **14**, 274–284.
(7) WANG, A. and RACK, H. J. (1991) Transition wear behavior of SiC-particulate and SiC-whisker-reinforced 7091 Al metal matrix composites, *Met. Sci. Engng* **147**, 211–224.

(8) NAKAZAWA, K., SUMITA, M., and MARUYAMA, N. (1992) Effect of contact pressure on fretting fatigue of high strength steel and titanium alloy, *ASTM STP 1159*, ASTM, Philadelphia, pp. 115–125.

(9) HIRANO, K. (1990) Fatigue crack growth characteristics of whisker reinforced aluminium alloys, Fatigue 90 Proceedings of the Fourth International Conference on Fatigue and Fatigue Thresholds, pp. 863–868.

(10) SUMITA, M., MARUYAMA, N., and NAKAZAWA, K. (1993) Role of second phase in fretting fatigue strength in a SiC-whisker-reinforced aluminium alloy composite, *J. Japan Inst. Metals*, **57**, 1141–1148 (in Japanese).

M. Okane*, T. Satoh*, Y. Mutoh*, and S. Suzuki†

Fretting Fatigue Behaviour of Silicon Nitride

REFERENCE Okane, M., Satoh, T., Mutoh, Y., and Suzuki, S., **Fretting fatigue behaviour of silicon nitride,** *Fretting Fatigue,* ESIS 18 (Edited by R. B. Waterhouse and T. C. Lindley) 1994, Mechanical Engineering Publications, London, pp. 363–371.

ABSTRACT A fretting fatigue test machine for ceramic materials was designed and fabricated. Fretting occurred at the contact surface between the rectangular specimen which was maintained under static fatigue condition and the cylindrical contact piece. Fretting fatigue strength decreased with increasing relative slip amplitude. Fretting damage was observed on the contact surface at very small relative slip amplitudes, which would never result in fretting damage in metallic materials. Since fretting cracks initiate at a very early stage of fatigue life, crack propagation life is dominant in fretting fatigue life.

Introduction

Silicon nitride, which possesses excellent high temperature strength, wear resistance, etc., is a major candidate material for potential structural components in high performance engine and gas turbine applications. In practical applications, contacts between ceramic and metallic structural components cannot be avoided in fits, joints, bearings, etc., where fretting damage is often induced.

The notch sensitivity of ceramic materials is considered to be significantly high compared to metallic materials. Therefore, when some damage or cracks are induced on the contact surface of the ceremic component due to fretting, reduction of fatigue strength may be remarkable compared to the case of metallic materials. No available report has been found on the fretting fatigue of ceramic materials, although there are some for fretting wear (1)(2).

In this study, as a first step to investigate the basis properties of fretting fatigue of ceramic materials, a fretting fatigue test machine for ceramic materials was designed and fabricated. Some preliminary fretting fatigue tests on silicon nitride were carried out. The effect of relative slip amplitude on fretting fatigue strength in silicon nitride is also discussed.

Experimental Procedure

Specimen

The material used was an HIP-sintered silicon nitride with additives of Al_2O_3 and Y_2O_3. The chemical composition and mechanical properties of the test

* Department of Mechanical Engineering, Nagaoka University of Technology, Nagaoka-shi 940–21, Japan.
† Isuzu Ceramics Research Institute, Fijisawa-shi 252, Japan.

Table 1　Chemical composition (a) and mechanical properties of the material used (b)

Al	Y	Ca	W	Fe
1.28 ∼ 1.75	3.59 ∼ 3.88	0.01 ∼ 0.055	0.378 ∼ 0.416	0.0072 ∼ 0.022

(a)

σ_B (MPa)	K_{IC} (MPa$\sqrt{m}$)	H_v (GPa)	E (GPa)	v	α (/°C)
1048	7.6	14.3	304	0.28	3.1×10^{-6}

(b)

σ_B: 4-point bending strength; K_{IC}: Fracture toughness; H_v: Hardness; E: Young's modulus; v: Poison's ratio; α: Coefficient of thermal expansion

material are summarized in Tables 1(a) and (b), respectively. The fatigue specimen is a rectangular bar with dimensions of $3 \times 4 \times 36$ mm and the contact piece is a cylindrical bar with diameter of 2 mm and length of 12 mm.

Testing Procedure
The schematic illustrations of the fretting fatigue test apparatus are shown in Fig. 1(a). The oscillating motion, produced by an eccentric cam, was transmitted to the end of the cantilever beam and consequently to the specimen set in the test section on the cantilever beam. The slip amplitude $S_{a,b}$, which is not coincident with relative slip amplitude S_a between the specimen and contact piece, was achieved by the oscillating amplitude of the beam at the point at which the test section was fixed. The cylindrical contact piece was pressed on the rectangular specimen by the dead weight, as shown in Fig. 1(b). Therefore, fretting occurred at the line contact surface between the specimen and the contact piece. In the fretting fatigue test, the specimen was maintained under a constant static fatigue stress in four-point bending (supporting span: 30 mm,

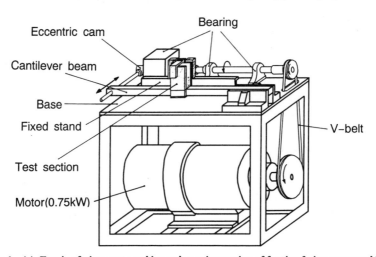

Fig 1　(a) Fretting fatigue test machine; schematic overview of fretting fatigue test machine

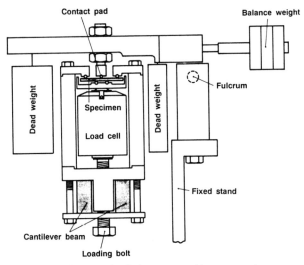

Fig 1(b)   Fretting fatigue test machine; test section

loading span: 10 mm). The bending force was applied by using the loading bolt. The frictional force and the relative slip amplitude were measured during tests using strain gauges attached to the contact rod and the extensometer, respectively, as shown in Fig. 1(c).

Fretting fatigue tests were carried out under slip amplitudes $S_{a,b}$, of 4 and 20 μm, frequencies of 5 and 10 Hz and contact load of 43 N. For comparison, two kinds of static fatigue tests were carried out: (1) static fatigue tests without fretting contact and (2) static fatigue test with static contact. All the tests were

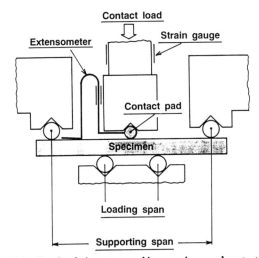

Fig 1(c)   Fretting fatigue test machine; specimen and contact piece

Table 2   Test conditions

	Static fatigue with static contact	Fretting fatigue (static fatigue with fretting)	
Amplitude of cantilever $S_{a,b}$, μm	0	4	20
Frequency, Hz	—	5, 10	5, 10
Contact load P, N	43	43	43

conducted at room temperature in laboratory air. The test conditions are summarized in Table 2.

Contact surfaces were observed in detail on the specimens tested up to certain fretting cycles using a low-vacuum scanning electron microscope with a Robinson-type reflect-electron detector which enabled direct observation of ceramics without conductive film coatings.

Results and discussions

Fretting fatigue strength

The relationship between applied stress normalized by bending strength σ_a/σ_B and time to failure is shown in Fig. 2. In the figure, data points represented by arrows show that no failure occurred up to the test time of 2×10^5 s. Static fatigue strength and fretting fatigue strength were defined as the stress at which the specimen survived over the testing time of 2×10^5. S–T curves were horizontal with relatively large scatter. Fretting significantly reduced

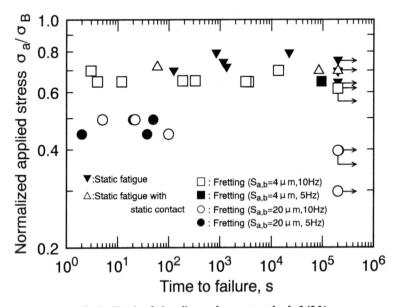

Fig 2   Fretting fatigue lives under a contact load of 43 N

fatigue strength of silicon nitride. No significant effect of frequency on fretting fatigue strength was found in the present range of frequency. Fracture points of the specimen in the fretting fatigue test were in the contact area, while those in the static fatigue test without fretting contact were dispersed in the loading span. The fatigue strength decreased with increasing relative slip amplitude: $\sigma_a/\sigma_B = 0.75$ for static fatigue without fretting (▼), $\sigma_a/\sigma_B = 0.62$ for fretting fatigue under slip amplitude of 4 μm (□, ■), and $\sigma_a/\sigma_B = 0.4$ for fretting fatigue under slip amplitude of 20 μm (○, ●). Application of static contact without fretting motion gave no influence on static fatigue strength (△). Fracture points were not always in the contact area but dispersed in the loading span. Therefore, fretting action with relative slip is essential in reaching the fatigue strength.

Tangential force coefficient and relative slip amplitude

Variations of tangential force coefficient and relative slip amplitude against number of fretting cycles is shown in Figs 3 and 4, respectively. The tangential force coefficient was defined as the ratio of frictional force F and contact load P, F/P. The results of two specimens for each test condition are plotted in the figures. By decreasing the slip amplitude $S_{a, b}$, the tangential force coefficient i.e., frictional force and the relative slip amplitude were reduced. A sudden increase in tangential force coefficient was observed in the specimens tested under $\sigma_a/\sigma_B = 0.3$ and slip amplitude of 20 μm. The attained higher values of tangential force coefficient coincided with that for the specimens tested under the higher applied stress of $\sigma_a/\sigma_B = 0.4$. This transition behaviour of the tangential force coefficient may have resulted from the change of wear surface

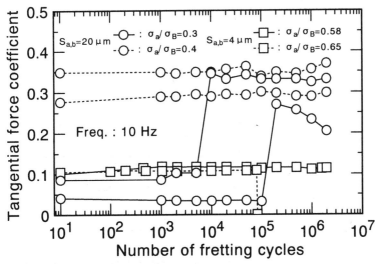

Fig 3   Variation of tangential force coefficient under fretting fatigue tests with $S_{a, b} = 4$ μm and 20 μm

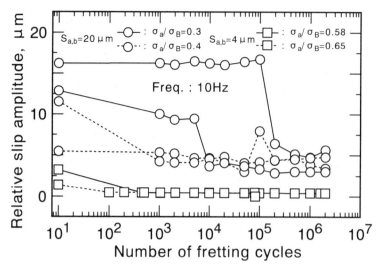

Fig 4   Variation of relative slip amplitude under fretting fatigue tests with $S_{a,b}$ = 4 µm and 20 µm

morphology. The fretting surface for specimens with a low tangential force coefficient was slightly worn with drop-out of silicon nitride particles, while that for specimens with a high tangential force coefficient was severely worn with wear debris.

Based on the measurements of maximum surface roughness or wear depth in the fretted surface of metallic materials (3)(4), no surface-damage-induced fretting is observed when the relative slip amplitude is lower than a certain level: 4 µm for quenched 0.35 percent C steel (Hv = 500) (5) and 7.5 µm for spring steel (Hv = 498) (6). On the other hand, in silicon nitride, the surface was damaged by fretting even under the much smaller relative slip amplitude of 0.5 µm, as shown below. Therefore, ceramic materials such as silicon nitride seem to be more sensitive to fretting, compared with metallic materials.

SEM observations of the fretted surface

SEM observations of the fretted surface are shown in Fig. 5. It is clear that the extent and width of fretting damage under the slip amplitude $S_{a,b}$ of 20 µm are more severe and wider than those under the slip amplitude $S_{a,b}$ of 4 µm. Considerable wear debris was observed for a slip amplitude $S_{a,b}$ of 20 µm, while no debris was found for a slip amplitude $S_{a,b}$ of 4 µm. Slight contact marks are observed in the specimen tested under static fatigue with static contact, as shown in Fig. 6, where the observation was at 45 degrees to the specimen surface to make the marks clear. The fracture points under this test condition were dispersed in the loading span and not always in the contact area. Therefore, although the contact load at the present level damaged the contact surface slightly, it did not seriously affect the static fatigue strength.

←Sliding direction→   ←Sliding direction→

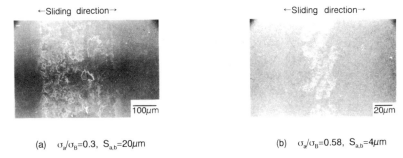

(a) σ_a/σ_B=0.3, $S_{a,b}$=20μm   (b) σ_a/σ_B=0.58, $S_{a,b}$=4μm

Fig 5   SEM observations of the fretted surface tested up to 2 × 106 cycles. (a) $\sigma_a/\sigma_B = 0.3$, $S_{a, b} = 20$ μm; (b) $\sigma_a/\sigma_B = 0.58$, $S_{a, b} = 4$ μm

SEM observations of the fretted surface of the specimens tested up to 4×10^4 fretting cycles under slip amplitude $S_{a, b}$ of 4 μm are shown in Fig. 7. In the moderately worn region of the contact area, micro surface cracks induced by fretting were found, as shown in Fig. 7(a). In the severely worn region of the contact area, traces of the drop-out silicon nitride particles and micro surface cracks were observed, as shown in Fig. 7(b). Based on the measurement of tangential force coefficient and relative slip amplitude, it is considered that surface damage induced by fretting develops and stable morphology of the fretted surface is attained in the very early stage of fatigue life, i.e, in one hundred fretting cycles. The surface damage includes surface cracks and traces of drop-out silicon nitride particles which may be nuclei for cracks. Therefore, the fretting fatigue fracture process is as follows: fretting cracks initiate in the very early stage of fatigue life and propagate rapidly under the influence of the fretting action, until, subsequently, final unstable fracture occurs.

Effect of relative slip amplitude on fretting fatigue strength

The relationship between the relative slip amplitude and normalized fatigue strength σ_w/σ_B is shown in Fig. 8. In metallic materials, it is known that fretting fatigue strength decreases with increasing relative slip amplitude and

←Direction of specimen thickness→

←Contact line

σ_a/σ_B=0.63, T=2×10⁵s

Fig 6   SEM observation of the contact surface under the static fatigue condition with static contact

(a)   Micro cracks

(b)   Traces of drop-out silicon nitride particles

Fig 7   SEM observations of the fretted surface of the specimen tested up to 4×10^4 cycles under $\sigma_a/\sigma_B = 0.65$ and $S_{a, b} = 4$ µm. (a) micro cracks, (b) traces of drop-out silicon nitride particles

attains a constant value when the relative slip amplitude become larger than a critical value (7)(8). This is because the frictional force between specimen and contact piece increases with increasing relative slip amplitude, and becomes constant above the critical value of relative slip amplitude. In the case of the

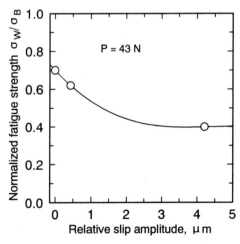

Fig 8   Relationship between normalized fatigue strength and relative slip amplitude

present material, the normalized fretting fatigue strength σ_w/σ_B decreased with increasing relative slip amplitude, as shown in Fig. 8.

Conclusions

Fretting fatigue tests on silicon nitride were carried out using a fretting fatigue test machine fabricated for ceramic materials. The main results obtainedd are summarized as follows:

(1) Fretting significantly reduces the static fatigue strength of silicon nitride: the fretting fatigue strength decreases as the relative slip amplitude increases.
(2) Fretting damage occurs on the contact surface under small relative slip amplitude which does not induce fretting damage in metallic materials.
(3) Application of static contact without fretting motion does not influence static fatigue strength. Therefore, fretting action, which induces surface wear and frictional force, is essential to the degradation of fatigue strength.
(4) The fretting fatigue fracture process in silicon nitride is as follows. Fretting cracks initiate at a very early stage of fatigue life and propagate with increasing speed until final unstable fracture occurs.

References

(1) HORN, D. R., WATERHOUSE, R.B., and PEARSON, B. R., (1986) Fretting wear of sintered alumina, *Wear*, **113**, 225.
(2) KLAFFKE, D. (1989) Fretting of ceramics, cfi/Ber, DKG 66, No. 1/2 **18**.
(3) NISHIOKA, K. and HIRAKAWA, K. (1968) Fundamental investigation of fretting fatigue. Part 5 – the effect of relative slip amplitude, *Trans. Japan Soc. mech. Engrs*, **34**, 2068.
(4) TANAKA K., MUTOH, Y., and SAKODA, S. (1985) Effect of contact materials on fretting fatigue in a spring steel, *Trans. Japan Soc. mech. Engrs*, **51**, 1200.
(5) NISHIOKA, K. and HIRAKAWA, K. (1968) Fundamental investigation of fretting fatigue. Part 2 – fretting fatigue testing machine and some test results, *Trans. Japan Soc. mech. Engrs*, **34**, 1183.
(6) EDWARDS, P. R. (1981) The application of fracture mechanics to predicting fretting fatigue, *Fretting fatigue*, (Edited by R. B. Waterhouse), Applied Science Publishers, p. 67.
(7) MUTOH, Y., NISHIDA, T., and SAKAMOTO, I. (1988) Effect of relative slip amplitude and contact pressure on fretting fatigue strength, *J. Soc. Mater. Sci. Japan*, **37**, 649.

ENVIRONMENTAL EFFECTS

*D. E. Taylor**

Environmental Fretting Fatigue

REFERENCE Taylor, D. E., **Environmental fretting fatigue**, *Fretting Fatigue*, ESIS 18 (Edited by R. B. Waterhouse and T. C. Lindley) 1994, Mechanical Engineering Publications, London, pp. 375–389.

ABSTRACT The environment in which fretting takes place can have a profound effect on the nature and degree of the resultant surface damage and fatigue crack generation. The effect is increasingly significant in environments more chemically reactive than air for example in the hostile aqueous environment provided by seawater or even that of body fluids. Examples of the many practical situations in which components are required to function in such environments are given including wire ropes, ship propeller systems, and orthopaedic implants. An overview of recent research investigations into the nature and level of fretting fatigue performance of metallic alloys in a variety of ambient temperature environments is presented. The relative roles of the mechanical and chemical contributions to the fretting fatigue process and the ways in which these and other factors affect crack generation and growth are discussed.

Introduction

Fretting may result in surface damage and may involve the disruption of surface films, abrasion by debris particles, and the making and breaking of welds between contacting asperities by direct shear of fatigue. Thus fretting can occasion two particular difficulties: wear, with the associated production of debris leading to loss of fit between previously contacting surfaces and, perhaps more seriously, the nucleation of cracks which in components subject to cyclic stress can lead to significant and marked reduction in fatigue life.

The environment in which fretting takes place can have a profound effect on the nature and degree of the resultant damage. The effect is increasingly significant when the environment is more chemically reactive than air as, for example, in the hostile oxygenated aqueous environment which seawater provides. In aqueous electrolytes the mechanical action of fretting disrupts surface films (often abrasive oxides) in the contact zone and results in continued and possibly accelerated dissolution of the underlying metal (**1**). The fretting induced vibration of the electrolyte is accompanied by the disturbance of diffuse layers adjacent to the contacting surfaces. In addition the fretting contact often provides a situation in which crevice corrosion is possible due to differential aeration effects. The lubricant properties of the aqueous solution must also be considered; the liquid may act as a lubricant reducing friction or separating the surfaces, and hence reducing wear rate or impairment by fretting fatigue. Thus the conjoint action of mechanical and electrochemical factions is likely to play a major part in fretting fatigue in aqueous environments.

Occurrence of environmental fretting fatigue

There are many practical applications in which components are required to function in aqueous environments. The press or shrink fit of a hub or bearing

* School of Engineering and Advanced Technology, University of Sunderland SR1 3SD, UK.

housing on to a loaded rotating shaft is in itself a frequent source of fatigue cracks initiated by fretting, but the situation is further aggravated if it is operating, as in many marine applications, in salt water or salt spray (2). One particular example concerns that of a ship's propeller shrunk on to a tail shaft. The incidence of fretting corrosion fatigue failures in such circumstances remains a problem. In a study (3) of 2500 shafts in large single-screw ocean-going vessels, approximately 10 percent of those removed for inspection were found to contain fatigue cracks, most of which occurred under the propeller hub or under the bronze liner fitted on to the shaft or in the keyway – all common fretting situations. A further similar application involves mechanical seal rings on ship propeller shafts. A small degree of misalignment or vibration of the shaft may result in axial oscillations of the seal ring, which on being transmitted to the shaft packing results in fretting of the shaft surface (4).

The assembly of wires to form flexible cables or ropes presents a situation that is particularly prone to fretting. It is clear that the flexure of the cable or rope must result in the sliding of the individual wires against their neighbours. There are many such interwire contacts which can become sites for fretting damage when the cables are cyclically stressed. Steel ropes widely used in marine environments for mooring and as hawsers, and overhead power cables made up of wound aluminium wires as conductors, are both multi-component systems. Wave or vessel motion in the case of mooring ropes and air flow past power cables results in fluctuating tension leading to the possibility of fretting and cracking occurring at the numerous interwire contacts. In these circumstances the prevailing environmental conditions can exert a considerable influence.

A further, very different, application where fretting has been seen to contribute to fatigue failure in an aqueous corrosive environment concerns surgical implants, for use as fracture fixation devices, or for hip replacement. Metal bone-plates usually contain a series of screw holes and are attached to the bone via metal screws so as to hold the two faces of a fracture together whilst it heals. Metal-to-metal contact occurs between the bevelled underside of the screw head and the countersink surrounding the hole. As the limb is exercised in the body fluid, effectively an oxygenated 1 percent sodium chloride solution, small movements occur between the contact surfaces, and fatigue cracks may initiate by fretting leading to failure at the hole (2). Fretting fatigue fractures have also been associated with the femoral stems of hip prostheses. The operation involves replacement of the hip's ball and socket joint with a metal ball normally riding in a high density polyethylene socket. In order to insert the stemmed femoral component some surgeons removed the greater trochanter a protruding portion of bone and subsequently reattached it with a stainless steel wire anchored by winding it around the prosthesis stem. Movement was possible between the wire and stem producing fretting and consequent fatigue failure of the stem at the point of attachment of the wire (5). The above surgical practice is no longer encouraged.

The examples given merely serve to illustrate the diverse situations in which fretting in various environments can be a factor in fatigue failure.

The environment and fretting fatigue

Fretting in the main is concerned with the generation of fatigue cracks whilst corrosion involves both generation by attack of emerging slip bands, and propagation by affecting the tip of the growing crack (2). It is of interest, therefore, to consider the possible interaction between fretting and corrosion at the crack initiation stage. It is possible that resultant corrosion products lead to a reduction in the coefficient of friction and act in the same way as an artificially applied coating. This has been found appropriate in high temperature oxidation, but does not generally appear to be the case in aqueous solutions. In practice in neutral solutions most metals develop a surface hydroxide or oxide film and the extent to which this grows depends on its ionic and electronic conductivity. In some cases, the film is not protective, as in the rusting of mild steel in sodium chloride solution, whilst in others the film is extremely thin and highly protective, e.g., the passive film formed on stainless steel and titanium alloys. The presence of such films and also the physical adsorption of various species can have a profound effect on the sliding and coefficient of friction of metal surfaces in contact.

Once the surface film has been disrupted, then some form of conjoint action between the fretting and the corrosive environment may occur. Waterhouse (2), divided materials into two groups with regard to their response to fretting fatigue in aqueous electrolytes; materials which are not corrosion resistant in the particular environment and materials which are. In the first group, e.g., mild steel in sodium chloride solution, when fretting and corrosion are both present, the behaviour is entirely dictated by the corrosive effect and fretting produces no further reduction in the fatigue strength. The second group, which rely on a protective surface film as the basis of their inactivity in a corrosive environment such as seawater, experience little or no reduction in fatigue strength due to corrosion alone, but a large reduction when fretting and corrosion are present even though the corrosive environment may be relatively mild, e.g., air. In this case the prime function of the fretting is the disruption of the protective surface oxide film. An example of the second group is shown in Fig. 1 which indicates the effect of fretting on the fatigue behaviour of a titanium alloy (6). In common with most titanium alloys this is little affected by corrosion fatigue but fretting reduces the fatigue strength by 55 percent whether the environment is air or 1 percent sodium chloride.

Investigations of the influence of environment on fretting fatigue behaviour

That the environment could have a significant effect on fretting fatigue behaviour was considered in early experiments carried out in, amongst other atmospheres, vacuum, inert gases, oxygen, and dry and moist air. Initially

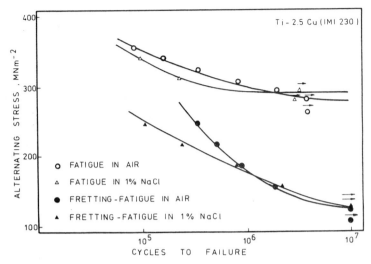

Fig 1 **Effect of fretting on the fatigue behaviour of the titanium alloy Ti–2.5Cu tested in rotating bending** (6)

investigations were largely concerned with the effect of oxygen. In all cases the reduction factor based on the fatigue strength in air was less for fretting fatigue in an oxygen free atmosphere than in air (7). Two stage tests by Fenner and Field (8) showed that the number of fretting cycles to initiate a propagating fatigue crack in an aged Al–4Cu alloy at a stress level of 193 ± 40 MNm^{-2} was 240000 in air but exceeded 2×10^6 cycles in vacuum. They concluded that both the number of cycles to initiate a crack and that to propagate it were greater in vacuum than in air. It was determined that the corrosive agent in air was water vapour and its effect was concerned both with crack initiation and growth. Exclusion of water vapour increased the life at a particular alternating stress by eight times.

Early results of Reeves and Hoeppner (9) suggested that the fretting fatigue strength of a 0.4–0.5 percent C steel in air and in vacuum was largely determined by the extent of mechanical damage rather than by environmental effects.

More recent work by Poon and Hoeppner (10) involved a randomized block fretting fatigue programme designed to statistically determine the relative roles of the mechanical and chemical factors in the mechanism of fretting fatigue. Tests were performed on the aluminium alloy 7075 in both laboratory air and in vacuum. It was found that the number of cycles to failure in vacuum was between 10 and 20 times longer than that in laboratory air having a relative humidity between 40 and 60 percent. It was established that the three parameter Weibull distribution best fitted the fretting fatigue data for both test environments and that the difference in the mean lives of the specimens tested in vacuum and in laboratory air was significant with 95 percent confidence.

They therefore accepted that the chemical factor played the dominant role in reducing specimen life when fretting fatigue occurs. Fractrographic analysis of the wear surfaces of specimens tested in air showed that pitting was very common whilst that on those tested in vacuum indicated that rewelding of metal was a frequent feature. This was interpreted as supporting the above hypothesis concerning the dominant chemical role.

Endo and Goto (11)(12) have also found in reversed bending that the presence of water vapour accelerated the initiation and propagation of fretting fatigue cracks in an aluminium alloy, whilst on carbon steel the environmental effect seemed minimal. S–N curves of fretting fatigue tests for the aluminium alloy indicated that the fatigue strength increased in the order room air, dry air (40–100 ppm), dry air (4–10 ppm). The S–N curve in argon (40–100 ppm) was similar to that in dry air (40–100 ppm). The results thus indicate that water vapour has a deleterious effect on fretting fatigue (12). The fretting fatigue strengths of a 0.34 percent C steel were the same in room air of relative humidity 50–80 percent and in dry air of humidity 4–10 ppm, indicating that ductile steels are not affected by water vapour in fretting fatigue. The crack propagation rate within the carbon steel was determined in two stage tests in argon in order to investigate the effect of oxygen. The crack rate was slowed down remarkably by the removal of oxygen compared to that in air as is shown in Fig. 2. This difference may be attributed to the adsorption effect of oxygen at the crack tip. Since the aluminium alloy was sensitive to humidity but very little affected by oxygen, Endo and Goto measured the crack propagation rate in dry air (Fig. 3). Not only the crack propagation but also crack initiation were accelerated considerably by a small amount of humidity, the environmental effect predominating over the mechanical effect.

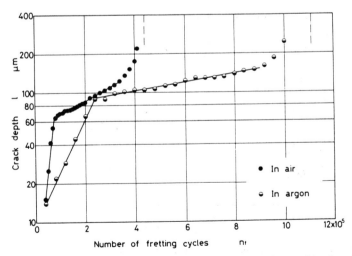

Fig 2   Fretting fatigue crack propagation curves of 0.34 percent C carbon steel in air and in argon (σ_b = 190 MNm^{-2}, p = 0.39 kNm^{-1}, f = 30Hz (11)

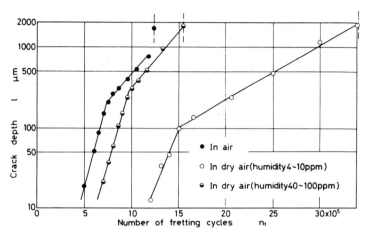

Fig 3   Fretting fatigue crack propagation curves of Al–4Zn–2Mg alloy in air of various humidities
        ($\sigma_o = 140\text{MN}^{-2}$, $p = 8\text{km}^{-1}$, $f = 20\text{Hz}$) (11)

The titanium alloy Ti–6Al–4V tested in fluctuating tension by Wharton and
Waterhouse (13) provided longer fretting fatigue lives in low-cycle high stress
fatigue but lower fretting fatigue strengths in high-cycle low stress fatigue in
dry argon (non-corrosive) as compared to humid argon (corrosive) (Fig. 4).
The increase in fatigue strength was attributed to the protective influence of
corrosion products. Thick layers of corrosion product formed within the
fretted region in humid argon (and indeed in 1 percent NaCl which was also
investigated) giving some protection to the surface and rendering the generation of fatigue cracks by fretting less likely, resulting in higher fatigue

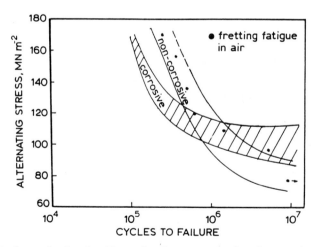

Fig 4   Effect of corrosive (e.g. humid argon) and non-corrosive (e.g. dry argon) atmospheres on
        fretting fatigue of Ti–6A1–4V showing that behaviour in air corresponds to that in non-
        corrosive atmospheres (13)

strengths at high cycle lives. At higher stresses and short lives the corrosion atmospheres were regarded as having a significant effect on crack propagation rates with a consequent reduction in fatigue life. It was noted that results in air were included and were seen to fit within the non-corrosive regime.

An analysis of the factors controlling fretting fatigue life has more recently (14) been carried out on the same titanium alloy as reported on above in a number of environments including pure water, synthetic seawater, CH_3OH, and $LiCl–CH_3OH$. The failure mechanism of fretting fatigue was related to both the friction forces and the environments. At a contact pressure of 50 MPa in air and argon 'elastic slip' occurred between the contacting surfaces, and the main crack was generated at the outer area of the contact. At the same contact pressure in the liquid environments 'macro-slip' arose and the main crack generated at the middle area of the contact. The fretting fatigue life in synthetic seawater or CH_3OH from which the effect of the friction force was eliminated was the same as that in air. However, the life in pure water was longer and that in $LiCl–CH_3OH$ was shorter, compared with the life in air. It seemed that the former was caused by the removal of micro cracks due to wear particles produced by the fretting, and the latter was caused by the stress concentration due to adhesion of the wear debris to the contacting surfaces.

The above investigations were carried out at frequencies of 1 and 20 Hz. In all environments except water the fretting fatigue lives were significantly higher at the higher frequency.

There have to date been few investigations on the effect of frequency on fretting fatigue. Figure 5 shows results obtained in Waterhouse's laboratory (2) on stainless steel in Hanks solution in fluctuating tension. These indicate as with those quoted above, that greater fatigue strength is associated with higher frequencies, Waterhouse postulated that such differences may be due to the effect of frequency on corrosion fatigue.

In testing cold drawn 0.64 C steel wires Takeuchi and Waterhouse (15), verified that fretting was a significant factor in the reduction of fatigue strength in seawater and that the effect was not solely that of corrosion fatigue. Fatigue curves for the wire with and without fretting clearly indicated that fretting caused an additional reduction of the fatigue life in seawater, thus conflicting with Waterhouse's earlier definition of performance of non-corrosion resistant materials (2). They then embarked on an investigation to further assess the effect of seawater on the fretting fatigue of steel wires and to study the beneficial effects of cathodic protection. Specimens were tested in fluctuating tension at a frequency of 5 Hz and stress ratio of 0.3, the centre of the specimen being surrounded by a cell through which artificial seawater was continuously pumped. The cell was also equipped with a three electrode system allowing potentiostatic control to be applied. To investigate the generation and propagation of fatigue cracks tests were run for limited numbers of cycles below the failure limit at three alternating stresses. The specimens were then sectioned and by careful repeated grinding and polishing, the depths of cracks

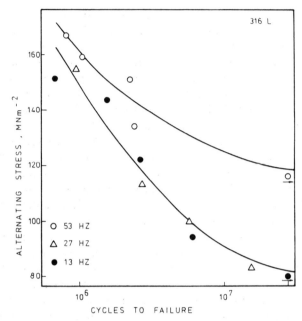

Fig 5 **Effect of frequency on the fretting fatigue behaviour of an austenitic stainless steel 316L in Hanks solution tested in push–pull mean stress 247 MNm^{-2} (2)**

(perpendicular to the surface) were measured with an optical microscope. From these observations the shape of the cracks were determined and curves of crack depth against numbers of cycles were plotted which in turn allowed crack growth rate (da/dN) versus crack depth curves to be constructed. Figure

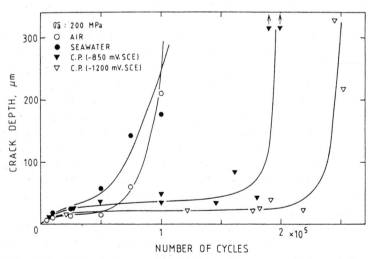

Fig 6 **Curves of crack length versus number of cycles in air, seawater, and with cathodic protection σ_a = 200MPa (15)**

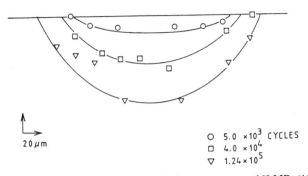

Fig 7   Crack profile of cracks growing in seawater. $\sigma_a = 160$ MPa (15).

6 shows a typical set of crack depth versus numbers of cycles curves based on sectioning and measurement of crack depths. Figure 7 shows the crack profile of typical cracks in seawater, indicating that they are of the thumbnail type, being elliptical in the early stages of fretting and then later becoming semi-circular. The cracks were more numerous in the specimens tested in seawater and were generated in the same low number of cycles as those in air. However, the subsequent growth rate was much accelerated in seawater. As seen in Fig. 8 the initial rate of crack growth is increased by an order of magnitude in seawater compared with air. After passing through a minimum which indicates that the crack is growing out of the region of influence of fretting, the crack growth rate in air is greater than that in seawater. In Endo's experiments (11) on steel in argon and air the growth rate increased by a factor of three, while Takeuchi and Waterhouse (15) reported a factor varying between 5 and 10 as the stress amplitude increased. This was to be expected in the much more

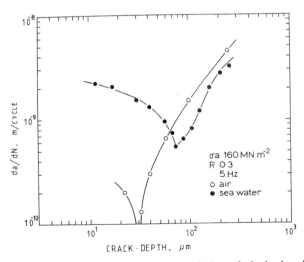

Fig 8   Crack propagation curves for fretting fatigue cracks in steel wire in air and seawater (15)

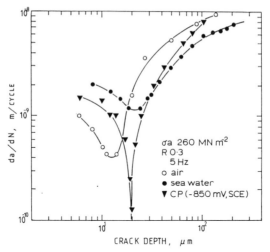

Fig 9   Retardation of fretting fatigue crack in steel wire by application of cathodic protection (15)

corrosive seawater, evidence for which was provided in tests carried out with cathodic protection which suppressed the corrosion reaction. Figure 9 shows 9 crack growth rate curve determined under the influence of applied potential, showing that the initial rates of crack growth are reduced.

An unexpected feature of Figs 8 and 9 is the slower growth rate in seawater at the longer crack lengths where the fretting ceases to influence the crack. This occurs because many more cracks initiate in seawater than in air and thus the stress intensity factor at an individual crack is smaller at a given crack length and load level. This is a well known feature of corrosion fatigue which also produces many cracks (16).

The application of impressed cathodic protection to a high strength low alloy weathering steel, Corten A, fretting fatigued in seawater (17) has identified electrochemical dissolution processes as having a significant influence on fatigue life. Corrosion fatigue in the absence of fretting produced a 60 percent reduction in fatigue life compared to that in air whilst the imposition of fretting led to a further 24 percent reduction in fatigue life. The restoration of the fretting fatigue life in seawater to a level similar to that observed in air through the use of impressed cathodic protection indicated that in this case the mechanical action played a minor role in the life reduction process. In further work by Takeuchi et al. (18) an attempt was made to predict fretting fatigue lives on the basis of fracture mechanics using a form of the Paris–Erdogan equation and plotting log da/dN against log 'K' curves. The results of the life predictions are shown in Fig. 10. The predicted fatigue life in air agreed with the experimental results. However, in seawater the predicted values, particularly at lower alternating stresses, were much greater than the experimental values. In seawater the prediction defines a much higher fatigue

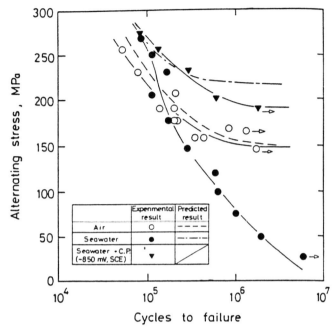

Fig 10 Fretting fatigue curves for steel wire in air, seawater, and seawater with cathodic protection at −850mV (SCE) (18)

limit than that for the experimental S–N curve, the difference being attributed to the fact that no account was taken of the contributory electrochemical effects. The analysis reveals that techniques such as cathode protection, which remove the deleterious electrochemical factor in the generation and propagation of fatigue cracks, significantly improve fretting fatigue performance in seawater.

Recent Japanese two stage test (19)(20) carried out at frequencies between 0.1 and 20 Hz in seawater under freely corroding conditions on high strength steels showed that fretting damage reached a limit, i.e., saturated beyond a certain number of fretting cycles. Lowering test frequency usually leads to lower fatigue life for a given fretting condition and number of cycles, indicating that a chemical (rate controlled) factor is important (21). In the Japanese work fretting fatigue lives in seawater at the lower frequencies were considerably lower than at higher frequencies whilst lives in air were consistent and at the level found in seawater at the higher frequencies. The work led to the definition of two groups; those in which the smallest numbers of fretting cycles to cause damage saturation was less than 0.1 percent of the fretting fatigue life (group 1) and those (group 2) in which it was more than 40–60 per cent. In group 1 cracks propagated normal to the alternating stress axis from the early stage of crack propagation. However, in group 2, cracks propagated at less than 90 degrees, but changed direction to become normal as they propagated beyond a certain length. It was suggested that the lowering in fretting fatigue lives of group 1 resulted from the decrease in crack initiation life caused by the acceleration of corrosion pit formation, and that of group 2 from crack initiation and growth caused by fretting. The same authors (22) also confirmed that the application of cathodic protection restored the fatigue strength to values greater than those in air as found by Takeuchi *et al.* (18). As mentioned earlier, in seawater the coefficient of friction is lower and the cooling effect is present, both of which have an influence on crack initiation (23). In addition cathodic protection reduces the propagation rate of the crack and the formation of calcareous deposits within the crack may prevent crack closure thereby reducing the effective value of K.

Conclusion

An overview of research investigation into the nature and level of fretting fatigue behaviour of metal alloys in a variety of environments leads to the following general conclusions.

(1) Although the role of the environment in fretting fatigue is not, as yet, clearly established there have been significant recent developments involving the measurement of crack growth rates and the effect of the environment upon them.

(2) The mechanism of fretting fatigue includes chemical and mechanical factors, the observed damage commonly resulting from both. The factor which dominates is dependent on the particular circumstances prevailing. Most researchers favour the chemical factor as playing the major role in reducing fretting fatigue life with the mechanical damage serving to disrupt surface films and expose underlying chemically reactive sites. Cathodic protection which removes deleterious electrochemical effects on both the generation and propagation of fatigue cracks greatly improves fretting fatigue performance, usually returning values to or above those found in air.

(3) The fretting fatigue behaviour can be significantly affected by the nature of the corrosion products forming within the fretted region. Thick layers of such product may reduce the coefficient of friction between contacting surfaces and also provide some protection to the surfaces, thereby delaying the generation of fatigue cracks.

(4) Reducing the frequency of stressing normally leads to a reduction in fretting fatigue life.

References

(1) WATERHOUSE, R. B. (1970), Tribology and electrochemistry, *Tribology* 3, 501.
(2) WATERHOUSE, R. B. (1981) Fretting fatigue in aqueous electrolytes, Fretting Fatigue (Edited by R. B. Waterhouse) Applied Science Publishers, pp. 159–175.

(3) HECK, J. W., and BAKER, E., (1963) *J.Soc. Naval Architects and Marine Engnrs*, **71**, 327–346.
(4) BATTILANA, R. E. (1971) *Chem. Engng*, **78**, 130–132.
(5) SMETHURST, E. and WATERHOUSE, R. B. (1977), *J. Mater. Sci.*, **12**, 1781–1792.
(6) WATERHOUSE, R. B. and DUTTA, M. K. (1973) The fretting fatigue of titanium and some titanium alloys in a corrosive environment, *Wear*, **25**, 171–175.
(7) WATERHOUSE, R. B. (1972) *Fretting corrosion*, Pergamon, Oxford.
(8) FENNER, A. J. and FIELD, J. E. (1960) *Trans. N E Coast Instn. Engrs. Shipbuilders*, **76**, 184.
(9) REEVES, R. K. and HOEPPNER, D. W. (1978) Microstructural and environmental effects on fretting fatigue, *Wear*, **47**, 221–229.
(10) POON, C., and HOEPPNER, D. W. (1979) The effect of environment on the mechanism of fretting fatigue, *Wear*, **52**, 175–191.
(11) ENDO, K. (1981) Practical observation of initiation and propagation of fretting fatigue cracks, *Fretting fatigue* (Edited by R. B. Waterhouse) Applied Science Publishers, pp. 127–141.
(12) ENDO, K. and GOTO, H. (1978) Effects of environment on fretting fatigue, *Wear*, **48**, 347–367.
(13) WHARTON, M. H. and WATERHOUSE, R. B. (1980) Environmental effects in the fretting fatigue of Ti–6 A1–4V., *Wear*, **62**, 287–297.
(14) MARUYAMA, N., SUMITA, M., and NAKAZAWA, K., (1991) Effect of testing environments on fretting fatigue strength of Ti–6 A1–4V, *J. Iron Steel Inst. Japan*, **77**, 290–297.
(15) TAKEUCHI, M. and WATERHOUSE, R. B. (1990) The initiation and propagation of fatigue cracks under the influence of fretting in 0.64 C roping steel wires in air and seawater, *Environment Assisted Fatigue*, EGF7 (Edited by P. Scott), Mechanical Engineering Publications, London, pp. 367–379.
(16) WATERHOUSE, R. B. (1976) *Corrosion*, (Edited by L.L. Shrier) Vol. 1, (Second edition), Newnes–Butterworth, London, pp. 96–113.
(17) PRICE, S. and TAYLOR, D. E. (1992) The application of electrochemical techniques to evaluate the role of corrosion in fretting fatigue of a high strength low alloy steel, *Standardisation of fretting fatigue test methods and equipment*, (Edited by H. Attia and R. B. Waterhouse) ASTM, Philadelphia, pp. 217–228.
(18) TAKEUCHI, M., WATERHOUSE, R. B., MUTOH, Y. and SATOH, T. (1991) The behaviour of fatigue crack growth in the fretting-corrosion-fatigue of high tensile roping steel in air and seawater, *Fatigue Fracture. Engng. Mater. Structures*, **14**, 69–77.
(19) NAKAZAWA, K., SUMITA, M., and MARUYAMA, N. (1990) Saturation of damage in fretting fatigue of high strength steels in seawater, *J. Iron Steel Inst. Japan*, **76**, 917–923.
(20) NAKAZAWA, K., SUMITA, M., MARUYAMA, N., and KAWABE, Y. (1989) Fretting fatigue of high strength for chain cables in seawater, *ISIJ International*, **29**, 781–787.
(21) TAYLOR, D. E. and WATERHOUSE, R. B. (1989), Wear, fretting and fretting fatigue, *Metal behaviour and surface engineering* (Edited by S. Curioni, R. B. Waterhouse, and D. Kirk) IITT International, pp. 13–35.
(22) NAKAZAWA, K., SUMITA, M., and MARUYUMA, N. (1990) Effect of cathodic protection of fretting fatigue of high strength steels in seawater, *J. Iron Steel Inst. Japan*, **76**, 1552–1558.
(23) PEARSON, B. R. and WATERHOUSE, R. B. (1984) The fretting corrosion in seawater of materials used in off-shore structures, Proceedings of the Ninth International Congress on Metallic Corrosion, National Research Council, Vol. **2**, pp. 334–341.

Y. Mutoh* and T. Satoh*

High Temperature Fretting Fatigue

REFERENCE Mutoh, Y. and Satoh, T., **High temperature fretting fatigue,** *Fretting Fatigue,* ESIS 18 (Edited by R. B. Waterhouse and T. C. Lindley) 1994, Mechanical Engineering Publications, London, pp. 389–404.

ABSTRACT Dominant factors of fretting fatigue are the fretting wear and the frictional force on the contact surface, which induce the early initiation of cracks and the acceleration of crack growth rate, respectively. Formation of glaze oxide layers at elevated temperatures promote a low coefficient of friction and consequently a high fretting fatigue strength, in nickel-based alloys and some kinds of titanium alloys. The Goodman relationship for predicting the mean stress effect and the modified Miner's rule for estimating the effect of variable loading are successfully applied to the case of fretting fatigue at both room and elevated temperatures. Shot peening is effective in improving fretting fatigue strength, even at elevated temperatures.

Introduction

Fretting fatigue has become a serious problem in the steam and gas turbines which operate at elevated temperatures under increasingly severe conditions, brought about by higher demands in power and efficiency. However, little work has been carried out on the effect of fretting on high temperature fatigue, whereas it has been studied as a wear problem. The studies on fretting fatigue at elevated temperatures have been reviewed by Taylor **(1)** mainly from a metallurgical point of view. In the review paper by Taylor, fretting fatigue behaviour at elevated temperatures is widely discussed with reference to titanium-based alloys, nickel-based alloys, and aluminium-based alloys, which are for gas turbine and aircraft applications. In the present paper, fretting fatigue behaviour at elevated temperatures is discussed mainly on steam turbine steels from a mechanical point of view.

Fretting fatigue process

In plain fatigue, crack initiation may account for 50–90 percent of total fatigue life, whereas in fretting, the crack initiation can occur in only 5 percent, or less, of fatigue life **(2)**. A typical example of a crack propagation curve in a steam turbine steel in high temperature fretting fatigue is shown in Fig. 1 **(3)**. The fretting crack, which means a crack influenced by fretting, initiates in 3 percent or less of fatigue life and propagates up to 150 μm in depth in only 5 percent of the fatigue life. Therefore, almost the whole life of fretting fatigue is spent propagating the crack. The fretting crack propagates under fretting action, which accelerates the crack growth rate. The relationship between crack growth rate da/dN and apparent stress intensity range ΔK for a steam turbine steel at elevated temperature is shown in Fig. 2 **(3)**. The crack growth rate in fretting fatigue is significantly accelerated in the low ΔK region, i.e., in the

* Department of Mechanical Engineering, Nagaoka University of Technology, Nagaoka-shi 940 − 21, Japan.

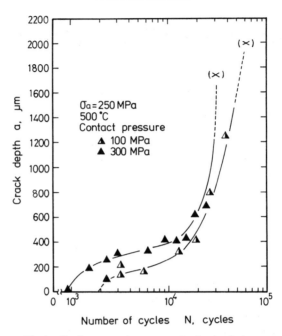

Fig 1   Crack propagation curve under fretting fatigue

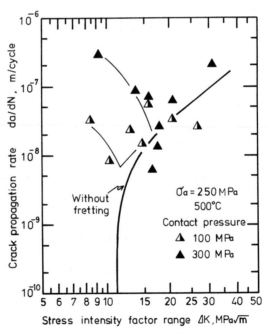

Fig 2   Relationship between stress intensity factor range ΔK and crack propagation rate da/dN

early stage of crack growth, due to fretting action. Similar 'V' shaped crack growth curves for fretting fatigue have been reported **(4)–(6)**. When the frictional force acting on the contact surface is taken into account in calculating the stress intensity range, the 'V' shaped crack growth curve is reduced to a monotonic crack growth curve which coincides with that for plain fatigue without fretting **(7)**. The main reasons for reducing fatigue strength by fretting are, therefore, the early initiation of a crack due to fretting wear and the acceleration of crack growth rate due to frictional force.

Oxidation

The fatigue properties of metallic materials are generally reduced as the temperature is raised because of a decrease in the yield strength, which results in the degradation of crack initiation resistance, and also an increase in the crack growth rate. Corrosion processes are also enhanced. In fretting fatigue, the coefficient of friction (the frictional force) is one of the dominant factors controlling fretting fatigue strength. A lower frictional force is beneficial. The coefficient of friction is generally reduced as the temperature is raised, as shown in Fig. 3. Therefore, the fretting fatigue strength at elevated temperatures is in the competitive relationship between the reduction in plain fatigue strength and the lowering of the coefficient of friction. The formation of a 'glaze' oxide at elevated temperatures induces a large reduction in the coefficient of friction, which was observed in nickel-based alloys **(9)(19)**, as shown in Fig. 3. This

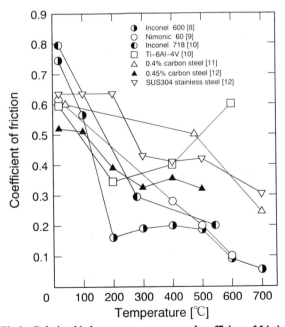

Fig 3   Relationship between temperature and coefficient of friction

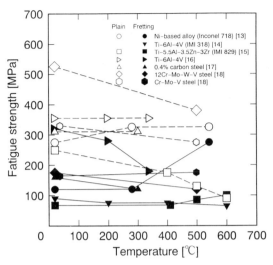

Fig 4   Relationship between temperature and fatigue strength

extreme reduction in the coefficient of friction gives rise to an increase in fretting fatigue strength, as illustrated in Fig. 4. In titanium alloys, the formation
of a glaze oxide depends on the material, fretting condition, and environment
(20), and consequently the fretting fatigue strength at elevated temperature is
also dependent on these conditions, examples of which are shown in Fig. 4.
The fretting fatigue strength of Ti–5.5Al–3.5Zn–3Zr (IMI829) alloy increases
with increasing temperature, while that of Ti–6Al–4V alloy decreases. In steam
turbine steels (12Cr–Mo–W–V and Cr–Mo–V steels in Fig. 4), the glaze oxide
was not observed on the contact surface **(18)**. The coefficient of friction at
elevated temperature was almost coincident with that at room temperature.
Therefore, the fretting fatigue strength of a 12Cr–Mo–W–V blade steel was
reduced at elevated temperatures, as indicated in Fig. 4. In a Cr–Mo–V rotor
steel, the oxide particles trapped in the crack mouth at an elevated temperature were observed, as schematically illustrated in Fig. 5. The oxide par-

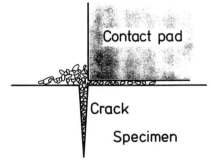

Fig 5   Schematic illustration of oxide particles trapped in the crack mouth.

ticles filled the crack mouth inducing a high level of crack closure: the measured crack opening ratios U ($= \Delta K_{eff}/\Delta K$) of the crack with trapped oxide particles were approximately 0.28, while those of the crack without trapped particles were around 0.4–0.5 **(18)**. Consequently higher fretting fatigue strength was obtained at elevated temperatures, as shown in Fig. 4.

Variable loading

Practical components suffer variable loading with different mean load levels. In plain fatigue, the following relationships are known to predict the effect of mean stress **(21)**

$$\sigma_w = \sigma_{w0}\{1 - (\sigma_{mean}/\sigma_B)^n\}$$

where σ_w is the fatigue strength at a given mean stress, σ_{w0} is the fatigue strength at zero mean stress, σ_{mean} is the mean stress, σ_B is the tensile strength, $n = 1$ is the modified Goodman relationship, and $n = 2$ is the Gerber relationship.

Figure 6 shows the S–N curves for a series of values of mean stresses at room and elevated temperatures in 12Cr–Mo–W–V turbine steel **(22)**. The relationship between the tangential force coefficient, which is defined as the ratio of tangential force (frictional force) and contact load, and the stress amplitude is independent of mean stress, as shown in Fig. 7. From the S–N curves shown in Fig. 6, the relationship between fatigue limit (fatigue strength) and mean stress is plotted in Fig. 8. The modified Goodman relationship, which is represented by solid linear lines in the figure, successfully predicts the effect of mean stress for fretting fatigue as well as for plain fatigue, regardless of temperature. Figure 9 shows the relationship between number of cycles to failure N_f and the stress amplitude σ_{a0} at zero mean stress, which is reduced from the stress amplitude at a particular mean stress by using the equation, $\sigma_{a0} = \sigma_a/(1 - \sigma_{mean}/\sigma_B)$. The modified Goodman relationship can also be applied to predict not only fatigue limit, but also fatigue life, i.e., the S–N curve.

In generating plants, the components of a steam turbine, such as blade and rotor joints, are subjected to variable loads caused by generating power regulation with the change of demands of electric power. Figure 10 shows the typical stress variation for the certain steam turbine component **(23)**. Figures 10(a) and (b) show the start-up and shutdown of a steam turbine with the periodical (every year and every month, respectively) inspections. Figure 10(c) shows the stress variation caused by the generating power regulation with the change of demands of electric power, and Fig. 10(d) illustrates the details of each block in (c), where the accidental peak value is induced by a failed nozzle. Although it is hard to reproduce exactly the actual service load in the laboratory fatigue test, the actual service load is simulated by the nine-step block load as shown in Fig. 10(e), where the stress in each step represents the peak value due to the failed nozzle and the total number of cycles in one block

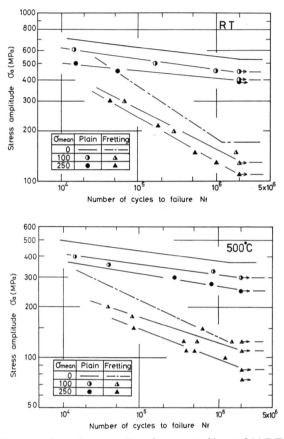

Fig 6 *S–N* **curves under various mean stress for steam turbine steel (a) R T; (b) 500°C**

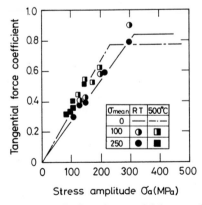

Fig 7 **Relationship between stress amplitude and tangential force coefficient under various mean stress**

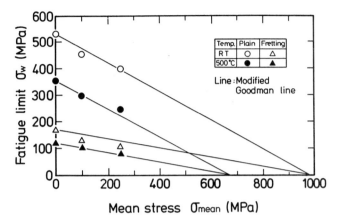

Fig 8   Relationship between mean stress and fatigue limit

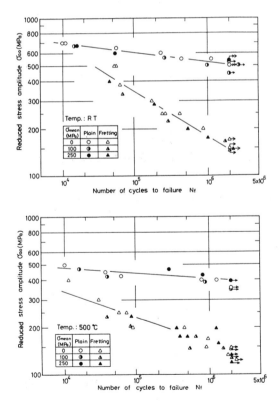

Fig 9   Relationship between number of cycles to failure and reduced stress amplitude at zero mean
stress. (a) R T; (b) 500°C

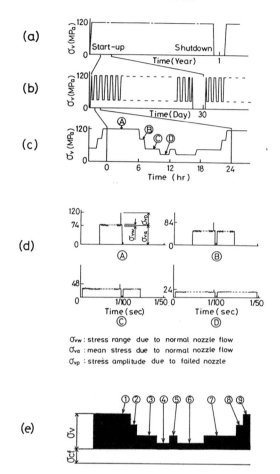

Fig 10 Simulated service load for steam turbine: (a) service load in a scale of year; (b) service load in a scale of day, (c) service load in a scale of hour; (d) peak load due to a failed nozzle; (e) simulated service load

is 103. Since the components of a steam turbine are subjected to a tensile mean stress due to the centrifugal force, the stress σ_{cf} due to the centrifugal force is also taken into account in the simulated block load. Another typical variable load will be random load, an example of which (Gaussian quasi-random load) is shown in Fig. 11.

$\sigma_{max} = 1.85\sigma_{rms}$

Fig 11   Random load

Figure 12 shows the relationship between the equivalent stress amplitude, σ_{eq}, and number of cycles to failure, N_f, for fretting fatigue tests under the simulated block load (Fig. 10(e)). The equivalent stress amplitude is evaluated as follows. The stress amplitude under block load is reduced to that at zero mean stress, according to the Goodman relationship. The number of blocks N_b at the cumulative damage $D = 1$ is calculated according to the modified Miner's rule (21), $D = [\Sigma(n_i/N_i)] \times N_b$, where n_i is the number of cycles in the block at the stress level of σ_i and N_i is the number of cycles to failure at σ_i. The equivalent stress amplitude, which gives the same number of cycles to failure, $(\Sigma n_i) \times N_b$, is obtained from the S–N curve for fretting fatigue under constant amplitude load. The S–N curves under block loading, where fretting fatigue life is related to the equivalent stress amplitude, are identical to those under constant amplitude load at both room and elevated temperatures.

The relationships between the root-mean-square (rms) value of peak stress, σ_{rms}, and N_f for plain and fretting fatigue under the random load shown in Fig. 11 are also shown in Fig. 12 (23). From this figure, the linear relationship between σ_{rms} and N_f is observed under random load as well as under constant amplitude load. The fatigue lives under random load related to σ_{rms} are shorter and the fatigue limits are lower than those under constant amplitude load in plain fatigue. On the other hand, the S–N curves for fretting fatigue under random load are identical to those under constant amplitude load at both room and elevated temperatures. In addition, for reference, the S–N curve under random load related to the equivalent stress amplitude is almost identical to that related to σ_{rms}.

The tangential force coefficients during the fretting fatigue test under block load are obtained in each of the steps of the block load. For fretting fatigue tests under random load, the rms value of tangential force coefficient is obtained using the rms value of frictional force. The maximum value of

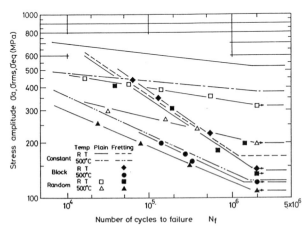

Fig 12 *S–N* curves under simulated block load and random load for a steam turbine steel

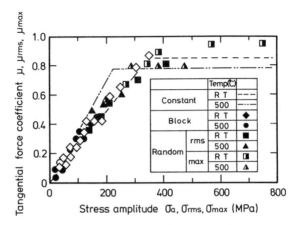

Fig 13　Relationship between stress amplitude and tangential force coefficient

tangential force coefficient during fretting fatigue tests under random load is also obtained using the maximum value of frictional force. Figure 13 shows the relationship between stress amplitude and tangential force coefficient at both room and elevated temperatures (23). The relationships under constant amplitude load are also indicated in this figure. As can be seen, the relationships under block load and random load are almost in agreement with those under constant amplitude load at elevated temperature as well as at room temperature.

In general, many stress amplitudes σ_i are applied under variable load. The Miner's linear cumulative damage rule is a well known method for estimating fatigue life under variable load (21). The modified Miner's rule where the S–N curve is extrapolated linearly in a lower stress region than the fatigue limit, is also proposed. The fretting fatigue lives under the two kinds of variable loads mentioned above are predicted on the basis of the modified Miner's rule. When estimating fatigue lives under block load with the mean stress, the stress amplitude is reduced to the stress amplitude with zero mean stress according to the Goodman relationship. The relationship between the predicted lives and the experimental results is shown in Fig. 14 (23). From the figure, the modified Miner's rule is effective for fretting fatigue under both block and random loads. On the other hand, for the plain fatigue under random load, the predicted lives tend to give a risky estimation. Since the fretting fatigue crack initiates at a very early stage of life, the fretting fatigue life is almost equal to the crack propagation life. Therefore, it seems that the linear cumulative damage rule is successfully applicable to fretting fatigue where the single process of crack propagation is dominant.

Similar results for the effectiveness of the modified Miner's rule in fretting fatigue have been also reported in an aluminium alloy under Gaussian random and programmed random corresponding to an aircraft gust spectrum

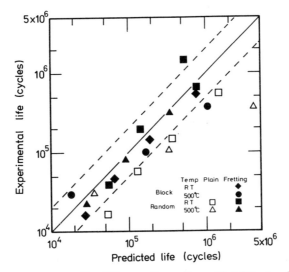

Fig 14 Relationship between predicted life according to the modified Miner's role and experimental life

(24), 0.55 percent carbon steel (spring steel) under Gaussian random (25), 0.45 percent carbon steel under two-step block loading (26), and high carbon bearing steel under a programmed load corresponding to an aircraft bearing spectrum (27).

Improvement of fretting fatigue strength

Methods for improving fretting fatigue strength can identified as follows.

(1) Designing to reduce contact pressure and relative slip amplitude.

(2) Combination of materials, where a soft contact material is beneficial.

(3) Surface treatment, which includes heat treatment, coating, shot peening, etc. Induced low tangential force coefficient and residual compressive stress are significantly beneficial.

The effectiveness of these improvement methods have been reported at room temperature: Shot peening (28)(29), coatings (17)(30)(31) and soft shims (32). At elevated temperature these methods may be still applicable. However, it is also known that residual stress is relieved by heating at elevated temperature. Therefore, improvement of fretting fatigue strength due to residual compressive stress depends on the combination of particular material and service temperature.

Fretting fatigue tests at elevated temperature were carried out using shot-peened specimens to improve the high temperature fretting fatigue strength of 12Cr–Mo–W–V steam turbine steel (33). In the tests, specimens shot-peened and subsequently exposed at elevated temperatures for long periods were pre-

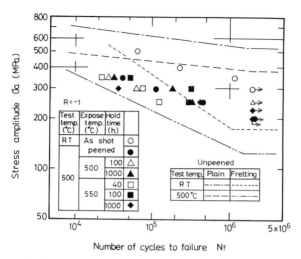

Fig 15 *S–N* curves for fretting fatigue of the as-peened specimens and the specimens shot-peened
and subsequently exposed at temperatures of 500°C or 550°C

pared to investigate the effectiveness of shot peening for long-term practical
use. The results of fretting fatigue tests of shot-peened specimens at both room
and elevated temperatures are shown in Fig. 15. For reference, the *S–N* curves
for plain and fretting fatigue of the unpeened specimens at room and elevated
temperatures are also shown in the figure. The fretting fatigue strengths of the
shot-peened specimens at room and elevated temperatures are higher than 300
MPa and 200 MPa, respectively. The shot peening improves the fretting
fatigue strength by a factor of around 1.8 at elevated temperature as well as at
room temperature. The fretting fatigue lives of the specimens shot-peened and
subsequently exposed at 500°C are almost identical, regardless of the exposure
conditions. Even after 1000h at 550°C, which corresponds to an exposure time
of 30 000 h at 500°C according to the Larson–Miller approach (34), the fretting fatigue strength is as high as 225 MPa. This value of 225 MPa means
improvement of the fretting fatigue strength by a factor of 1.8. Therefore, it is
suggested that shot peening is effective in improving the fretting fatigue
strength of 12Cr–Mo–W–V turbine steel, even for long-term practical use at
the service temperature of 500°C.

The relationships between stress amplitude and tangential force coefficient
for the unpeened and shot-peened specimens are shown in Fig. 16. At room
temperature, the tangential force coefficient of the shot-peened specimen
increases with increasing stress amplitude, and attains a constant value of 0.8.
At 500°C, the tangential force coefficients are almost identical, regardless of
the exposed conditions. These relationships for shot-peened and unpeened
specimens are almost identical at both room and elevated temperatures. The
oxidation of the material (12Cr–Mo–W–V turbine steel) is not significant at

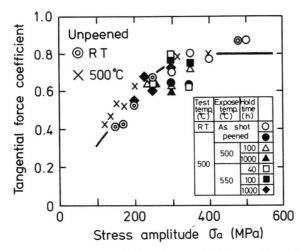

Fig 16 Relationship between stress amplitude and tangential force coefficient for the as-peened
specimens and the specimens peened and subsequently exposed at 500°C or 550°C

the test temperatures, and the oxidized surface is removed by fretting action.
The surface roughness for unpeened and peened specimens are almost identical during and after the fretting fatigue test. In addition, the hardening of the
peened surface is not significant. This particular surface behaviour of the
material used seems to result in the unique relationship shown in Fig. 16,
which is independent of the test conditions.

Examples of the compressive residual stress distribution in the depth direction for shot-peened 12Cr–Mo–W–V steel are shown in Fig. 17. For the as-
shot-peened specimen, the compressive residual stress at the specimen surface
is almost 600 MPa, and it slightly increases with an increase in depth. At a
depth of about 150 μm, the compressive residual stress attains a maximum

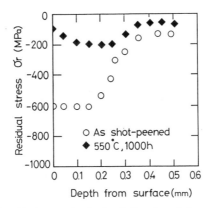

Fig 17 Distributions of residual stress in the depth direction of the as-peened specimen and the
specimen peened and subsequently exposed at 550°C for 1000h

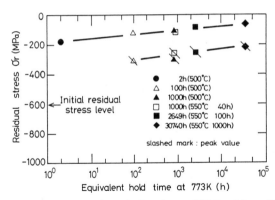

value. Beyond this point, the compressive residual stress decreases with an increase in depth. On the other hand, for the specimens exposed to elevated temperatures, the compressive residual stress increases with an increase in depth. It attains a maximum at 200 to 300 μm. Then it decreases with an increase in depth. The relationship between the compressive residual stress and the equivalent holding time at 500°C is shown in Fig. 18. Both the residual stress on the specimen surface and the maximum residual stress in the depth direction are shown in the figure. The exposure time of 40h at 550°C can be reduced to 1000h at 500°C according to the Larson–Miller relationship (34). As can be seen from Fig. 18, the compressive residual stress of the specimen exposed at 550°C for 40 h agreed well with that of the specimen exposed to 500°C for 1000 h. Therefore, the reduction in time of exposure according to the Larson–Miller relationship seems to be valid. The compressive residual stress in the specimen surface induced by shot peening process, which is initially about 600 MPa, is significantly decreased by exposure for 2 h at 500°C. However, the subsequent relaxation of compressive residual stress with time of exposure is not significant as shown in Fig. 18. Even after exposing for 30 000 h at 500°C, a peak compressive residual stress of 200 MPa, which is still effective in improving fretting the fatigue strength, remains.

The effectiveness of shot peening was also found in Ti–6Al–4V alloy at 350°C and Ti–5Al–5Zr alloy at 500°C (28)(29), Ti–6Al–4V alloy and 4340 steel at 400°C (35), and nickel-based alloy (D979) at 480°C (36).

Summary

Dominant factors for reducing fatigue strength by fretting are early crack initiation due to fretting wear and acceleration of crack growth rate due to frictional force. At elevated temperatures, fretting fatigue strength is in the competitive relationship between the reduction in plain fatigue strength and the lowering of the coefficient of friction (frictional force). Formation of a glaze oxide layer on the fretting surface brings about a large reduction in the coefficient of friction and consequently results in higher fretting fatigue strength at elevated temperature in nickel-based alloys. Oxide particles trapped in the crack mouth induce higher level of crack closure which results in the reduction of crack growth rate and consequently in higher fretting fatigue strength. Although the formation of glaze oxide and the oxide particles trapped are beneficial, they significantly depend on material, temperature, etc.

The mean stress effect in fretting fatigue may be estimated according to the Goodman relationship, as well as in the case of plain fatigue. The modified Miner's rule (the linear cumulative damage rule) is successfully applicable to predict the effect of variable loading in fretting fatigue.

The shot peening is effective in improving fretting fatigue strength even at elevated temperatures. The effect will depend on the combination of particular material and service temperature. Coating techniques which bring about both a low coefficient of friction and residual compressive stresses may be useful in improving the fretting fatigue strength at both room and elevated temperatures, but there is a need for further study.

References

(1) TAYLOR, D. E. (1981) Fretting fatigue in high temperature oxidising gas, *Fretting fatigue*, (Edited by R.B. Waterhouse), Applied Science Publishers, pp. 177–202.
(2) WATERHOUSE, R. B. (1992) Fretting fatigue, *Int. Mater. Rev.*, 37–2, 77–97.
(3) SATOH, T., MUTOH, Y. et al.(1993) Effect of contact pressure on high temperature fretting fatigue, *J. Soc. Mater. Sci., Japan*, **42**, 78–84.
(4) TAKEUCHI, M. and WATERHOUSE, R. B. (1987). *Proceedings of the Tenth Congress on Metallic Corrosion*, Vol 5, IBH Publishes, pp. 1959–1966.
(5) TAKEUCHI, M., WATERHOUSE, R. B., MUTOH, Y., and SATOH, T. (1991) The behaviour of fatigue crack growth in the fretting-corrosion-fatigue of high tensile roping steel in air and seawater, *Fatigue Fractures Engng Mater. Structures*, **14**, 69–77.
(6) SATOH, K., FUJII, H., and KODAMA, S. (1986) Crack propagation behaviour in fretting fatigue, *Wear*, **107**, 245–262.
(7) SATOH T. and MUTOH Y. (1994) Effect of contact pressure on fretting fatigue crack growth behaviour at elevated temperature, *Fretting Fatigue*, ESIS 18 (Edited by R. B. Waterhouse and T. C. Lindley), Mechanical Engineering Publications, London, pp. 405–416.
(8) KAYABA, T., IWABUCHI, A. and KATO, K. (1984) Fretting wear of Ni–Cr alloys at high temperatures, *J. Japn. Soc. Lubr. Engng. Int.* **5**, 47.
(9) WATERHOUSE, R. B. (1986) The fretting wear of nitrogen-bearing austenitic stainless steel at temperature to 600°C, *J. Tribology*, **108**, 359–362.
(10) HAMDY, M. M. and WATERHOUSE, R. B. (1981) The fretting wear of Ti–6Al–4V and aged inconel 718 at elevated temperatures, *Wear*, **71**, 237–248.
(11) HARRIS, S. J., OVERS, M. P., and GOULD, A. J. (1985) The use of coating to control fretting wear at ambient and elevated temperatures, *Wear*, **106**, 35–52.
(12) KAYABA, T. and IWABUCHI, A. (1981–82) The fretting wear of 0.45%C steel and austenitic stainless steel from 20 to 650°C in air, *Wear*, **74**, 229–245.
(13) HAMDY, M. M. and WATERHOUSE, R. B. (1979) The fretting fatigue behaviour of a nickel-based alloy (INCONEL 718) at elevated temperatures. *Proceedings of the International Conferance on Wear of Materials*, ASME, pp. 351–355.
(14) HAMDY, M. M. and WATERHOUSE, R. B. (1979) The fretting fatigue behaviour of Ti–6Al–4V at temperatures up to 600°C, *Wear*, **56**, 1–8.
(15) HAMDY, M. M. and WATERHOUSE, R. B. (1982) The fretting fatigue behaviour of the titanium alloy IMI829 at temperatures up to 600°C, *Fatigue Fracture Engng Mater. Structures*, 5–4, 267–274.

(16) BETTS, R. K. (1971) AFML–TR–71–212, Wright-Patterson Air Force Base, Ohio.

(17) OVERS, M. P., HARRIS, S. J., and WATERHOUSE, R. B. (1979) The fretting wear of sprayed molybdenum coatings at temperatures up to 300°C, *Proceedings of the International Conference on Wear of Materials*, ASME, pp. 379–387.

(18) MUTOH, Y., SATOH, T., TANAKA, K., and TSUNODA, E. (1989) Fretting fatigue at elevated temperatures in two steam turbine steels, *Fatigue Fracture Engng Mater. Structure*, 12–5, 409–421.

(19) STOTT, F. H., LIN, D. S., and WOOD, G.C. (1973) The structure and mechanism of formation of the 'glaze' oxide layers produced on nickel-based alloys during wear at high temperature, *Corrosion Sci.*, 13, 449–469.

(20) WATERHOUSE, R. B. and IWABUCHI, A. (1985) The composition and properties of surface films formed during the high temperature fretting of titanium alloys, *Proceedings of JSLE International Tribology Conference*, Vol. 1, Elsevier, pp. 53–58.

(21) FROST, N. E., MARSH, K. J., and POOK, L. P. (1974) *Metal fatigue*, Oxford University Press.

(22) SATOH, T., MUTOH, Y., and TSUNODA, E. (1991) Effect of mean stress on fretting fatigue strength at elevated temperature, *Trans. Japan. Soc. Mech. Engrs*, 57–534, 262–267.

(23) SATOH, T., MUTOH, Y., and TSUNODA, E. (1990) High temperature fretting fatigue under a simulated service load in a steam turbine steel. *Fatigue 90*, 1, 621–626.

(24) EDWARDS, P. R. and RYMAN, R. J. (1975) Studies in fretting fatigue under variable amplitude loading conditions, RAE Technology, Rep. 73132, Royal Aircraft Establishment.

(25) MUTOH, Y., TANAKA, K., and KANDOH, M. (1989) Fretting fatigue in SUP9 spring steel under random loading, *JSME Int. J.* 32, 274–281.

(26) MUTOH, Y., TANAKA, K., and KONDOH, M. (1987) Fretting fatigue in JIS S45C steel under two-step block loading, *JSME Int. J*, 30, 386–393.

(27) MUTOH, Y., TANAKA, K., and ITOH, S. (1988) Fretting fatigue properties of high strength stainless steel, *J. Soc. Mater Sci., Japan.*, 37, 643–648.

(28) HARRIS, W. J. (1969) Rolls-Royce Research Report R245/52/69.

(29) SYERS, G. (1978) Rolls-Royce Research Report LLR 10245/SF/78.

(30) GABEL, M. K. and BETHK, J. J. (1979) Coatings for fretting prevention, *Wear*, 46, 81–96.

(31) VARDIMAN, R. G., CREIGHTON, D., SALIVAR, G., EFFATIAN, A., and RATH, B. B. (1982) Effect of ion implantation on fretting fatigue in Ti–6Al–4V alloy, *ASTM STP780*, ASTM, Philadelphia, pp. 138–149.

(32) TANAKA, K., MUTOH, Y., and SAKODA, S. (1985) Effect of contact materials on fretting fatigue in a spring steel, *Trans. Japan, Soc., Mech., Engrs.*, 51, 1200–1207.

(33) MUTOH, Y., SATOH, T., and TSUNODA E. (1992) Improving fretting fatigue strength at elevated temperatures by shot peening in steam turbine steel. *ASTM STP1159*, ASTM, Philadelphia, pp. 199–209.

(34) LARSON, F. R. and MILLER, J. (1952) A time–temperature relationship for rupture and creep stresses, *Trans. ASME*, 74, 765.

(35) PAVLISCAK, T. J. (1968) PhD Thesis, Ohio State University.

(36) FIELDER, L. J. (1976) Materials and process technology laboratories, Report MR0326, Avco Lycoming Division, Connecticut.

T. Satoh* and Y. Mutoh*

Effect of Contact Pressure on Fretting Fatigue Crack Growth Behaviour at Elevated Temperature

REFERENCE Satoh, T. and Mutoh, Y. **Effect of contact pressure on fretting fatigue crack growth behaviour at elevated temperature,** *Fretting Fatigue,* ESIS 18 (Edited by R. B. Waterhouse and T. C. Lindley) 1994, Mechanical Engineering Publications, London, pp. 405–416.

ABSTRACT Fretting fatigue tests at elevated temperatures were carried out using 12Cr–Mo–W–V steam turbine steel, in order to investigate the influence of mean contact pressure (clamping pressure) on fretting fatigue crack growth behaviour. It was shown that the high temperature fretting fatigue cracks initiated at an early stage of fatigue life (less than a few percent of the fretting fatigue life), and grew at stress intensity factors lower than the threshold value for a long fatigue crack. Moreover, the crack growth rate for a fretting fatigue crack was significantly accelerated compared to that for a naturally initiated short fatigue crack. These effects are due to the fretting action, in particular the frictional force between the specimen and the contact pad. The fatigue crack growth curve presented in terms of the stress intensity factor range ΔK, in which the frictional force was taken into consideration, was in good agreement with that for a long fatigue crack without fretting.

Introduction

Fretting fatigue is one of the most serious problems in the fatigue fracture of joints in structures such as blade and rotor joints (1) and force fitted parts of a shaft (2). In the fretted region, many fatigue cracks initiate at an early stage of life (2) and their growth rates are accelerated by fretting action, consequently the fatigue strength and life of the joining parts are significantly reduced. It is well known that there are many factors affecting fretting fatigue (3), for example mean stress, friction coefficient, mean contact pressure, and relative slip amplitude. It is of importance to understand the effect of these relevant factors on fretting fatigue crack growth behaviour from a machine design point of view.

In this study, in order to investigate the effect of the mean contact pressure on high temperature fretting fatigue crack growth behaviour, two-step fretting fatigue tests with two levels of mean contact pressure (100 MPa and 300 MPa) were carried out at an elevated temperature of 500°C using 12Cr–Mo–W–V steam turbine steel. A fracture mechanics analysis was also carried out to evaluate the stress intensity factor for the fretting fatigue crack, where the frictional force acting on the contact surface is taken into consideration.

* Department of Mechanical Engineering, Nagaoka University of Technology, Nagaoka-shi 940 – 21, Japan.

Experimental Procedure

In the present study, three kinds of fatigue crack growth tests were carried out: fatigue crack growth test under fretting fatigue conditions; fatigue crack growth tests using long through-thickness crack; and tests with naturally initiated short fatigue crack. All the tests were carried out at the temperature of 500°C, which was the service condition of the steam turbine.

Material

Two kinds of steam turbine steels, 12Cr–Mo–W–V steel and 11Cr–Mo–V–Nb steel were used in the present study. The steel 12Cr–Mo–W–V was used for the fatigue test specimens and 11Cr–Mo–V–Nb steel for the contact pad in the fretting fatigue tests. The chemical compositions of the materials used are shown in Table 1. Heat treatment conditions and mechanical properties at room temperature and 500°C are given in Table 2. The strengths of the two materials were almost identical at both temperatures.

Fretting fatigue crack growth test

Two-step fretting fatigue tests using multiple specimens **(4)** were carried out to investigate the fretting fatigue crack growth behaviour. The two-step test is as follows: the contact pads are removed from the specimen after a certain number of cycles in fretting fatigue and subsequently a plain fatigue test of the same specimen follows. The shape and dimensions of the specimen and the contact pad used are shown in Fig. 1(a). Fretting was induced by pressing a pair of flat contact pads on the gauge part of the specimen that had machined flats. In order to investigate the effect of mean contact pressure on high temperature fretting fatigue crack growth behaviour, the two-step tests with two levels of mean contact pressure, 100 MPa and 300 MPa, which were controlled to be constant during the test, were carried out. The crack length and shape were measured by fracture surface observation by means of the scanning electron microscope. These tests were performed under a load-controlled tension–compression condition with stress ratio $R = -1$ and at the applied stress amplitude of 250 MPa. The details of fretting fatigue test procedures and frictional force measurement method are give in reference **(5)**.

Fatigue crack growth test

A fatigue crack growth test of a long through-thickness crack was carried out under a stress ratio of $R = -1$ using the thin-walled cylindrical specimen shown in Fig. 1(b). At the centre of the specimen gauge section, a through notch was introduced by electric spark machining. The length of the fatigue

Table 1 Chemical compositions of the materials used

Material	C	Si	Mn	P	S	Ni	Cr	Mo	V	W	Co	Sn	Al	Ti	Nb	N
12Cr–Mo–W–V Steel	0.23	0.49	0.70	0.021	0.001	0.67	11.21	0.93	0.23	0.95	0.02	0.01	0.01	0.01		
11Cr–Mo–V–Nb Steel	0.17	0.39	0.66	0.019	0.007	0.35	11.44	0.92	0.21	0.02	0.01	Tr	Tr	0.01	0.47	0.04

Table 2 Heat treatment conditions and mechanical properties

Material (°C)	Heat treatment (°C)	Temp.	Yield strength σ_{ys} (MPa)	Tensile strength σ_B (MPa)	Elongation ψ (%)	Reduction in area ϕ (%)	Young's modulus E (GPa)	Vickers hardness Hv
12Cr–Mo–W–V Steel	1052 OQ 650 AC 635 AC	RT 500	852 651	987 676	17.2 27.0	52.0 77.7	209 160	346 —
11Cr–Mo–V–Nb Steel	1095 OQ 650 AC 625 AC	RT 500	874 —	1000 700	18.9 22.5	54.8 69.9	206 —	342 —

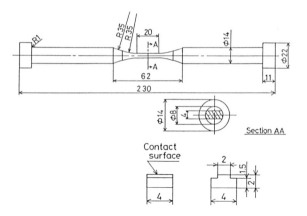

(a) Fretting fatigue crack

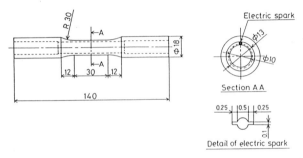

(b) Long fatigue crack

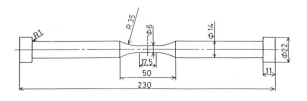

(c) Short fatigue crack

Fig 1   Shapes and dimension of specimen; (a) fretting fatigue crack growth, (b) long fatigue crack growth and (c) short fatigue crack growth test

crack was measured using a travelling microscope with an accuracy of 0.01 mm. The stress intensity factor range ΔK was calculated using the equation for a CCT specimen given in reference (6), where the measured crack length was reduced to the circumferential length. The threshold stress intensity factor range ΔK_{th} was determined based on the K-decreasing method.

Short fatigue crack growth test

In order to investigate the fatigue crack growth properties of a naturally initi-ated short crack, a plain fatigue test using the smooth specimen shown in Fig. 1(c) was carried out under a stress ratio of $R = -1$ at applied stress amplitude of 350 MPa. Crack lengths at the specimen surface were measured by means of the plastic replica films.

Results and discussion

Fretting fatigue crack growth behaviour

It is well known that the fretting fatigue cracks initiate at an early stage of life, not only in air (7)(8) but also in corrosive environments such as seawater (9). High temperature fretting fatigue cracks also initiate within 1 percent of fatigue life at the edge of the contact region, as shown in Fig. 2. Since the contact pad is removed from the specimen after a certain number of fretting fatigue cycles, the fretting fatigue crack geometry can be observed, as shown in Fig. 3. From these observations, assuming that the fretting fatigue crack is a semi-elliptical surface crack, the crack depth, *a*, and the semi-width, *c*, were measured. Relationships between the fretting fatigue crack length and the number of cycles normalized by the fretting fatigue life are shown in Fig. 4. The similar relationship for the naturally-initiated fatigue crack is also indi-cated in this figure. In plain fatigue, more than 45 percent of the total fatigue life is required for crack initiation to take place. On the other hand, in fretting fatigue a crack initiates in less than a few percent of the fatigue life regardless of the mean contact pressure level in the range of the present study and propa-

Fig 2   **Fretting fatigue crack initiated in early stage of life (mean contact pressure = 300 MPa, $N/N_f = 1$ percent)**

Fig 3   Fretting fatigue crack geometry (mean contact pressure = 100 MPa, N/N_f = 20 percent)

gates up to 150 μm in depth in only 5 percent of the fatigue life. Therefore, almost all of the whole life of fretting fatigue is spent in propagation of the fatigue crack. Variations of the aspect ratio of the fretting fatigue crack with crack depth are indicated in Fig. 5, showing that fretting fatigue crack geometry changed from flat-shape to semi-circular shape with crack growth. The crack geometry for the high mean contact pressure of 300 MPa is flatter than that of the low mean contact pressure of 100 MPa.

Fretting crack growth curve estimated by the apparent stress intensity factor

The relationship between crack growth da/dN and apparent stress intensity factor range ΔK, where only the applied stress amplitude is taken into account, is shown in Fig. 6. The stress intensity factor range at the deepest point for the fretting crack was calculated using the Newman–Raju equation

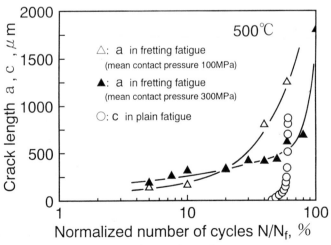

Fig 4   Fatigue crack growth curves

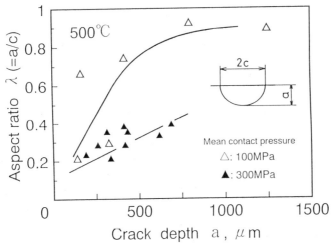

Fig 5   Variations of the aspect ratio of fretting fatigue crack with crack growth

for a semi-elliptical surface crack (10) considering the variation of aspect ratio with crack growth mentioned above. In the figure, crack growth curves for a long fatigue crack and naturally initiated short fatigue crack under plain fatigue condition are also indicated. The calculation of the stress intensity factor range for a short fatigue crack was also conducted using the Newman–Raju equation for a semi-circular surface crack, i.e., aspect ration $\lambda = 1$, which

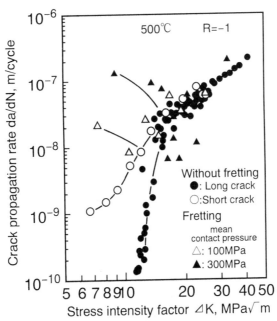

Fig 6   Relationship between apparent stress intensity factor range ΔK and crack growth rate $\mathrm{d}a/\mathrm{d}N$

is based on the observations of the fracture surface. The crack growth curve for a long fatigue crack can be expressed in the following Paris equation **(11)** which is modified by the threshold stress intensity factor range ΔK_{th}

$$\frac{da}{dN} = C_0(\Delta K^m - \Delta K_{th}^m)$$

where C_0 and m are material constants. The values of ΔK_{th}, C_0, and m for the material used were 11.0 MPa $\sqrt{m}$, 3.27×10^{-12}, and 3.0, respectively.

Both the fretting fatigue crack and the naturally-initiated short fatigue crack grow at stress intensity factor ranges lower than the threshold value ΔK_{th} for a long crack. The fretting crack growth rates were significantly accelerated in the lower ΔK region, i.e., in the early stage of crack growth, compared not only to that for a long fatigue crack without fretting but also to that for the short fatigue crack. Regardless of the mean contact pressure, the crack growth rate first decreased with increasing apparent stress intensity factor, i.e., with extension of fretting fatigue crack. The minimum crack growth rate occurred at the meeting point with the crack growth curve for a long fatigue crack, and the crack growth rate increased with increasing stress intensity factor along the same curve as for a long crack without fretting. Similar behaviour of fretting fatigue crack growth has been observed at room temperature **(9)(12)** and in seawater **(9)**.

The crack growth rate in the early stage of life for the mean contact pressure level of 300 MPa was faster than those for the mean contact pressure level of 100 MPa. The frictional force between the specimen and contact pad for the mean contact pressure of 300 MPa and 100 MPa for the stress amplitude of 250 MPa were 672 N and 520 N, respectively as shown in Fig. 7. Then the higher crack growth rate for the mean contact pressure of 300 MPa resulted

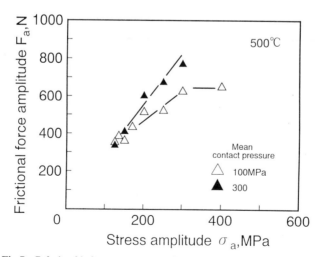

Fig 7   Relationship between stress amplitude and frictional force amplitude

from the higher frictional force between the specimen and the contact pad compared to that for 100 MPa. The lower aspect ratio of fretting crack shape for the mean contact pressure of 300 MPa also contributed to the higher crack growth rate. The stress intensity factor at the deepest point for the crack with the lower aspect ratio was higher than that for the crack with higher aspect ration.

Fretting crack growth curve estimated by the effective stress intensity factor

The fretting fatigue crack growth rate was significantly accelerated in the early stage of life due to the frictional force mentioned above. The effective stress intensity factor for the fretting fatigue crack, in which the frictional force was taken into consideration, was estimated on the basis of elastic–plastic fracture mechanics analysis **(5)(13)**.

According to the observations of the fretted surface, the maximum adhesion occurred in the vicinity of the edge of contact pad feet, and fretting fatigue cracks initiated at the edge of the contact region, as shown in Fig. 2. A distribution of the frictional force would not be uniform along the contact lengths of the fretted pad. When calculating the stress intensity factor range due to the frictional force, the distribution of frictional force was assumed as follows: some fraction of the frictional force, $\alpha F_a l$, is concentrated at the edge of the contact region and the remaining frictional force, $(1 - \alpha)F_a l$, is uniformly distributed along the contact region, where α is the frictional force concentration factor at the contact edge and l the length of contact region. The stress intensity factors of a through-thickness crack with the concentrated frictional force $\alpha F_a l$ at the edge of the contact surface and with the uniformly distributed frictional force $(1 - \alpha)F_a l$ along the contact surface in a semi-infinite body are given as follows **(6)**

$$K_{F1} = 1.29\alpha F_a l\left(\frac{1}{\pi a}\right)^{1/2}$$

$$K_{F2} = 1.29(1 - \alpha)F_a(\pi a)^{1/2}\left[\frac{3}{2\pi}\ln\left\{\frac{l + (l^2 + a^2)^{1/2}}{a}\right\} - \left(\frac{1}{2\pi}\right)\frac{l}{(l^2 + a^2)^{1/2}}\right]$$

Consequently, the stress intensity factor range ΔK under fretting conditions can be calculated as

$$\frac{\Delta K}{2} = K_T + K_{F1} + K_{F2}$$

where K_T is the stress intensity factor due to the applied stress amplitude. The frictional force concentration factor at the contact edge for the material used is 0.17 **(5)**. The actual shape of the fretting crack was semi-elliptical, as mentioned above. However, the shape changed from flat to semi-circular as the crack was growing, as shown in Fig. 5. The K-value of a semi-elliptical surface crack with surface frictional force has not yet been evaluated. Therefore, considering the results of the two-step fretting fatigue tests (14), the problem was simplified as follows: when $N/N_f < 0.4$, the crack is assumed to be a through crack with surface frictional force due to fretting. When $N/N_f > 0.4$, a semi-elliptical crack is assumed to propagate without frictional force due to fretting. The aspect ratio of the semi-elliptical crack is given in Fig. 5. The stress intensity factor of the semi-elliptical crack was evaluated by using the Newman–Raju equation (10). The relationship between the fretting fatigue crack growth rate and the effective stress intensity range at the deepest points of the crack is shown in Fig. 8. As can be seen from the figure, the 'V' shaped crack growth curves shown in Fig. 6 are reduced to a single crack growth curve which coincides with that for a long crack without fretting.

Prediction of fretting fatigue strength

Prediction of fretting fatigue strength was also carried out on the basis of the fracture mechanics technique mentioned above. The fretting fatigue strength was defined as the stress which makes the effective stress intensity factor range for fretting fatigue crack equal to or lower than the threshold value for long fatigue crack. The predicted fretting fatigue strengths and the experimental results (14) are shown in Fig. 9. The experimental fretting fatigue strengths

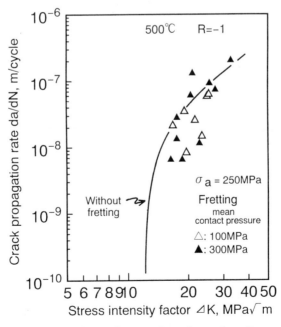

Fig 8   Correlation of fretting crack growth rate and actual stress intensity range, considering frictional force

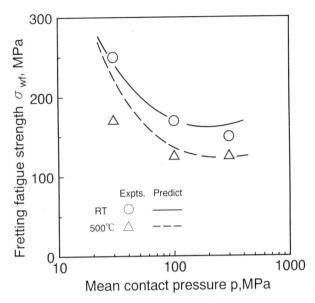

Fig 9   Prediction of fretting fatigue strength under several kinds of mean contact pressure level

decreased with increasing mean contact pressure, and then attained a constant value when the mean contact pressure was higher than 100 MPa. The predicted fretting fatigue strengths were in good agreement with experimental results.

Conclusions

The main results obtained are summarized below.

(1) High temperature fretting fatigue cracks initiate within a few percent of fretting fatigue life. This grows at a stress intensity factor range lower than the threshold value, and significant acceleration of crack growth rate is observed.
(2) The fretting crack growth rate for higher mean contact pressure level is higher than that for lower mean contact pressure level. This is mainly due to the higher frictional force between the specimen and the contact pad.
(3) When the frictional force acting on the contact surface is taken into account when calculating the stress intensity factor, the fretting fatigue crack growth curves are reduced to a single crack growth which coincides with that for a long crack without fretting.
(4) The predicted fretting fatigue strength based on elastic–plastic fracture mechanics analysis decreases with increasing mean contact pressure and attains a constant value when the mean contact pressure is higher than a certain value; these are in good agreement with experimental results.

References

(1) RUIZ, C., BODDINGTON, P. H. B., and CHEN, K. C. (1984) An investigation of fatigue and fretting in a dovetail joint, *Expl Mech.* **24**, 208–217.
(2) WATERHOUSE, R. B. (1972) *Fretting corrosion*, Pergamon Press, Oxford.
(3) WATERHOUSE, R. B. (1992) Fretting fatigue, *Int. Mater. Rev.*, **37**, 77–79.
(4) WHARTON, M. H., TAYLOR, D. E., and WATERHOUSE, R. B. (1973) Metallurgical factors in the fretting-fatigue behavior of 70/30 brass and 0.7% carbon steel, *Wear*, **23**, 251–260.
(5) MUTOH, Y., SATOH, T., TANAKA, K., and TSUNODA, E. (1989) Fretting fatigue at elevated temperatures in two steam turbine steels, *Fatigue Fracture Engng Mater. Structures*, **12**, 409–421.
(6) MURAKAMI, Y. (1987) *Stress intensity factor handbook*, Pergamon Press, Oxford.
(7) ENDO, K. and GOTO, H. (1976) Initiation and propagation of fretting fatigue crack, *Wear*, **38**, 311–324.
(8) ALIC, J. A. and HAWLEY, A. L. (1979) On the early growth of fretting fatigue crack, *Wear*, **56**, 377–389.
(9) TAKEUCHI, M., WATERHOUSE, R. B., MUTOH, Y., and SATOH, T. (1991) The behaviour of fatigue crack growth in the fretting-corrosion-fatigue of high tensile roping steel in air and seawater, *Fatigue Fracture Engng Mater. Structures*, **14**, 69–77.
(10) NEWMAN, J. C. Jr. and RAJU, I. S. (1981) An empirical stress-intensity factor equation for the surface crack, *Engng Fracture Mech*, **13**, 185–192.
(11) KLESNIL, M. and LUKÁŠ, P. (1972) Influence of strength and stress history on growth and stabilization of fatigue cracks, *Engng Fracture Mech.*, **4**, 77–92.
(12) SATO, K., FUJII, H., and KODAMA, S. (1986) Crack propagation behaviour in fretting fatigue, *Wear*, **107**, 245–262.
(13) TANAKA, K., MUTOH, Y., SAKODA, S., and LEADBEATER, G. (1985) Fretting fatigue in 0.55C spring steel and 0.45C carbon steel, *Fatigue Fracture Engng Mater. Structures*, **8**, 129–142.
(14) SATOH, T., MUTOH, Y., YADA, T., TAKANO, A., and TSUNODA, E. (1992) Effect of contact pressure on high temperature fretting fatigue, *J. Soc. Mater. Sci.*, **42**, 78–84.

PREVENTATIVE METHODS

*J. Beard**

Palliatives for Fretting Fatigue

REFERENCE Beard, J., **Palliatives for fretting fatigue**, *Fretting Fatigue*, ESIS 18 (Edited by R. B. Waterhouse and T. C. Lindley) 1994, Mechanical Engineering Publications, London, pp. 419–436.

ABSTRACT There are many palliatives suggested in the literature for in-service fretting fatigue problems. However, there is little guidance for selecting the right type of palliative for a specific application. In order to make a successful selection, it is essential to understand the mechanisms of fretting and how key fretting parameters influence crack nucleation processes. A palliative needs to be selected with the objective of modifying the fretting process in some deliberate and controlled way. Classifying fretting problems into displacement and force controlled categories can help determine the principal characteristics required of the fretting palliative. A number of diverse types of palliatives are considered and guidance given to which situations they may be most effective in.

Introduction

Under oscillating load conditions, load transfer between contacting components may induce differential strains that can result in small amplitude relative sliding at the surfaces. This fretting process can bring about a reduction of fatigue strength and is, therefore, called fretting fatigue.

A well known example in the diesel engine industry of this type of problem is the result of fretting on the back of connecting rod large-end shell bearings. Most medium speed and large diesel engines suffer some degree of fretting in this area and the author is aware of several manufacturers who have experienced fatigue failures as a result.

For problems such as this and others, fretting fatigue palliatives can be found in the literature by the score. However, guidance to making a correct selection for a specific application is scarce. Until recently there was no attempt at standardizing test methods for quantifying the effectiveness of a palliative. One result of this situation is that the majority of experimental data in the published literature is produced under vastly differing conditions and the results cannot easily be compared.

There is poor understanding of the mechanisms of fretting fatigue and few workers in the field have investigated the true mode by which a palliative influences the fretting damage process. It is, thus, perhaps not surprising to find that the reported effectiveness of palliatives treatments are often contradictory or misleading.

To the uninitiated engineer looking for a solution to a fretting problem the subject becomes totally bewildering and confusing. Palliative selection is generally made in desperation, on a completely arbitrary basis with the consequent high risk of failure. In order to make a rational palliative selection, it is

* National Centre of Tribology, AEA Technology, Warrington, Cheshire, UK.

essential to understand the mechanisms of fretting and how key parameters
influence the crack nucleation process.

It is also necessary to appreciate the limitations of the data available in the
published literature, and how this can influence selection. Such an understanding is only possible by knowing how the palliative operates, but few investigators provide this insight. Only when such information is available is it feasible
to select with confidence a palliative for a specific application. Even then selection may still have to be generalized to a generic class simply because of the
difficulty in obtaining more refined data.

The situation is difficult but not hopeless and a lot can be achieved even in
the present situation. The palliative options available fall into two classes:
those relying on design changes to directly influence the most important fretting parameters, and those which involve surface engineering. The latter may
operate by improving resistance to crack nucleation or by influencing the principal fretting parameters.

The effect of wear and the influence of slip amplitude

Fretting fatigue and fretting wear are generally treated separately and there
are few papers that consider whether any interrelationship exists. It is the
author's view that in many instances the two processes can have common
mechanisms. Moreover, in fretting fatigue there are occasions when fretting
wear can control whether fatigue failure occurs.

Wear influences fretting fatigue in a number of ways. One of the most
important is the modification it produces in the contact stress brought about
by changes of geometry. Within the last fifteen years contact stress has come
to be recognized as an important factor in the initiation of fretting fatigue
crack nuclei.

To the experimentalist trying to quantify the effects of fretting on fatigue
strength, wear represents one of the most difficult factors to cope with. Wear
inevitably produces conformance and thus the real possibility of a reduction in
the stress concentration at the contacting geometry. In practice, therefore, this
effect can be beneficial.

However, under test, unless the influence of the changing contact geometry
is known precisely it is difficult to be certain whether the success of a palliative's measure is due to a favourable fretting fatigue performance or simply
decreased wear resistance.

Wear considerably complicates the picture with regard to the effect of slip
amplitude on fretting fatigue strength since slip can change wear rates by up to
two orders of magnitude, as shown in Fig. 1. Bill (1) reports that fretting
fatigue life increases with increasing slip amplitude. Beard (2) also found a
marked increase in fretting fatigue strength with increasing slip amplitude.

Nishioka and Hirakawa (3) report a somewhat different finding. From the
results of their investigations they derived the following empirical formula for

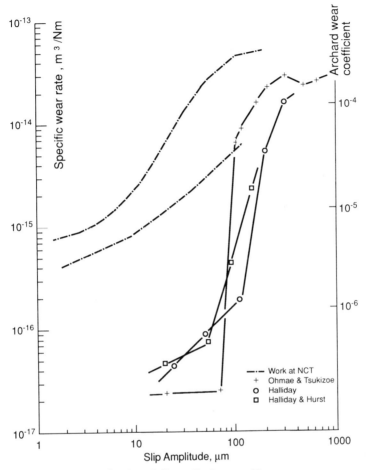

Fig 1   The effect of slip amplitude on specific wear rate

fretting fatigue strength

$$\sigma_{fwl} = \sigma_{wl} - 2\mu P_0 \left\{ 1 - \exp\left(-\frac{s}{k}\right) \right\}$$

where

σ_{wl} is the normal fatigue life,
P_0 is the maximum Hertzian contact pressure,
s is the slip amplitude,
μ is the friction coefficient,
k is a constant for the material.

Thus, decreasing slip amplitude increases fretting fatigue strength. This result is obtained after compensating for the effect of wear on contact geometry. If

the influence of wear is ignored the net result is that the stress to initiate
cracks becomes larger as relative slip increases. Earlier work by Funk (4) also
reported the reduction in fatigue strength with increasing slip. The fretting
fatigue strength initially dropped off rapidly, but then the rate of decrease
reduced to have virtually no effect above 20 μm.

Field and Waters (5) also found that below a critical value increasing slip
reduced fretting fatigue strength. However, at slip levels higher than the critical value fretting fatigue strength increased. The effect was attributed to the
wearing away of fatigue crack nuclei before they could reach a minimum
threshold size.

There may, however, be an alternative explanation which better unifies the
various research findings. In work carried out by the author the initiation of
crack nuclei developed very early in the fatigue life. In the tests, the crack
propagation stage was short and, therefore, the fatigue life was largely determined by the initiation of the fretting crack. The results shown in Fig. 2 indicate that the critical contact pressure initiating fretting fatigue failure is
determined by the level of slip.

Careful metallographic examination of interrupted tests showed that the slip
parameter exerted its influence on the fretting fatigue strength by controlling
crack propagation direction. With increased slip there is a greater tendency for
the crack, at a few microns in length, to turn and propagate parallel to the
surface. The crack eventually breaks back to the surface releasing a large wear
particle and leaving behind a pit rather than producing a propagating fatigue
crack, see Fig. 3. The mechanism is in many ways similar to the delamination

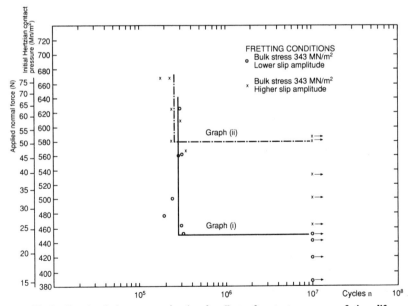

Fig 2   **Fretting fatigue curves showing the effects of contact pressure on fatigue life**

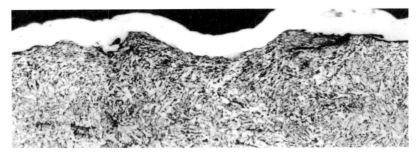

Fig 3   Pit formed by fretting crack propagation

theory of wear but cracks are initiated at the surface. At a critical point in the life of the crack the outcome of either wear or fatigue failure is delicately balanced. Under the right conditions both fatigue failure and wear particle formation may occur through crack branching. The possible directions of crack propagation are summarized in Fig. 4.

Why the crack propagation direction should be so influenced by slip amplitude is yet to be understood. The influence of slip amplitude on the direction of crack propagation is an area ripe for investigation both experimentally and

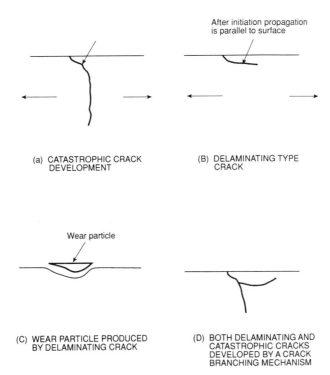

Fig 4   Fretting crack propagation

analytically. Unfortunately, from the stand-point of palliative selection it is not yet possible to predict analytically at what level increasing slip amplitude will start to become beneficial. This is currently a major area of difficulty when trying to make a rational palliative selection.

Microstructural changes at the surface

Many investigators of the subject see the mechanism of fretting fatigue as being entirely explained in terms of the generated contact stress. However, this mechanism, whilst no doubt being important, is certainly not the only one. Fatigue crack nucleation as a consequence of the formation of white etching layers (WEL) is a little recognized fretting fatigue mechanism.

Transformation of the surface material to form an exceptionally hard non-etching layer is little known outside of the diesel engine and wire rope industries. Nevertheless, in practice it is probably more common than might be expected, but simply not recognized.

The structure of white etching layers has not yet been unambiguously identified. However, the majority of investigations suggest that a WEL is either a martensite or supersaturated solution of carbon in ferrite; a minority suggest an agglomeration of carbide. In addition to its non-etch properties, which cause it to look white under the microscope, white etching material exhibits extreme hardness. Hardness values in excess of 1200Hv are frequently recorded.

The difficulty of predicting or measuring a significant temperature rise in fretting leads to the currently favoured stress induced mechanism of formation (6). A zone of intense shear deformation appears to precede the advance of the layer, when the WEL is formed from the substrate material. There also appears to be a mechanism whereby material transfer from one surface to the other undergoes a structural change to a WEL; a layer formed in such a way is shown in Fig. 5.

The number and size of white etching formations appear to increase with both increasing slip and load. This might be anticipated with a stress induced mechanism since both parameters effectively increase the intensity of surface shear. White etching layers can be formed when no bulk fatigue stress is present i.e., under fretting wear conditions.

WEL appear only to form with steels and so the selection of a non-ferrous coating is potentially a good palliative. However, if the crack nucleation process is to be inhibited, it is important to establish if material transfer occurs to ensure the palliative coating is applied to the appropriate surface.

Designing out the problem

Fretting fatigue is not often identified at the design stage; therefore, it is common to find that in practical situations several palliatives are tried simultaneously in the hope of eliminating the problem. This fire-fighting situation could often easily be avoided if full consideration of the consequences of fret-

Fig 5   White etching layer produced during fretting

ting was made at the design stage. Each contact interface needs to be reviewed to assess the likelihood of fretting and the consequences of a significant reduction in fatigue strength.

Slip is caused by a concentration of shear stress at the contact interface, for a Hertzian contact this has been elegantly shown by Mindlin (7). Simple palliative design changes can often be introduced to reduce the propensity for generating a concentration of shear stress. Other than for Hertzian contacts an analytical solution of contact conditions, even for relatively simple geometries, is not available. Analysis of the potential for fretting of real components must, therefore, be carried out by numerical methods. Finite element analysis is an effective and well established method by which the interface between component assemblies can be realistically assessed at the design stage.

A simple two-dimensional elastic analysis can be particularly effective at highlighting an area susceptible to fretting. A complex but real example is the analysis of the big-end of a diesel engine connecting rod. The two-dimensional analysis provides considerable information in relation to both fretting at the joint and also between the housing and shell bearing. A pressure is exerted around the circumference of the bore due to the bearing nip and the crank pin load. The ratio of the shear stress in the plane of the surface and the stress normal to the surface indicate the minimum value of friction coefficient required to prevent slip.

As a result of this analysis action can be taken to either stiffen the housing, increase the bearing nip pressure or reduce the firing and inertia load on the connecting rod until fretting is no longer predicted.

Such an analysis can be used to optimize the position of the joint. It is usual to position the joint horizontally (perpendicular to the rod axis) unless for maintainability the connecting rod is too wide to be extracted through the bore. In such cases the joint is often inclined to the horizontal. Fretting wear and fatigue of connecting rod joints are another well known problem area. Horizontally split joints tend to fret towards the edge nearest the bearing i.e., the inboard side. There are reported cases of a wedge of material being removed so that either bolt preload is lost and fatigue failure of the bolts occurs or excessive nip is introduced when the shell bearing is replaced.

Inclined split joints tend to fret at the outside edge and the bearing nip problem does not arise. An FE analysis of the design can be used to optimize the angle of the split to minimize the extent of fretting.

In certain cases, optimizing the design of an assembly on the basis of other essential design criteria may mean that fretting becomes inevitable. For this situation a palliative may be the only option available to minimize fretting damage. In such instances a detailed analysis of the fretting interface can be useful to classify the type of fretting. This in turn can assist with the selection of a suitable palliative.

Classifying fretting by the source of movement

Controlling slip appears to be a valid method of improving fretting fatigue strength. It is, therefore, worth examining how the relative movement between surfaces is generated. Understanding the source of the fretting movement can also be useful in identifying the potential effects a palliative may have on the fretting fatigue performance. If the slip amplitude is fixed and effectively independent of normal load then the problem is described as being displacement controlled.

If the slip amplitude varies with the applied normal load fretting it is said to be force or stress controlled. Identifying which case applies can be helpful in the selection of a palliative based on modifying the surfaces in contact.

There is an important distinction between systems in which the relative movement comes from an alternating force or stress applied to elements of a machine, and ones in which the movement is caused by defined displacements. This distinction must be made as the effectiveness of any palliative is often critically dependent upon the source of the motion.

In force or stress excited systems, the degree of slip at a surface is often a non-linear function of the applied system force or stress. This is influenced by damping which is a function of normal load, and the static and dynamic friction coefficient. The success of a palliative such as a low shear stress lubricating layer which may reduce the contact stress cannot be guaranteed if perhaps the system response increases as a result of the low friction.

The connecting rod large end shell/housing fretting problem previously described is an example of such a response found by the author during investigation of the fretting fatigue problem. Friction was reduced between shell and housing by a surface nitrocarburizing treatment known as 'Sulfinuz'. The effect of the palliative was to increase the relative slip. Unfortunately, fretting fatigue crack nucleation was due to white etching layer formation. The increase in slip produced by the reduction in friction encouraged white etching layer formation which led to an increased density of crack nuclei. Fortunately, in this instance, the increase in fatigue strength produced by the Sulfinuz treatment prevented the WEL cracks from penetrating the substrate. Nevertheless, the example does illustrate that in a force controlled situation it is particularly important to consider the likelihood of forming white etching layers before considering reducing friction.

Amplitude controlled slip

There are some mechanical components which are subjected to a fixed amplitude of movement irrespective of loading, for example, gear couplings or crown splines where the amplitude of movement is related to the degree of angular misalignment and not to the transmitted torque. Other examples are in business machines such as impact printers where the print head is designed to move through a defined travel. Often the characteristic of these systems is that movements are intentional and little can be done to reduce or limit the relative slip at the surface without affecting the function of the machine. Reducing contact pressure and lowering friction is usually the best approach to alleviating fretting fatigue.

Some fretting situations are effectively amplitude driven if the clamping pressure cannot be controlled or increased. Fretting between the stands of a wire rope is an example where the applied normal load cannot be externally controlled and fretting is effectively amplitude driven by rope stretch. Another example is the fretting on the lands of a fitted bolt which has been caused by small amounts of joint movement, see Fig. 6.

Surface engineering

When all the design options have been exhausted, engineering the surfaces to provide resistance to fretting fatigue is usually the way forward. Surfaces can be modified either directly or by coating with metallic or non-metallic materials. The former can be subdivided into three groups: thermochemical treatments, thermal treatments, and mechanical working.

Within the coatings groups, there are basically six generic categories of processes by which a coating can be applied. These are electrochemical, chemical, welding, spraying, chemical vapour deposition (CVD), and physical vapour deposition (PVD), see Fig. 7. Within some of these groups, there are several distinct variants and so without even considering the choice of material the scope for selection of a surface engineered solution to a fretting fatigue problem is immense.

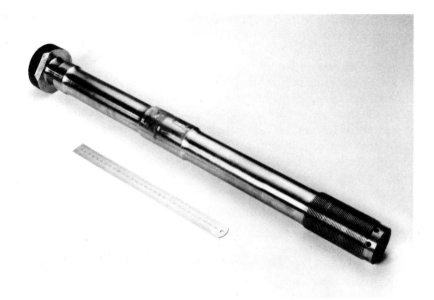

Fig 6   Displacement controlled fretting of a fitted bolt

There are a large number of factors influencing coating material or process selection without even considering the effects on fretting fatigue performance. The importance of these factors must be assessed according to the requirements of the individual application, but generally of primary consideration are substrate material, cost, size, deposition rate, geometry, process temperature, thickness, and adhesion. From the point of view of alleviating fretting fatigue, selection is based on the effect the palliative has on friction coefficient, normal fatigue stress, wear, and durability.

Thermochemical treatments

The surface chemistry of ferrous materials can be altered in several ways. Thermochemical treatments involve the diffusion of carbon, nitrogen and, less usually, chromium, boron, aluminium, or silicon into the surface. Treatments fall into two main categories: those carried out at high temperature with the steels in the austenitic conditions, and those at low temperature (below 600°C) with the steel in the ferritic condition.

By virtue of the compressive stresses developed through the diffusion of elements in the surface layer the fatigue strength of steel is normally enhanced by these treatments. As a general principal, any process which increases the normal fatigue strength of the steel, providing it is not accompanied by a significant change in friction coefficient, should improve the fretting fatigue performance. The process giving rise to the greatest increase in normal fatigue strength may not always provide the largest improvement in fretting performance.

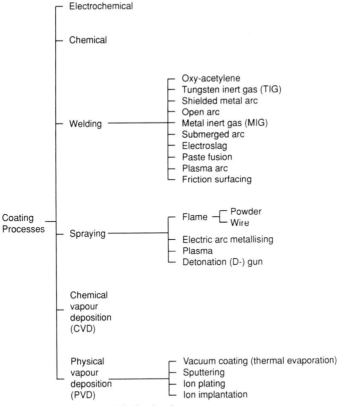

Fig 7   Coating processes

The most common thermochemical treatments are carburizing, carbo-nitriding, nitrocarburizing, and gas nitriding. This category of treatments increases surface hardness and produces high levels of residual compressive stress at the surface.

Carburizing is a high temperature process (also known as case-hardening) in which carbon is diffused into the surface from either solid, liquid, or gaseous carbonaceous media at a temperature of between 825°C and 925°C. The component is quenched and tempered to produce a hard, tough case with a depth of several millimetres. For precision fitting components it is important to take into account the dimensional changes (typically 0.1 percent) which occur.

Boronizing, chromizing, aluminizing, and siliconizing are pack processes and, like carbonizing, are all carried out at high temperature. Unfortunately there appears to be nothing in the literature to indicate their performance as a palliative against fretting fatigue.

Nitriding is a lower temperature process, less than 600°C. Components show less distortion but the steels must contain nitride forming elements like chromium or aluminium. In many cases, nitriding forms a thin surface layer of iron nitrides which is often referred to as a compound or white layer. This is usually detrimental to the fatigue properties and, therefore, is removed by finish grinding.

As a fretting fatigue palliative, nitriding is reported by Kreitner (8) to be significantly better than carburizing. In this work, carburizing gave a larger increase in normal fatigue strength but under fretting conditions the fatigue strength suffered a large reduction of 45 percent. By contrast, after nitriding the fretting fatigue strength was more than 90 percent of the normal fatigue strength.

Diffusion treatments can have a considerable effect on friction. The Sulfinuz treatment, which is a salt bath nitrocarburizing process, on En32 steel is reported to reduce the friction coefficient at low sliding velocities from 0.8 down to 0.2 (9). Taylor and Waterhouse (10) found the process to be beneficial in fretting fatigue and this, at least in part, seems to be due to the reduction in contact stress resulting from lower friction. In displacement controlled fretting, reduced frictional traction will enhance the effectiveness of diffusion treatments. However, in a force controlled situation the reduction in friction may increase slip amplitude and negate the improvements in strength.

Unlubricated, carburizing, and nitrocarburizing surfaces have been found to have only moderate fretting wear resistance (11)(12). If wear can be tolerated in the application then a more conforming geometry with lower contact stress may well have benefits. Similarly, the production of wear debris will prevent severe adhesive welding of the surfaces, and this too will be beneficial.

Coating processes

Electrochemical and chemical
In previous investigations of palliatives, both hard and soft coatings have been employed to mitigate against fretting fatigue. However, it is probably true to say that there are more conflicting opinions in the literature on the performance of individual coatings than for any other class of palliative. The reasons for the disagreement on coating performance are not always clear. However, sometimes a successful coating performance can be dependent on the characteristics of the test rig or the application.

For example, in a rig producing low amplitude force controlled fretting, such as that described by Budinski (13), coatings having high coefficients of adhesion may bring about seizure of the contacts. A palliative which is given a favourable assessment by such a rig may in fact only be effective in force controlled situations where seizure can be brought about. Sikorskii (14) has correlated high coefficients of adhesion with soft coatings such as silver, indium, and lead. It is perhaps, therefore, no surprise to find that these are precisely the coatings which are said to be beneficial in fretting (13).

It has also been suggested that soft metal coatings work by absorbing the fretting movement (15). However, the amount of movement which can be absorbed elastically is so low that it does not seem a valid explanation.

Waterhouse (16) found that electro-deposited coatings such as lead, zinc, silver, tin, copper, and nickel results in a higher fretting fatigue strength with increased coating thickness. A correlation was reported between the fretting fatigue limit and the square of the coating thickness. In this work all the coatings, except for nickel, showed heavy plastic deformation but the surface damage was not as severe as on the uncoated steel. Copper and silver came out to be the best materials but silver tended to be extruded out of the contact zone whereas copper did not. Although force controlled problems may benefit from this type of palliative, the poor durability of soft coatings is likely to be the main factor limiting their use. Coating porosity, residual stress, and hydrogen embrittlement can all reduce normal fatigue strength and must be taken into account.

Of the harder surface coatings that can be applied, electroplate chromium is probably the best known for wear resistance. However, hard chrome plate is highly stressed and extensively cracked. Under lubricated conditions the fissures and cracks formed in the plate usefully act as reservoirs to hold the oil. Under fatigue conditions unfortunately they also act as stress raising notches which seriously reduce the normal fatigue strength, in some cases by up to 80 percent. To regain some of the lost fatigue strength, controlled shot peening is usually recommended.

Notwithstanding the effect on fatigue strength, chrome plate has been reported by Wise and Burdon (17) to improve fretting fatigue strength. This work does seem to be at odds with other investigations (16)(18) which find hard chrome to be of little benefit. On the present evidence, chrome plating would be low on the list of potential palliatives for most unlubricated fretting fatigue problems.

Electroless nickel, another well established wear resistant coating, can be viewed in much the same way as hard chromium. It has a detrimental effect on normal fatigue and has not been found to be beneficial (16). Electroless-nickel impregnated with PTFE potentially may reduce the friction coefficient. In displacement controlled situations it might be beneficial in reducing contact stress, but there is no reported data available. However, work at the National Centre of Tribology carried out to establish the wear resistance of this coating has shown it to have poor performance under high contact load. With typical high loads encountered in fretting fatigue, the coating is unlikely to have sufficient durability.

Phosphating is extensively used in the automotive industry to give corrosion protection and as a pre-treatment to improve the adhesion of paint. It is a low cost, low temperature chemical conversion coating. It has a high degree of porosity and so holds lubricant well which gives it good anti-scuffing properties in film lubricated appplications such as cams and tappets. However, under dry unlubricated conditions the coating is friable. The thickness of the coatings is only a few microns and, therefore, the durability under fretting fatigue conditions is likely to be poor. It also has a detrimental effect on the unfretted fatigue strength (19). Although recommended in previous work (17) the author is sceptical about its use in the majority of fretting fatigue problems.

For aluminium, hard anodizing is effective in reducing fretting wear. The cellular porosity inherent in all anodized coatings can be utilized to allow the controlled ingress of low friction dry lubricants such as PTFE. In displacement controlled applications they may be beneficial but the current limited evidence does not support any improvement in fretting fatigue strength (20)(21).

Spraying

Hard coatings deposited by spraying can provide high wear resistance but there is limited information on their application in fretting fatigue. Most of these types of coatings have been employed to inhibit fretting wear rather than fretting fatigue. There are numerous coatings commercially available and deposition can be achieved by several methods.

Sprayed hard coatings inevitably reduce the fatigue strength. The lowering of fatigue strength occurs when the sprayed molten particles strike the relatively cold substrate. The molten particles rapidly solidify and the contraction leaves the surface in tension. Nevertheless, Syers (22) concluded that 'D' gun coatings are generally effective and with titanium alloys the fretting fatigue strength was improved by 33 percent, but again normal fatigue was badly affected and reduced by 40 to 50 percent. Harris (23) also found tungsten carbide (WC) deposited by plasma spraying and 'D' gun processes to be beneficial.

Sprayed molybdenum is probably best known for its use in the automotive industry. The coating is used to impart wear resistance to gear synchronization discs and piston rings. Molybdenum is chosen for its high melting point, hard oxide film, and good oil retaining characteristics in marginal lubricated applications. It has been reported to double the fretting fatigue life (10) although there is a substantial reduction in normal substrate fatigue properties.

The general benefit of hard sprayed coatings may be through a reduction in the coefficient of friction and so they may be best suited to displacement controlled problems. Alternatively, their often layered structure may encourage cracks to propagate within the coating and not penetrate the substrate.

Other palliatives

There are several other palliatives to consider in fretting fatigue problems including inserts, lubricants, and the unusual one of grooving the surface. The potential benefits and disadvantages of these are considered.

Lubricants

Lubricants can obviously produce a large decrease in the coefficient of friction. In force controlled situations this effect is likely to result in an increase in slip which may be unsatisfactory. With clamped assemblies the load transmission across the joint may be modified and bolts or fasteners can become overloaded. However, lubricants can prove very useful in displacement controlled fretting fatigue. Application of a lubricant will lower the level of shear traction and very effectively reduce the fretting fatigue contact stress. The main difficulty with lubricants lies in their durability. Unless continuously replenished, the self cleaning action of fretting can quickly remove any boundary lubricating film.

Gaining access to the inner regions of the fretting contact without separating the surfaces is difficult for oils of high viscosity and greases. A thin, low viscosity penetrating oil is, therefore, often recommended in the literature (24)(25). Work at NCT has, however, shown that under certain conditions high solid content pastes can be more durable than oil or grease.

Lubricants are better suited to low cycle applications, and where only a grease can be used, it must be replenished at regular intervals. The choice of thickener appears to have an important effect on durability (26)(27)(28). Polyurea grease seems to be the best choice according to Schlobohm (29), with lithium based grease next, but some way behind. Again additives such as ZDDP have been shown to be beneficial, (30).

An alternative to oil and grease is dry lubricant such as molybdenum disulphide or zinc oxide. Dry lubricants are again very effective at reducing fretting damage but over a limited life (31). If the problem involves high numbers of fretting cycles ($> 2 \times 10^5$), then dry lubricants such as these are unlikely to be satisfactory without re-application.

Most work with dry lubricants has concentrated on MoS_2 (23)(32)(33)(35) with mixed results. Fluorocarbon coatings, particularly PTFE loaded resins, have been reported to give longer endurance. Certainly in fretting wear, PTFE is one of the best performing polymers. There are numerous proprietary formulations of dry lubricant on the market. The effectiveness in fretting fatigue is likely to be determined by the strength of adhesion to the substrate. Few of these coatings will survive with nominal contact pressures above 50MPa. Work carried out at NCT has suggested that the epoxy based formulations have the best adhesion and, therefore, give the best prospects for success in alleviating displacement controlled fretting fatigue. Little work has so far been carried out on this type of palliative.

Interfacial layer

It has been shown that micro-slip can be prevented when a thin layer of flexible material (for example rubber or terylene) is interposed between the fretting surfaces (34). With a layer of the right compliance and thickness, the shear stress concentration at the edge of contact giving rise to the slip can be eliminated. Although this palliative can be very effective at preventing relative sliding there are design limitations. The palliative is usually only suitable for relatively small slip amplitudes. The increase in compliance produced by the palliative may also be unacceptable to the overall performance of the design. The durability of the interfacial layer under the alternating loading may also be another limiting factor.

Notwithstanding these limitations, the ability of interfacial layers to eliminate the shear concentration can prove very effective in many fretting fatigue problems involving bolted or riveted joints. Sandifer (35) has carried out practical testing of interfacial layers for use in the aircraft industry and found the palliative under limited conditions to be a success.

De-stressing notches

This palliative has been suggested by Kreitner (8) who tested a flat fatigue specimen machined with closely spaced de-stressing notches (0.4 mm deep) running laterally and longitudinally over the contact area. The tests showed that with the de-stressing notches fretting was prevented from influencing the fatigue strength of the specimen. Moreover, although there was a reduction in the fatigue limit caused by the notches, the improvement in fretting fatigue strength was still in excess of 100 percent.

The effectiveness of this palliative has been confirmed by Bramhall (36) who also obtained a significant increase in fretting fatigue strength using de-stressing grooves cut in a lateral direction only.

The advantage of this palliative is that the fatigue strength is governed purely by the notch effect of the de-stressing notches, it is, therefore, more predictable. Also, the palliative is applicable to both displacement and force controlled problems.

Surface cold working by shot peening

Shot peening is one of a number of treatments which can be used to produce surface cold working. This type of treatment increases the surface hardness and induces a compressive stress in the surface layers. If the process is carried out correctly the normal fatigue strength should be increased substantially.

Shot peening also produces a roughened surface texture, and rough surfaces are thought to be more resistant to fretting fatigue (37). This may be because a rough surface has, to a limited extent, the same de-stressing effect as the notches described previously. A shot peened surface will also retain lubricant better and enhance the effectiveness of an oil lubricant.

Conclusion

The selection of fretting palliatives at the current time is difficult and still very subjective. To achieve any chance of success, selection must be made only after careful consideration of the application requirements.

A good understanding of the effect of slip amplitude on fretting fatigue strength is important to achieve the best palliative selection. In the future, more research needs to be directed at understanding this parameter.

References

(1) BILL, R. C. (1982) Review of factors that influence fretting wear, *ASTMSTP 780*, ASTM, Philadelphia, pp. 165–182.
(2) BEARD, J. (1982) *An investigation into the mechanisms of fretting fatigue*, PhD Thesis, University of Salford.
(3) NISHIOKA, K. and HIRAKAWA, K. (1972) Fundamental investigation of fretting fatigue – Part 5, *Bull JSME*, **12**, 692–697.
(4) FUNK, W. (1968) A test process for investigating the effect of fretting corrosion on fatigue strength, *Metalloberflache* **22**, 362–367.
(5) FIELD, J. E. and WATERS, D. H. (1967) Fretting fatigue strength of En26 steel, NEL Report No 275 Min Tech.
(6) NEWCOMBE, S. B. and STOBBS, W. M. (1984) A transmission electron microscope study of white-etching layers on a rail head, *Mater. Sci. Engng*, 195–204.
(7) MINDLIN, R. D. (1949) Compliance of elastic bodies in contact, *J. appl. Mech*, **16**, 259–268.
(8) KREITNER, L. (1979) The effect of false brinelling and fretting fatigue on the fatigue life of assembled machine components, Forschungshefts Forschungskuratorium Maschinenbau e.V., Part 56 (1976): RAE Trans 1998.
(9) The 'Cassel' Sulfinuz Process, ICI Mond Division.
(10) TAYLOR, D. E. and WATERHOUSE, R. B. (1972) Sprayed molybdenum coatings as a protection against fretting fatigue, *Wear*, **20**, 401–407.
(11) BILL, R. C. (1978) Fretting of AISI 93100 steel and selected fretting resistance surface treatments, ASLE Trans. 21 No. 3 pp 236–242.
(12) BILL, R. C. (1979) Fretting wear of iron, nickel and titanium under varied environmental conditions. NASA TM–78972.
(13) BUDINSKI, K. G. (1979) Control of fretting corrosion, *Thin solid films*, **64**, 359–363.
(14) SIKORSKII, M. E. (1963) Correlation of the coefficient of adhesion with various physical and mechanical properties of metals, *J. bas. Engng*, **85**, 279–285.
(15) WATERHOUSE, R. B. (1972) Fretting corrosion, *International series of monographs on materials science and technology* (Edited by D. W. Hopkins) Vol. 10, Pergamon, Press, Oxford.
(16) WATERHOUSE, R. B., BROOK, P. A., and LEE, M. C. (1962) The effect of electro-deposited metals on the fatigue behaviour of mild steel under conditions of fretting corrosion, *Wear*, **5**, 235–244.
(17) WISE, S. and BURDON, E. S. (1964–65) The dual roles of non-metallic coatings on the fretting corrosion of mild steel, *Inst Loco Eng.*, **54**, 298.
(18) MCDOWELL, J. R. (1977) Fretting corrosion tendencies of several combinations of materials, *ASTMSTP144*, ASTM, Philadelphia.
(19) WATERHOUSE, R. B. and ALLERY, M. (1965) The effect of non-metallic coatings on the fretting corrosion of mild steel, *Wear*, **8** 112–120.
(20) HARRIS, W. J. (1972) The influence of fretting on fatigue. Part III, AGARD Advis. Rep. Agard-AR–45.
(21) WATERHOUSE, R. B. (1965) The formation, structure and wear properties of certain non-metallic coatings on metals, *Wear*, **8** 421–447.
(22) SYERS, G. The protection of certain titanium alloys against fretting fatigue. Rolls Royce Lab. Research Report 10245/FS/78118.
(23) HARRIS, W. J. (1975) The influence of fretting on fatigue. AGARD Conf. Proc. AGARD-CP–161, Paper 7.
(24) NEYMAN, A. (1992) The influence of oil properties on the fretting wear of mild steel, *Wear*, **152**, 171–181.
(25) SATO, J., SHIMA, M., SUGAWARA, T., and TAHARA, A. (1988) Effects of lubricant on fretting wear of steel, *Wear*, **125**, 83–95.
(26) SCHLOBOHM, R. T. (1982) Formulating grease to minimise fretting corrosion, *NLGI Spokesman*, **46**, 334–338.

(27) WUNSH, F. (1977) Relationship between the chemical structure of a lubricant and fretting, *Tribology Int.* 147–151.

(28) MISHIMA, M., KINOSHITA, H., and SEKLYA, M. (1990) Prevention of fretting corrosion to wheel bearings by Urea Grease, *NLGI Spokesman*, 496–503.

(29) SCHLOBOHM, R. T. (1982) Formulating grease to minimise fretting corrosion, *NLGI Spokesman*, **46**, 334–338.

(30) SATO, J. SHIMA, M., SUGAWARA, T., and TAHARA, A. (1988) Effects of lubricant on fretting wear of steel, *Wear*, **125**, 83–95.

(31) GODFREY, D. and BISSON, E. E. (1950) Effectiveness of molybdenum disulfide as a fretting corrosion inhibitor, NACA Tech Note No. 2180.

(32) GABEL, M. B. K. and BETHKE, J. J. (1978) Coatings for fretting prevention *Wear*, **46**, 81–96.

(33) BOWERS, J. E., FINCH, N. J., and GOREHAM, A. R. (1968) The prevention of fretting fatigue in aluminium alloys, *Proc. Inst. mech. Engrs.* **182**, 703–708.

(34) JOHNSON, K. L. and O'CONNOR, J. J. (1963–64) Mechanics of fretting, *Proc. Inst. mech Engrs.*, **178**, Paper 11.

(35) SANDIFER, J. P. Evaluation of methods of reducing fretting fatigue in 2024–T3 aluminium lap joints, *Wear*, **26**, 405–412

(36) BRAMHALL, R. (1973) Studies in fretting fatigue. D Phil. University of Oxford.

(37) LEADBEATER, G., NOBLE, B., and WATERHOUSE, R. B. (1984) The fatigue of aluminium alloy produced by fretting on a shot peened surface, *Advances in fracture research, Proceedings of the Sixth International Conference on Fracture*, Pergamon Press, Oxford, pp. 2125–2132.

*T. Hattori**

Fretting Fatigue Problems In Structural Design

REFERENCE Hattori, T., **Fretting fatigue problems in structural design,** *Fretting Fatigue,* ESIS 18 (Edited by R. B. Waterhouse and T. C. Lindley) 1994, Mechanical Engineering Publications, London, pp 437–451.

ABSTRACT This paper briefly discusses three fretting fatigue strength evaluation methods. The maximum stress method is used for optimizing shrink fitted shaft coupling and blade-dovetail structures. The fracture mechanics method is used to optimize groove shape. Finally , a new method is presented that uses two stress singularity parameters to estimate the fretting crack initiation stress amplitude. This method is applied to optimize wedge angles and pad materials.

Introduction

Fretting fatigue results when a pair of structural elements are in contact and cyclic stress and relative displacement occur along their contact surface. These fretting conditions can be seen in bolted and rivetted joints, shrink-fitted shaft couplings, the blade-dovetail regions of turbo machinery, and the coil wedges of turbine generator rotors, as shown in Fig. 1. Under these fretting conditions, fatigue strength can decrease to less than one-third the value found under non-fretting conditions.

In this paper the fretting fatigue problems are explained in each of these structures and three fretting fatigue strength evaluation methods are presented.

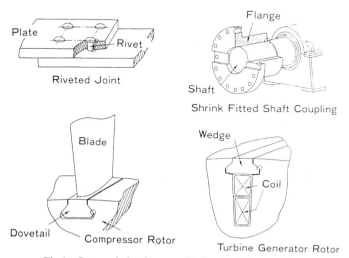

Fig 1 Contact design features with fretting fatigue problems

* Mechanical Engineering Research Laboratory, Hitachi Limited, Tsuchiura-City, Ibaraki, Japan.

Evaluating fretting fatigue strength using experimental results

Fretting fatigue problems have been studied by many researchers, including Wöhler (1), Eden (2), Kühnel (3), Thum (4), Horger (5), and Nagashima (6). These early researchers used fretting fatigue strength evaluation methods based mainly on experimental results, using fretting specimens under each contact condition for the structures shown in Fig. 2 (7), since stress analysis near the contact edge for each contact condition is very difficult.

Evaluating fretting fatigue strength using maximum stress

Along with the development of photo-elasticity (8) and finite element method techniques (9) under contact conditions, fretting fatigue strength evaluation methods using the stress concentration on contact surfaces have also been developed (10)(11). For instance, the author has analysed the fretting fatigue strength of shrink-fitted shaft couplings under repeated torsional loads (10), and fretting fatigue strength and mechanical damping of blade-dovetail structures (11) using finite element method analysis under contact conditions. Figure 3 shows the finite element mesh used for stress analysis of shrink-fitted shaft couplings. Figure 4 shows the estimated and experimental results of fretting fatigue strength. The finite element mesh used for stress and mechanical damping analyses of blade-dovetail assemblies is shown in Fig. 5. The estimated stress at dovetail corners and the mechanical damping constant are shown in Figs. 6 and 7, respectively. Using these results, the optimization of blade-dovetail structures is discussed.

Analysing the fretting fatigue threshold using fracture mechanics

Generally, fretting cracks are initially small and propagate very slowly. Therefore the fretting fatigue threshold was analysed by comparing the stress

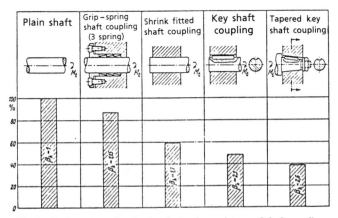

Fig 2   Fatigue strength reduction factor for each type of shaft coupling

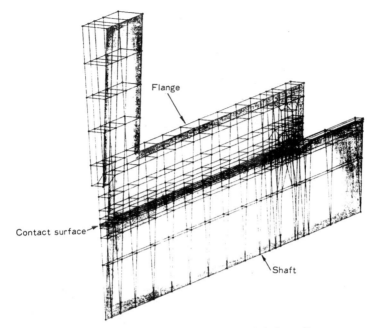

Fig 3   Finite element model of shrink-fitted shaft coupling

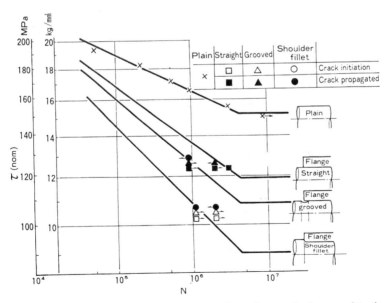

Fig 4   *S–N* curves of each shrink-fitted shaft coupling under completely reversed torsion

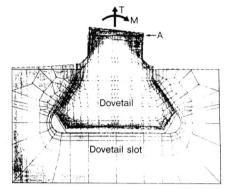

Fig 5   Finite element model of blade-dovetail structure

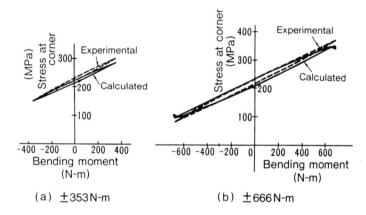

(a) ±353 N-m   (b) ±666 N-m

Fig 6   Calculated stresses at dovetail slot corners

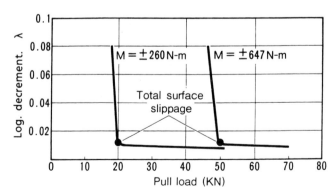

Fig 7   Calculated results of root damping

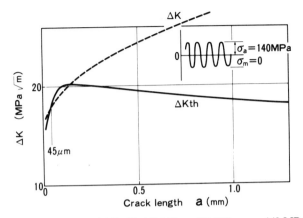

Fig 8   Comparison of ΔK with ΔKth (P_0 = 196 MPa, σ_a = 140 MPa)

intensity factor range of small fretting cracks with those initiated at the contact edge and threshold stress intensity factor range of the material, as shown in Figs 8–10 **(12)**. In these figures, the mean of the axial nominal stress, σ_m, is 0 and the amplitudes of the axial nominal stress, σ_a, are 140 MPa, 142 MPa, and 143 MPa, respectively. As shown in Fig. 8 (σ_a = 140 MPa), the small crack initiated by the fretting damage arrested at a length of 45 μm and, as shown in Fig. 10 (σ_a = 143 MPa), the crack initiated by the fretting damage grew slowly, and finally the specimen broke. Figure 9 shows the critical condition where the crack arrested at the length of 85 μm; however, above this stress amplitude (σ_a = 142 MPa), the crack grew continuously. This critical condition is the fretting fatigue limit. These estimated fretting fatigue thresholds coincide well with the experimental results, as shown in Fig. 11. The observed results of a fretting-damaged surface by scanning electron micro-

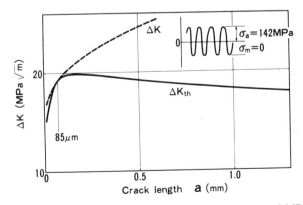

Fig 9   Comparison of ΔK with ΔKth (P_0 = 196 MPa, σ_a = 142 MPa)

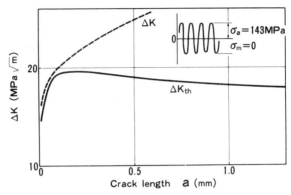

Fig 10   Comparison of ΔK with ΔKth ($P_0 = 196$ MPa,$\sigma_a = 143$ MPa)

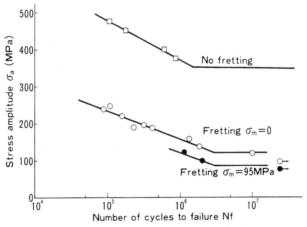

Fig 11   Estimated and experimental results of fretting fatigue strength

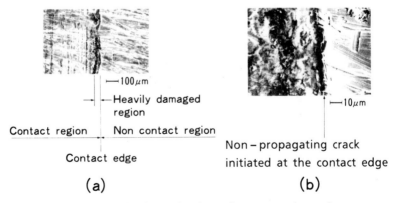

Fig 12   Fretting-damaged surface and non-propagating crack

scope are shown in Fig. 12. The loading condition of this surface was as follows:

Contact pressure $P_0 = 196$ MPa
Mean stress $\sigma_m = 0$
Stress amplitude $\sigma_a = 98$ MPa
Number of cycles $N = 2.4 \times 10^7$ (unbroken)

The heavily damaged region near the contact edge corresponds to the contact pressure and tangential stress concentrating regions (12). The non-propagating crack at the contact edge is shown in Fig. 12(b). A sectional view of these cracks observed by a scanning electron microscope is shown in Fig. 13. This crack length corresponds well with the estimated non-propagating crack length shown in Figs 8 and 9. Then, fretting fatigue crack propagation behaviour was estimated using fracture mechanics analysis. Figure 14 shows the crack propagation behaviour of the test material (N_i–M_o–V steel). The fretting crack propagation behaviour was calculated using this crack propagation rate curve and equivalent stress intensity factor range ΔK_e, defined as follows

$$\Delta K_e = \frac{\Delta K_{th}\,(R = 0, a = \infty)}{\Delta K_{th}\,(R, a)} \Delta K\,(R, a) \tag{1}$$

where R is the stress ratio and a is the crack length (mm). Calculated results of fretting fatigue crack propagation behaviour are shown in Fig. 15. The calculated crack propagation behaviour coincided well with that from experimental results as show in Fig. 16. Accordingly, the suitability of the estimation method for the fretting fatigue limit by using the ΔK_{th} of a small crack is confirmed. This fracture mechanics approach is then used to determine the optimimum groove shape, which is sometimes used to increase fretting fatigue

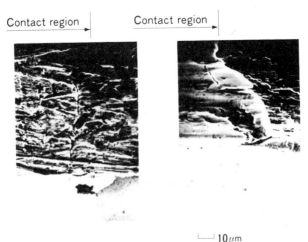

Contact region   Contact region

└──┘ 10 μm

Fig 13   Non-propagating crack length

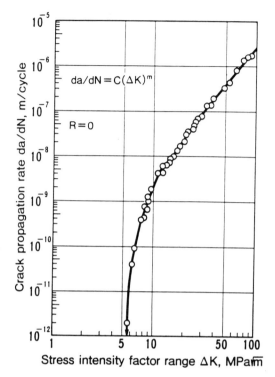

Fig 14   Crack propagation behaviour of Ni–Mo–V steel

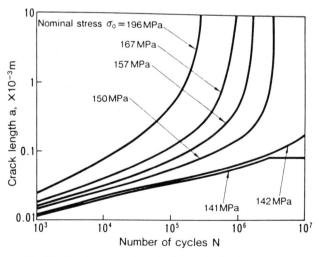

Fig 15   Crack propagation behaviour under fretting conditions

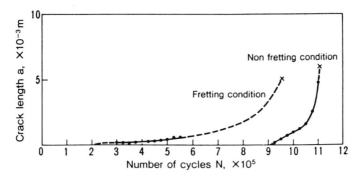

Fig 16   Crack propagation behaviour under fretting condition and non-fretting condition

strength. The calculated results of fatigue strength are shown in Fig. 17. From this we can see that fretting fatigue strength improves with increased groove depths. However, the fatigue strength of the groove bottom decreases with increased groove depths. Both the fretting fatigue limit and groove bottom fatigue limit must be considered for optimization of the groove shape. It is concluded that the optimum groove depth is about 1.5 mm. This estimated result is confirmed experimentally, as shown in Fig. 18. Fatigue fracture conditions of the grooved type specimens, near fatigue limit stress levels, are shown in Fig. 19. For a groove depth of $d = 1$ mm the fracture occurred at a contact edge and for a groove depth of $d = 2$ mm the fracture occurred at the groove bottom. These conditions coincided well with the estimated results shown in Fig. 17. However, for high stress amplitudes, fracture occurred at the groove bottom regardless of the groove depth. This is shown in Fig. 18 with the

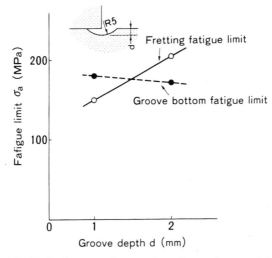

Fig 17   Fatigue strength comparison of grooved-type models

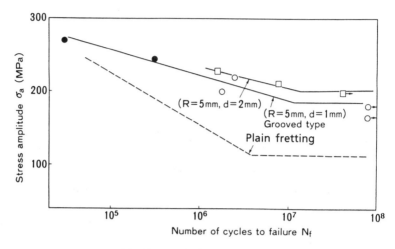

Fig 18   Fretting fatigue test results

symbol ● for a groove depth of 1mm. This optimized groove shape was
applied in the design of the turbine generator coil wedge.

Analysing fatigue crack initiation using stress singularity parameters.

In some cases, even the presence of small cracks cannot be permitted because
of reliability requirements. In these cases the initiation of fretting cracks has to
be estimated. The problem with fretting crack initiation evaluation is that the
stress and displacement fields near a contact edge show a singularity behaviour (14). Therefore, accurate strength evaluation cannot be made by using the
stress value alone. A new fretting crack initiation evaluation method was,
therefore, developed which uses two stress singularity parameters, K and λ,

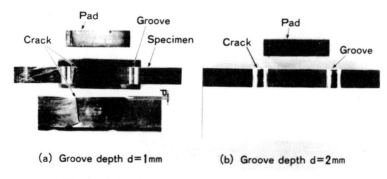

(a) Groove depth d=1mm   (b) Groove depth d=2mm

Fig 19   Fatigue fracture conditions of groove type specimens

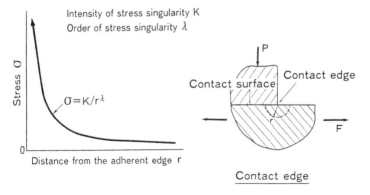

Fig 20   Stress distribution near a contact edge

which express the stress distributions near a contact edge, as follows (see Fig. 20)

$$\sigma(r) = K/r^\lambda \tag{2}$$

Where σ is the stress (MPa), r is the distance from contact edge (mm), K is the intensity of the stress singularity and λ is the order of the stress singularity. These two stress singularity parameters K and λ were calculated using stress distributions calculated by FEM. These stress singularity parameters were applied to study the influence of wedge angle and pad material stiffness on fretting fatigue crack initiation. The calculated results near contact edges are shown in Fig. 21 for each wedge angle and Fig. 22 for each pad material. Comparing these calculated parameters with the critical intensity value of Kc (see Fig. 23), the fretting fatigue crack initiation limit can be obtained in Fig. 24 for each wedge angle and in Fig. 25 for each pad material. These figures show that reducing the wedge angle to less than $\pi/3$ is a very effective means of avoiding fretting crack initiation, and using low stiffness pad material can considerably increase the fretting fatigue crack initiation limit. These results are currently being considered for the coil wedge design for a turbine generator rotor.

Conclusion

Three fretting fatigue strength evaluation methods have been briefly discussed. The maximum stress method is used for optimizing shrink-fitted shaft couplings and blade-dovetail structures. The fracture mechanics method is used to optimize the groove shape of a turbine generator coil wedge structure. The stress singularity parameters method makes it possible to estimate the fretting crack initiation stress amplitude, and permits this to be applied to optimize the wedge angle and stiffness of the pad material.

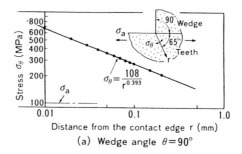

(a) Wedge angle $\theta = 90°$

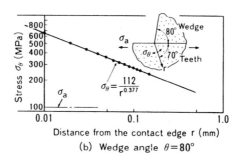

(b) Wedge angle $\theta = 80°$

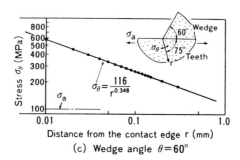

(c) Wedge angle $\theta = 60°$

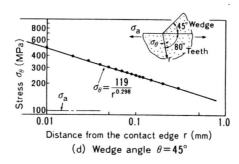

(d) Wedge angle $\theta = 45°$

Fig 21   Calculated stress distributions near a contact edge

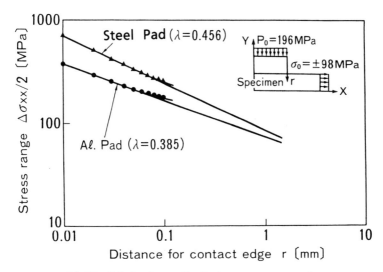

Fig 22   Calculated stress distributions near a contact edge

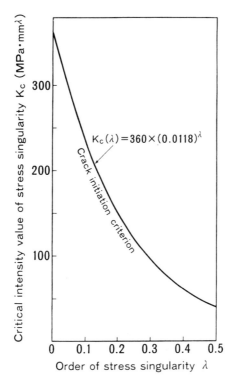

Fig 23   Fretting fatigue crack initiation criteria using stress singularity parameters

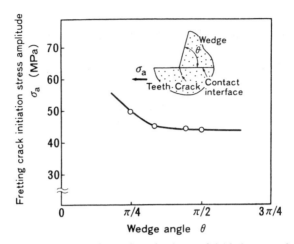

Fig 24   Effect of wedge angle on fretting crack initiation strength

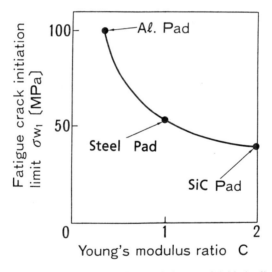

Fig 25   Estimated results of fretting fatigue crack initiation limit

References

(1) WÖHLER, A. (1960) Versuche über die relative Festigkeit von Eisen Stahl und Kupfer, *Zeit. Bauwesen*, **16**, 67.
(2) EDEN, E. M. *et al.* (1991) The endurance of metals, *Proc. IME*, 4, 839.
(3) KUHNEL, R. (1934) Grenzen der Werkstoffleistung-Dauerbrüche und ihre Ursachen, Glas. Ann , **115**,
(4) THUM, A. and WUNDERLICH F. (1933) Clamping pressure and fatigue bending tests, *Zeit VDI*, **77**, 851.

(5) HORGER O. J. and MAULBETSH, J.L. (1936) Increasing the fatigue strength of press-fitted axle assemblies by surface rolling, *J. Appl. Mech*, **3**, 91.

(6) NAGASHIMA, K. (1939) *Trans. Japan Soc mech. Engng*, **5**, 80 (in Japanese).

(7) CORNELIUS, E. A. and SCHMIDT, P. (1961) Das Ring federspannelement als Verbindung von Welle und Nabe, *Konstruction*, **13**, 91.

(8) NISHIOKA, K. *et al.* (1970) *Trans. Japan Soc. mech. Engng*, **36**, 1799 (in Japanese).

(9) OKAMOTO N. and NAKAZAWA, M. (1979) Finite element incremental contact analysis with various frictional conditions, *Int. J. Numer. Methods Engng*, **14**, 377.

(10) HATTORI, T. *et al.* (1981) Torsional fatigue strength of a shrink fitted shaft, *Bull. JSME*, **24**,1893.

(11) HATTORI, T. *et al.* (1984) Slipping behaviour and fretting fatigue in the disk/blade dovetail region, Proceedings, of the 1983 Tokyo International Gas Turbine Congress, Vol. **13** , 945.

(12) HATTORI, T. *et al.* (1988) Fretting fatigue analysis using fracture mechanics, *JSME Int. J, Ser. 1*, **31**, 100.

(13) HATTORI, T. *et al.* (1992), Fretting fatigue analysis of strength improvement models with grooving or knurling on a contact surface, *ASTM STP 1159*, ASTM, Philadelphia, p. 101.

(14) HATTORI T. *et al.* (1988), A stress singularity parameter approach for evaluating adhesive and fretting strength, Vol. 6, ASME, p. 43.

T. Hattori* and M. Nakamura*

Fretting Fatigue Evaluation using Stress Singularity Parameters at Contact Edges

Reference Hattori, T. and Nakamura, M., **Fretting fatigue evaluation using stress singularity parameters at contact edges**, *Fretting Fatigue*, ESIS 18 (Edited by R. B. Waterhouse and T. C. Lindley) 1994, Mechanical Engineering Publications London, pp. 453–460.

ABSTRACT The stress and displacement fields near contact edges show singularity behaviour. Consequently, methods of evaluating fretting fatigue strength using maximum stress are generally not valid. This paper present a new method for evaluating fretting fatigue strength. This method can be used to estimate the initiation of fretting fatigue cracks by using two stress singularity parameters, K and λ, which express stress distributions near a contact edge as follows

$$\sigma(r) = K/r^\lambda$$

where r is the distance from the contact edge.
   This method is used to analyse the influence of wedge angles and stiffness of pad material on the fretting fatigue crack initiation limit.

Introduction

The authors have previously reported a method for evaluating fretting fatigue (1)(2). This method uses stress analysis and fracture mechanics analysis of the contact structures. Using this method, the fretting fatigue limit can be estimated by comparing the calculated stress intensity factor of a crack which originates at the contact edge with the material's threshold stress intensity factor range. However, some heavy duty industrial machines, such as turbine machinery rotors, cannot be permitted to crack at all. In a turbine generator, contact interfaces between the rotor teeth and the wedge have contact edges, indicated by the circle (Fig. 1). This paper presents a new method for evaluating

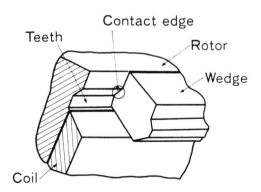

Fig 1   Contact edge in the turbine generator coil wedge

* Mechanical Engineering Research Laboratory, Hitachi Limited, Tsuchiura-City, Ibaraki, Japan.

453

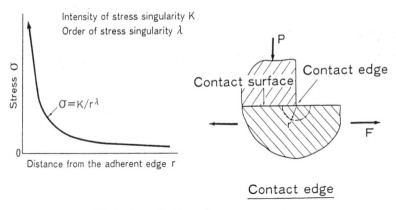

Fig 2 Stress distributions near a contact edge

the initiation of fretting fatigue cracks (3). This new method uses two stress singularity parameters, K and λ, which express stress distributions near a contact edge (Fig. 2). This method is used to analyse the influence of wedge angles and pad material rigidity on the crack initiation limit. From this it is shown that, using a small wedge angle and slightly rigid pad, the crack initiation limit can be considerably increased.

Stress singularity parameter approach for evaluating the initiation of fretting fatigue cracks

The stress fields near the contact edges show singularity behaviour (Fig. 2). These stress distributions near the contact edges can be expressed using the two stress singularity parameters, K and λ, a follows

$$\sigma(r) = K/r^{\lambda} \qquad (1)$$

where σ is stress (MPa), r is distance from singularity point (mm), K is intensity of stress singularity, and λ is order of stress singularity.

Parameter, λ and K are calculated as follows. The order of stress singularity (λ) for the contact edge is calculated analytically (4) using wedge angle θ_1, θ_2, Young's modulus E_1, E_2, Poisson's ratio $v1$, $v2$, and the frictional coefficient μ (Fig. 3). However, in this paper λ is calculated by best fitting equation (1) to the stress distributions near the contact edge. Stress distribution is calculated by numerical stress analysis, such as the finite element method, or the boundary element method. The intensity of stress singularity (K), is also calculated by best fitting equation (1) to the numerically analysed stress distribution. By comparing these calculated parameters (K and λ) with the crack initiation criterion Kc, fretting crack initiation conditions can be estimated for each wedge angle and pad stiffness. This criterion Kc is derived from the plain fatigue limit (σ_{wo}) and the threshold stress intensity factor range (ΔKth) of the rotor materials (Fig. 4).

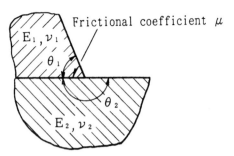

Fig 3 Geometry of the contact edge

Application results

Figure 5 shows an analytical model of the turbine-generator coil wedge region illustrated in Fig. 1. In this model, a constant contact pressure of 198 MPa is loaded on the upper surface of the wedge, and an alternating axial stress of ±98 MPa is loaded on the side surface of the teeth. The frictional coefficient

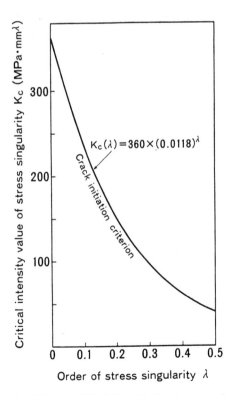

Fig 4 Fretting fatigue crack initiation criteria using stress singularity parameters

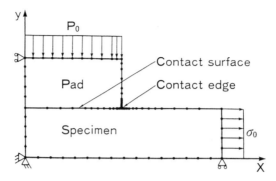

Fig 5   Boundary element mesh of simple fretting model

(μ) on the contact surface is set at 0.7 **(5)(6)**. In Fig. 6, distributions of the tangential stress range ($\Delta\sigma\theta$) are plotted for wedge angles of 90, 80, 60 and 45 degrees. These tangential stress ranges are used to evaluate fretting strength because they are thought to dominate crack initiation at the contact edge. Parameters λ and K are calculated using these stress distributions. The calculated λ for each wedge angle is plotted in Fig. 7. The solid line in this figure shows the order of stress singularity λ calculated theoretically under contact conditions of complete adhesion **(7)**. The calculated results of λ from stress distributions are slightly lower, because of micro-slippage near the contact edge. The crack initiation stress amplitude (σa) for each wedge angle is estimated by comparing calculated ΔK with Kc (Fig. 6). Figure 8 shows the estimated stress amplitude (σa) for each wedge angle. From these results it can be seen that the fretting crack initiates at a considerably lower stress amplitude compared with the fretting fatigue limit. Therefore, a non-propagating crack can often be observed in fretting conditions **(1)**. This figure also shows that reducing wedge angles to less than $\pi/3$ is a very effective means of avoiding fretting crack initiation.

These stress singularity parameters were applied to study the influence of pad material stiffness on fretting strength. Three kinds of pad materials were employed (steel, aluminium and SiC) and the same analytical model and loading conditions as shown in Fig. 5 were used. The elastic modulus of these materials is shown in Table 1. The calculated results of stress distributions near contact edges are shown in Fig. 9. Comparing these calculated parameters with the critical intensity value of Kc, the fretting fatigue crack initiation limit can be obtained (Fig. 10). This result shows that using low stiffness pad material can considerably increase the fretting fatigue crack initiation limit.

Table 1   Elastic constants of materials

	St. Pad	Al. Pad	SiC Pad
Young's modulus	206 GPa	71 GPa	410 GPa
Poisson's ratio	0.30	0.34	0.15

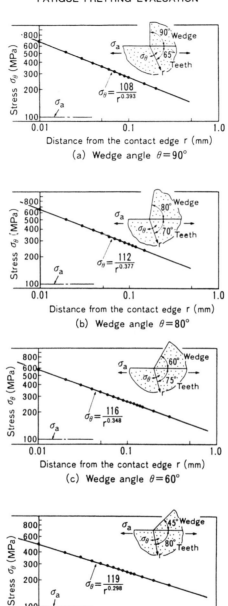

(a) Wedge angle $\theta = 90°$

(b) Wedge angle $\theta = 80°$

(c) Wedge angle $\theta = 60°$

(d) Wedge angle $\theta = 45°$

Fig 6 Calculated stress distributions near the contact edge

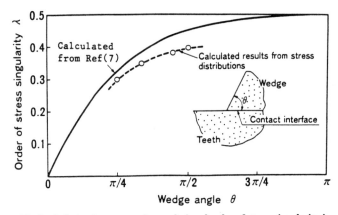

Fig 7   Relation between wedge angle θ and order of stress singularity λ

Conclusion

A new method of evaluating fretting fatigue strength has been developed, which uses two stress singularity parameters. The following conclusions were obtained.

(1) Fretting cracks initiate at considerably lower stress amplitudes compared with the fretting fatigue limit. Reducing the wedge angle to less than $\pi/3$ improves the fretting strength.
(2) The fretting fatigue crack initiation limit improves as pad stiffness is reduced.
(3) The fretting fatigue crack initiation limit, when using an Al pad, increases by about 1.9 times compared with using a steel pad, and decreases to about 74 percent when using an SiC pad.

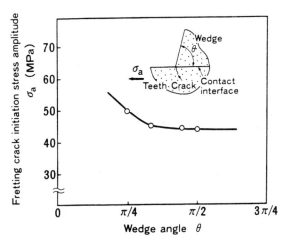

Fig 8   Effect of the wedge on the fretting crack initiation strength

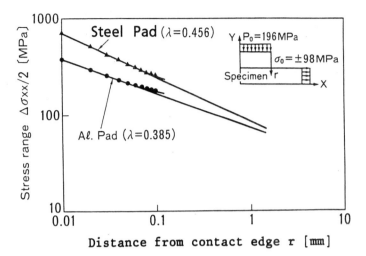

Fig 9   Stress distributions near a contact edge

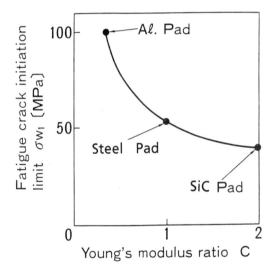

Fig 10   Estimated results of fretting fatigue crack initiation limit

References

(1) HATTORI, T., NAKAMURA, M., and WATANABE, T., (1984) Fretting Fatigue Analysis by using fracture mechanics, ASME Paper No. 84-WA/DE–10.
(2) SAKATA, H., HATTORI, T., and HATSUDA, T. (1987) An application of fracture mechanics to fretting fatigue analysis, *Role of fracture mechanics in modern technology*, Elsevier Science Publishers, p. 303.
(3) HATTORI, T., SAKATA, S., and WATANABE, T. (1988) A stress singularity parameter approach for evaluating adhesive and fretting strength, ASME Book No. G00485, MD–Vol. 6, p. 43.

(4) DEMPSAY, J. P. and SINCLAIR, G. B. (1981) On the singular behaviour at the vertices of a bi-material wedge, *J. Elasticity*, **11**, 317.
(5) HATTORI, T., KAWAI, S., OKAMOTO, N., and SONOBE, T. (1981) Torsional fatigue strength of a shrink fitted shaft, *Bull. JSME*, 24–197, 1893.
(6) HATTORI, T., NAKAMURA, M., SAKATA, H., and WATANABE, T. (1988) Fretting fatigue analysis using fracture mechanics, *JSME Int. J, Ser. 1*, **31**, 100.
(7) HEIN, V. L. and ERDOGAN, F. (1971) Stress singularities in a two-material wedge, *Int. J. Fracture Mech.* **7**, 317.

K. Hirakawa* and K. Toyama†

Influence of Surface Residual Stresses on the Fatigue Crack Initiation of Press-Fitted Axle Assemblies

REFERENCE Hirakawa, K. and Toyama, K. **Influence of surface residual stresses on the fatigue crack initiation of press-fitted axle assemblies,** *Fretting Fatigue,* ESIS 18 (Edited by R. B. Waterhouse and T. C. Lindley) 1994, Mechanical Engineering Publications, London, pp. 461–473.

ABSTRACT A series of experiments are conducted on press-fitted axle assemblies to investigate the influence of residual stresses of axles, introduced by induction hardening, on the fatigue strength regarding both initiation and propagation of fatigue cracks in the region of the fitted wheel.

After reviewing the influence of residual stress, it is apparent that some limited improvement in fatigue resistance can be obtained against initiation of fatigue cracks. Much greater improvement, however, can be obtained against fatigue crack propagation.

Notation

σ_n Nominal bending stress, MPa
σ_r Contact pressure at the fitted surface, MPa
τ Shear stress at the surface of axle, MPa
a Depth of surface crack, mm
c Half length of surface crack, mm
μ Coefficient of friction between mating surfaces
λ Depth below the surface of axle, mm
x Distance from contact edge in axial direction, mm
θ Parametric angle of the ellipse, degrees
Q Shape factor for elliptical crack
K_1 Mode I stress intensity factor, MPa $\cdot$ m$^{1/2}$
K_{res} Mode I stress intensity factor for the residual stress, MPa $\cdot$ m$^{1/2}$
a_0 Compressive residual stress zone size, mm
σ_{RO} Residual stress at the surface of axle, MPa
σ_{W1} Fatigue strength regarding crack initiation, MPa
σ_{W2} Fatigue strength regarding crack propagation, MPa

Introduction

Many fatigue failures of engineering structures and components are associated with fretting. Of these examples, it is frequently found that fatigue cracks initiate from mating surfaces which transfer alternating load due to the effect of

* Faculty of Engineering, Kyushu University, 6-10-1 Hakozaki, Fukuoka, Japan.
† Research and Development Centre, Sumitomo Metal Industries Limited, Fusocho 1–8, Amagasaki, Japan.

461

fretting. In situations involving shrink-fits on rotating components and load transfer situations between components attached with bolts, the presence of fretting fatigue should be considered.

It is well known that press-fitted axle assemblies exhibit low fatigue strength characteristics. This weakness arises from both fretting corrosion of mating surfaces and high stress concentration from shape effects (1–3). Press-fitted axle assemblies are widely used for railway wheel-sets. Recent trends toward greater railway car speed with increasing dynamic load, have attracted increasing attention to the fretting fatigue strength of press-fitted axles.

As a means of improving fatigue resistance, induction hardening of axles has been used in the axle assemblies for the high speed bullet train in Japan(4)(5). This paper discusses the influence of residual stresses and surface hardness on fatigue strength regarding both initiation and propagation of fatigue cracks in press-fitted axles.

Experimental procedure

Fatigue strength was determined in laboratory tests for rotating bending of press-fitted assemblies having axle diameters from 40 mm to 209 mm. Cantilever type fatigue testing machines were used. A typical press-fitted assembly for testing an axle diameter of 209 mm is shown in Fig. 1. Similar assemblies were used for smaller axles which were straight with no fillet (fillet radius = ∞). Plain carbon steel was used for the axle forging. Table 1 shows the chemical compositions and mechanical properties of the axles before induction hardening.

After heat-treatment of normalizing and induction hardening, axles were tempered at 200°C–230°C. The surface hardness of the induction hardened axle was Hv 520–580. Specimens of 52 mm axle diameter were additionally tempered at 400 and 600°C to reduce the residual stress levels. Axles were then ground and wheels were press-fitted with a interference selected to give a mean contact pressure of 70 MPa.

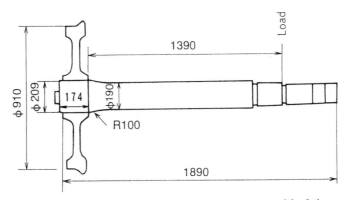

Fig 1   Press-fitted assembly having 209mm diameter axle prepared for fatigue testing

Table 1 Material composition and mechanical properties

No.	Axle diameter (mm)	Element %wt					Yield strength (MPa)	Tensile strength (MPa)	Elongation (%)	Reduction of area (%)
		C	Si	Mn	P	S				
1	40	0.39	0.25	0.68	0.010	0.008	350	610	31.0	52.0
2	52	0.35	0.25	0.79	0.011	0.025	338	541	31.0	51.8
3	100	0.36	0.22	0.59	0.040	0.037	343	559	30.0	52.0
4	150	0.37	0.26	0.72	0.010	0.010	357	600	32.0	51.0
5	209	0.36	0.24	0.67	0.015	0.010	346	609	29.3	55.8

Mechanical properties are before induction hardening.

All axles were loaded for 20 million (2×10^7) stress reversals, unless breakage developed earlier. The wheel was then pressed off and the fitted portion of the axle was examined in the electron microscope with a magnification of $400 \times$. A series of disconnected short hairline circumferential cracks of shallow depth in this region were found accompanied with wear pits, as illustrated in Fig. 2.

The crack length, $2c$, at the surface and its depth, a, observed in the induction hardened 209 mm in diameter axles are shown in Table 2. As seen in the table, the fatigue crack length $2c$ was below 1.0 mm, and the depth was below 0.1 mm. The aspect ratios, a/c, were scattered around 0.25. The crack observed was always formed by coalescence of two or three coplanar cracks as shown in Fig. 2; therefore, it can be considered that the actual aspect ratio of an individual small crack before coalescence is approximately 0.4–0.6.

The distribution and magnitude of residual stress were determined by the Sachs boring-out method for six different series of axles. For reference, sub-

Fig 2(a) Fretting fatigue cracks and fretting pits

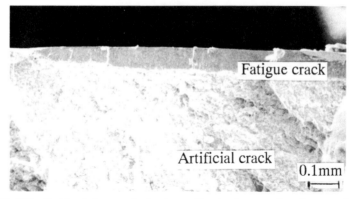

Fig 2(b)   Fretting fatigue cracks-fracture surface (coalescence of co-planar cracks)

critical quenched axle specimens of 52 mm in diameter were also tested, since sub-critical quenching (quenched below the critical transformation temperature) can introduce residual stresses without raising the hardness of the axles.

Stress calculation

The stress conditions of the press-fitted axles under bending load were calculated by an elastic finite element analysis method using the ABAQUS program. It was supposed in calculation, that slip was allowed to occur in the press-fitted contact surface where the shear stress, τ, was larger than the contact pressure, σ_r, multiplied by frictional coefficient, μ, as

$$\tau > \mu\sigma_r \qquad\qquad\qquad\qquad\qquad\qquad (1)$$

In the calculation, 0.6 was taken as the frictional coefficient, μ **(6)**.

From the stress conditions of the axles, stress intensity factors for an elliptical small surface crack in the fretted part of the axle were calculated. In the calculation of the stress intensity factors, it was assumed that the axle had an elliptical surface crack emanating perpendicular to the axle direction. The dis-

Table 2   Fatigue crack length and depth for 209 mm diameter axle

Bending stress (MPa)	Crack length $2c(\mu m)$	Crack depth $a(\mu m)$	Aspect ratio a/c
70	280	12	0.19
80	440	42	0.19
90	600	60	0.20
100	400	55	0.28
120	360	78	0.43
140	340	56	0.33
160	840	57	0.13

tances of the crack from the contact edge were taken as 0.8, 1.8, and 2.8 mm because small cracks perpendicular to the surface in these fretted part were always observed.

The stress distributions of the press-fitted axle calculated under the bending load, were applied to the axle which was assumed to have the above-mentioned surface cracks, and the mode I stress intensity factor was estimated from the crack opening displacement of the finite element meshes near the tip of the crack. In the above situation, crack growth was considered to occur in the tensile opening mode normal to the direction of maximum principal stress, and mode I stress intensity factor was taken as the governing parameter.

Stress states of press-fitted axle assembly

Axial bending stress distributions along the contact surface of an axle are shown in Fig. 3 as a function of applied nominal bending stress, σ_n, which is defined as (M/Z) where M is the bending moment at the edge of the press-fit and Z is the section modulus of the axle. Stress concentration can be seen in the figure; however, the concentration factors were not as high as expected from the shape effect produced when the axle and the hub are made from one block in which no slip occurs. This indicates that the stress concentration at the fitted part was relieved by the occurrence of the slip. It was noted that stress concentration factors, (1.37 for $\sigma_n = 150$ MPa, 1.50 for $\sigma_n = 100$ MPa, and 1.83 for $\sigma_n = 50$ MPa), depended on the applied stress, and the non-linear stress condition could be seen. These non-linear characteristics arose from the occurrence of slippage (fretting) near the edge of the contact surface, and also

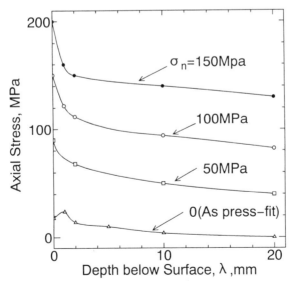

Fig 3   Axial stress distribution (1.8mm from the contact edge for 209mm diameter axle)

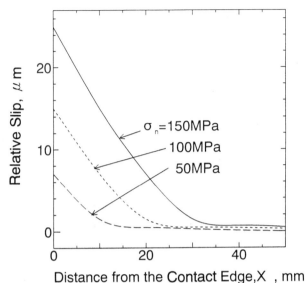

Fig 4   Distribution of slip amplitude along contact surface

the amplitude of the slip which depends on the applied stresses. The distributions of relative slip along the contact surface are shown in Fig. 4. The magnitude of the slip at the contact edge agreed well with the experimental values measured by a special strain gauge (7). Figure 5 shows the relaxation of contact pressure due to the applied bending stresses. As seen in Fig. 5, the relaxation of contact pressure increases with increasing applied stresses.

Stress intensity factor

Mode I stress intensity factors, K_I, for semi-elliptical cracks of aspect ratio, a/c, equal to 0.6, at the fretted part of axle, are shown in Table 3. From the table, it can be seen that the K_I for the deepest point of crack front, and K_I on

Table 3   Stress intensity factor for a semi-elliptical surface crack

Distance from contact edge (mm)	Nominal bending stress			
	100 MPa		150 MPa	
	$\theta = 1$	$\theta = 85$	$\theta = 1$	$\theta = 85$
	(degrees)		(degrees)	
0.8	9.15	9.30	11.90	12.34
1.8	9.18	9.52	11.62	12.22
2.8	8.90	9.55	11.75	12.23

Crack depth $a = 0.6$ mm, θ, parametric angle of the ellipse (degrees); crack length $c = 1.0$ mm, $x = 1.8$ mm

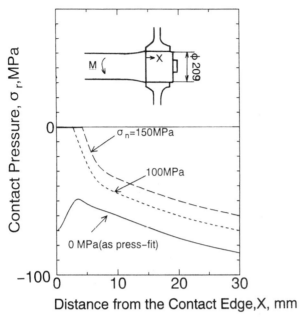

Fig 5 Contact pressure distribution along contact surface

the surface have nearly the same values. It can be assumed that the aspect ratio, a/c, of a small surface fatigue crack will be 0.6, if the fatigue crack propagates so as to maintain the K_1 along the crack front at a uniform value.

The calculated K_1 at the deepest point of the crack front is shown in Fig. 6. In the figure, K_1 is normalized by $\sigma_n\sqrt{(\pi a/Q)}$, where a is the crack depth and Q is the shape factor for an elliptical crack. $K_1/\sigma_n\sqrt{(\pi a/Q)}$ depends on the applied nominal stress σ_n in a similar way as the non-linear characteristic of the stress concentration factor. It is also noted that it increases rapidly at first with increasing crack depth and then peaks at a small crack depth. This could explain the characteristics of fretting fatigue where a non-propagating shallow fatigue crack is often observed.

To quantify residual stress effects on crack growth, the method of superposition of the respective stress intensity factors for the applied stresses and for the residual stresses was used. The magnitude and distribution of the residual stress field into which a crack is expected to grow were measured by the Sachs boring method. The K factor for the residual stresses, denoted as K_{res} here, was calculated by loading the crack faces with the residual stresses that exist normal to the plane of potential crack growth in the uncracked axle. The calculation was also made by the same finite element meshes.

A typical residual stress distribution of the induction hardened axles of 209 mm in diameter is approximated by straight lines and used for K_{res} calculation as shown in Fig. 7. The results of the finite element solution are plotted in Fig. 6.

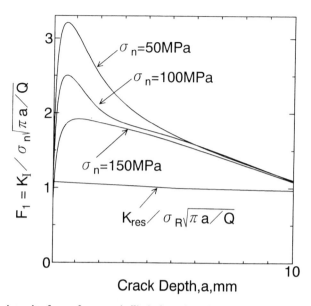

Fig 6   Stress intensity factor for a semi-elliptical crack at fretted axle surface (1.8 mm from the contact edge for 209mm diameter axle)

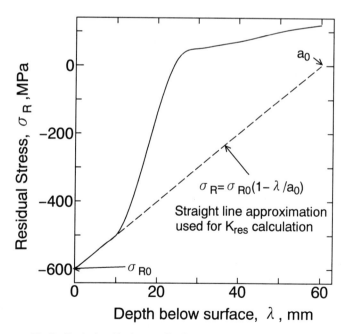

Fig 7   Typical residual stress distribution in 209mm diameter axle

Based on the Newman–Raju empirical equation for a surface crack in a plate under bending, **(8)(9)**, the following approximate equation for a small crack can be derived when the residual stress distribution is expressed as a linear equation in Fig. 7.

$$K_{res}/\sigma_{R0}\sqrt{(\pi a/Q)} = F\{1 - 0.61(a/a_0) - 0.06(a/c)(a/a_0)\} \qquad (2)$$

where

$$F = 1.13 - 0.09(a/c) \qquad (3)$$

For $(a/c) = 0.6$, equation (2) is rewritten as

$$K_{res}/\sigma_{R0}\sqrt{(\pi a/Q)} = 1.08\{1 - 0.65(a/a_0)\} \qquad (4)$$

Equation (4) agrees well with the FEM results for crack depth below 10 mm, Fig. 6. Therefore, equation (4) can be used for further discussion of the effect of residual stresses. For example, K_I for nominal bending stress $\sigma_n = 150$ MPa is compared with the K_{res} for the maximum surface residual stress $\sigma_{R0} = 350$ MPa and 250 MPa, where the compressive residual stress zone sizes, a_0, are varied. The result is shown in Fig. 8. This figure indicates that K_I is always larger than K_{res} for $\sigma_{R0} = 250$ MPa, but when $\sigma_{R0} = 350$ MPa it depends on a_0 whether K_I is larger than K_{res} or not. It can be noted that not only the maximum surface residual stress but also its distribution plays an important role in crack propagation.

Experimental results

Fatigue strengths regarding both initiation of the shallow cracks and complete fracture were determined after 2×10^7 stress reversals. The axles which did

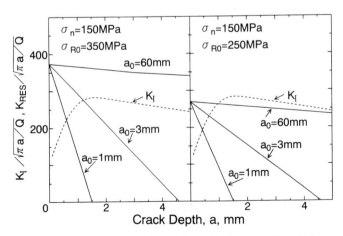

Fig 8 Comparison of the stress intensity factors for residual stress and applied stress (solid lines are for K_{res})

not fracture after 2×10^7 stress reversals, were withdrawn from the fitted hub. The fretted surfaces were then buffed and searched for the cracks by the electron microscope with the magnification up to $400 \times$. Most of the cracks larger than 0.1 mm in length could be detected. The minimum bending stresses which initiate the above cracks but not propagate to fracture are defined as the fatigue strength of crack initiation.

Figure 9 shows the fatigue strength as a function of the case depth. It is known that shallow cracks initiate at stresses far below the usual endurance limit determined from the fracture of axles.

In Fig. 10, the effects of surface residual stresses on fatigue strength are summarized. In the figure, data of the axles quenched below the transformation temperature are also plotted for comparison. Apparently, compressive residual stress has little influence on crack initiation, but the fatigue strength regarding crack propagation increases as the compressive residual stresses increase.

In Fig. 11, the size effect of the fatigue strength regarding crack propagation is shown. It seems that the fatigue strength decreases rapidly as the axle size increases. However, it can be noted that, as shown in Fig. 9, the ratio of the case depth to axle diameter for large diameter axles is restricted by the hardenability of the materials. It is, therefore, expected that the fatigue strength of the large diameter axle may be increased when the larger case depth, and consequently the larger residual stress, is introduced by using a higher hardenability material.

axle diameter (mm)	40	52	100	150	209
σw_1	△	□	◇	○	▽
σw_2	▲	■	◆	●	▼

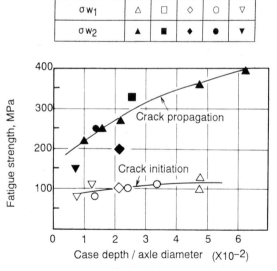

Fig 9   Effect of the case depth on the fatigue strength

axle diameter (mm)	40	52	100	150	209	Sub-critical quench
σ_{w1}	△	□	◇	○	▽	×
σ_{w2}	▲	■	◆	●	▼	+

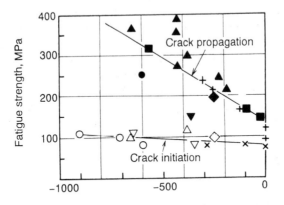

Fig 10   Effect of the residual stress on the fatigue strength

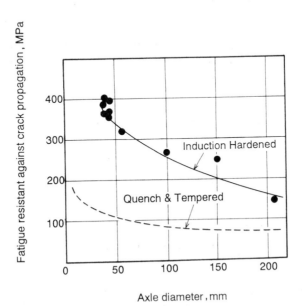

Fig 11   Size effect of the fatigue strength

Discussion

As stated above, the case depth as well as the residual stress have little influence upon the fatigue strength of crack initiation, σ_{w1}. This means that a crack may initiate from minute stress concentrators caused by localized plastic deformation on particular slip band or small wear pits created by fretting. This is because plastic deformation with permanent offset by slip is basically caused by reversed shear stresses, which are not influenced by residual static stress. At the stress levels of σ_{w1}, the crack depth is very small compared with the case depth, and hence it has little effect on the fatigue strength, σ_{w1}.

It seems that the case depth may have a large effect on the crack propagation, however. This influence is attributable to the residual stresses, because the residual stress has a tendency to increase with the increasing of case depth.

To utilize K_{res} factor in crack growth prediction, the effective stress intensities, K_{eff}, are defined as

$$K_{eff, max} = K_{max} + K_{res} \tag{5}$$

$$K_{eff, min} = K_{min} + K_{res} \tag{6}$$

where K_{min} and K_{max} are the stress intensity factors for the minimum and the maximum values of applied stress. Since $K_{eff, min}$ is less than zero, the effective stress intensity factor range becomes

$$\Delta K_{eff} = K_{max} + K_{res} \tag{7}$$

The condition for the crack not to propagate can be expressed as

$$\Delta K_{eff} \leqq \Delta K_{th} \tag{8}$$

where ΔK_{th} is the threshold stress intensity range for the crack not to propagate in the testing material. Since, σ_{w1} is the threshold stress for the crack to initiate but not to propagate

$$\Delta K_{th} = F_1 \sigma_{w1} \sqrt{(\pi a/Q)} \tag{9}$$

where F_1 is a dimensionless quantity account for the press-fitted axle which is already shown in Fig. 6.

From equations (7) and (9), equation (8) becomes

$$F_1 \sigma_{w2} \sqrt{(\pi a/Q)} + 1.08 \sigma_{R0}(1 - 0.65a/a_0)\sqrt{(\pi a/Q)} \leqq F_1 \sigma_{w1} \sqrt{(\pi a/Q)}$$

Then

$$F_1 \sigma_{w2} + 1.08 \sigma_{R0}(1 - 0.65a/a_0) \leqq F_1 \sigma_{w1} \tag{10}$$

when a/a_0 is negligibly small, equation (10) becomes

$$\sigma_{w2} = 1.08(\sigma_{R0}/F_1) \leqq \sigma_{w1} \tag{11}$$

From Fig. 6, the maximum value of F_1 is about 2.0–3.0 and, therefore, it can be seen that equation (11) expresses the tendency of the effect of σ_{R0} on the fatigue strength against crack propagation.

Concluding remarks

Fatigue strength regarding crack propagation can be greatly increased by the induction hardening of axles; however, crack initiation cannot be prevented by compressive residual stresses. It is, therefore, important to examine the behaviour of cracks once initiated in service conditions, where axles are subjected to variable amplitude stress reversals.

Acknowledgements

This report correlates and summarizes the work which has been carried out over approximately ten years. Dr H. Komatsu and Dr K. Tanaka are recognized for their help to obtain much of the data given in this paper.

References

(1) HORGER, O. J. (1952) Influence of fretting corrosion on the fatigue strength of fitted members, *ASTM STP 144*, ASTM, Philadelphia, pp. 40–51.
(2) NISHIOKA, K. and HIRAKAWA, K. (1978) Fracture mechanics approach to the strength of wheel-sets, Proceedings of the International Wheel-Sets Congress 6, pp. 2–4.
(3) WATERHOUSE, R. B. (1981) *Fretting fatigue*, Applied Science Publishing, London.
(4) ISHIZUKA, H. and SATO, Y. (1992) Fretting fatigue strength of axles for bullet train and their maintenance, Proceedings of the International Wheel-Sets Congress 10, pp. 195–199.
(5) TEZUKA, K. and TANAKA, S. (1988) Design method and service loads of car axle at high speed, Proceedings of the International Wheel-Sets Congress 9, pp. 341–348.
(6) NISHIOKA, K. and HIRAKAWA, K. (1968) Fundamental investigation of fretting fatigue – Part 2. Fretting fatigue testing machine and some test results, *Trans. Jap. Soc. Mech. Engrs*, **12**, 180–184.
(7) NISHIOKA, K. and HIRAKAWA, K. (1968) Fundamental investigation of fretting fatigue – Part 1. On the relative slip amplitude of press-fitted axle assemblies, *Trans. Jap. Soc. Mech. Engrs*, **11**, 437–443.
(8) NEWMAN, J. C. Jr. and RAJU, I. S. (1981) An empirical stress-intensity factor equation for the surface crack, *Engng. Fracture Mech.*, **15**, 185–192.
(9) KOPSOV, I.E. (1992) Stress intensity factor solution for a semielliptical crack in an arbitrarily distributed stress field, *Int. J. Fatigue*, **6**, 399–402.

*A. Bignonnet**

Some Observations of the Effect of Shot Peening on Fretting Fatigue

REFERENCE Bignonnet, A., **Some observations of the effect of shot peening on fretting fatigue,** *Fretting Fatigue,* ESIS 18 (Edited by R. B. Waterhouse and T. C. Lindley) 1994, Mechanical Engineering Publications, London, pp. 475–482.

Introduction

Fretting fatigue in mechanical assemblies significantly reduces the fatigue life. Palliatives can be used and shot peening appears to be one of the most effective methods of reducing fretting damage, **(1)**.

The main objective of this experimental work was to establish the influence of residual stresses induced by shot peening on the fretting fatigue of an alloyed steel. The experimental programme was devised to separate the effect of residual stresses and roughness. The influence on the results of using different fretting test rigs was also checked using two different systems.

Material

The material studied was a nickel-chromium steel, 35NCD16 according to the AFNOR standard, which had been quenched and tempered for 40HRC. The chemical composition and the mechanical characteristics of the steel are given in Tables 1 and 2.

Table 1 Chemical analysis of the 35NCD16 steel (%wt)

C	Si	Mn	Ni	Cr	Mo
0.39	0.33	0.40	3.7	1.7	0.31

Table 2 Mechanical characteristics of the 35NCD16 steel

UTS (MPa)	YS (MPa)	El (%)	RA (%)
1230	1095	14	55

* PSA–DRAS, Chemin de la Malmaison, 91570 Bièvres, France.

Specimens and testing devices
Specimens

Two types of flat specimens were chosen for repeated tensile fatigue tests, with and without fretting, with an R ratio $= 0.1$.

The gauge part of specimens is:

- 4 mm wide, 3 mm thick for the flexible beam fretting device, Fig. 1;
- 20 mm wide, 5 mm thick for the dynamometric ring fretting device, Fig. 2.

For each set of specimens three surface conditions were prepared.

(a) The first surface condition involved quenching and tempering at 560°C, followed by fine grinding.
(b) Shot peening formed the second surface condition. The almen intensity was 0.30–0.35 mm A [12–14 A for US standard], the shot was S170–52HRC, and coverage was 125 percent.
  The surface residual stresses were around − 600 MPa, decreasing to zero at a depth of 0.35 mm.
(c) The third surface condition was prepared by shot peening followed by stress relieving at 510°C for 1h. This treatment below the tempering temperature did not change the bulk properties. The remaining residual stresses at the surface were − 130 MPa.

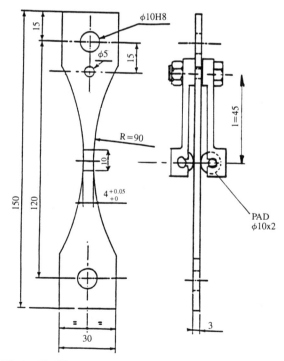

Fig 1   Flexible fretting device and corresponding test specimen

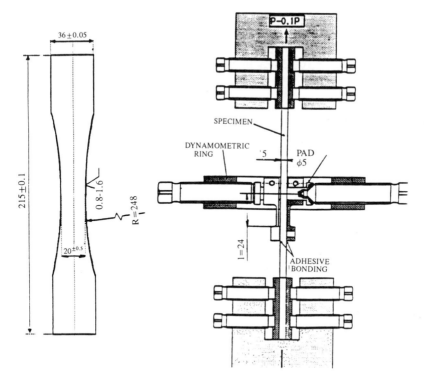

Fig 2   Dynamometric fretting device and corresponding test specimen

Testing devices

Two testing devices were used with fretting pads in 35NCD16 steel treated for 55HRC.

(a) The first device was a cylinder to plane contact using a 10 mm diameter, 2 mm wide cylinder. The normal force was applied by a flexible beam clamped to one specimen head, as shown in Fig. 1. The normal force was 170 N and the theoretical contact area was 0.280 mm^2.

(b) The second device was a plane to plane contact using a fretting pad of 5 mm in diameter. The normal force was applied with a dynamometric ring, as shown in Fig. 2. The normal pressure was 30 MPa.

In both of the above cases the displacement of the fretting area below the pad was proportional to the tensile forces

$$\delta = F/2Et \cdot \int_0^l \mathrm{d}x/w(x)$$

where δ is displacement, F is applied tensile force, t is thickness, w is width, E is Young's modulus, and l is free length of the fretting device. For the flexible beam device $1 = 45$ mm, and for the dynamometric ring device $1 = 24$ mm. Under these geometric conditions the same theoretical 'fretting displacement' was obtained for a given tensile stress in the gauge part of the specimen; typically a stress of 100 MPa in the gauge part is associated to a 'fretting displacement' of 11 μm.

Testing programme and parametric study

Wöhler curves

For both types of specimens the Wöhler curves were established with and without fretting for the following conditions:
– ground;
– shot peened;
– shot peened and stress relieved.

Testing with the dynamometric ring

Using the 20 mm wide specimens the fatigue behaviour of shot peened and ground specimens was characterized. Estimated fatigue limits ΔS_D, were, respectively, 800 MPa and 675 MPa.

When fretting was applied with a test frequency of 15 Hz for ground specimens the fatigue limit was lowered to $\Delta S_D = 370$ MPa, while for shot peened specimens the fatigue limit remained around $\Delta S_D = 750$ MPa. The results are presented in Fig. 3 for ground specimens and in Fig. 4 for shot peened specimens.

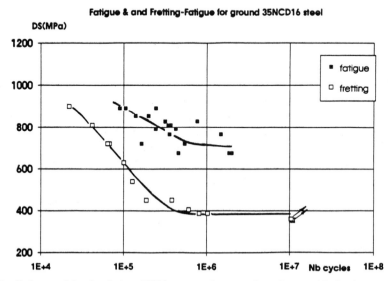

Fig 3 **Fatigue and fretting fatigue Wöhler curves for ground specimens with the dynamometric device**

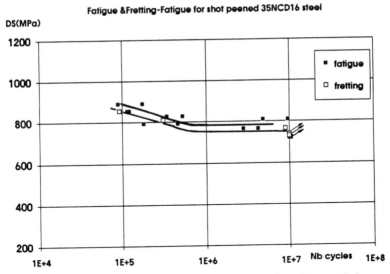

Fig 4 Fatigue and fretting fatigue Wöhler curves for shot peened specimens with dynamometric device

Testing with the flexible beam

For the testing conditions at a frequency of 35 Hz, the fatigue behaviour of shot peened and ground test pieces is very similar. Estimated fatigue limits, ΔS_D, are respectively 930 MPa and 900 MPa. These fatigue limits will be confirmed with more tests.

When fretting was applied the fatigue limit of ground specimens was lowered by a factor of about three, ΔS_D 300 MPa. In the case of shot peened specimens the fretting conditions applied had little effect on fatigue, and the fatigue limit remained above 700 MPa. When shot peening residual stresses were relieved, the fretting fatigue behaviour was similar to that observed with ground specimens.

The results are presented in Fig. 5 for ground specimens and in Fig. 6 for shot peened specimens.

Interrupted tests

Two types of interrupted tests were undertaken to evaluate the damage evolution: one to follow the development of cracks and one to follow the residual stresses below the contact. This latter part was not completed.

With fretting conditions, two stress levels were chosen for interrupted tests: one corresponding to a fatigue life of 10^5 cycles, the other one to 10^6 cycles. At each level five specimens were tested for destructive observations after, respectively, 10^2, 10^3, 10^4, 10^5, 10^6 cycles.

A first set of observations was available for fretting fatigue with the flexible

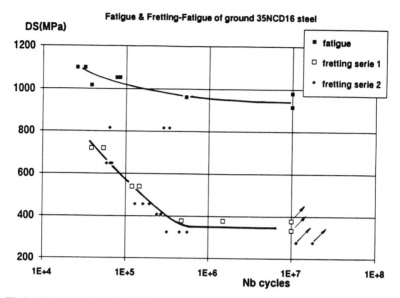

Fig 5   Fatigue and fretting Wöhler curves for ground specimens with flexible beam device

beam device on shot peened specimens. The two levels were $\Delta S = 920$ and 720 MPa. At the level 920 MPa, cracks and oxide debris were present on the contact surface, even after 100 cycles. At 720 MPa oxide debris was present in the contact after 100 cycles, but with a scanning electron microscope, no surface cracking was observed at this stage. Surface cracks were observable

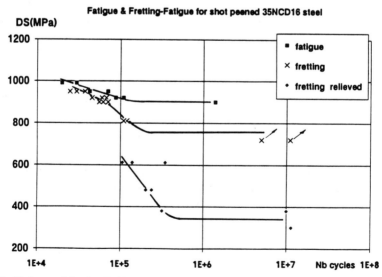

Fig 6   Fatigue and fretting Wöhler curves for shot peened specimens with the flexible beam device

after 1000 cycles. As often observed (see for example, (2) and (3), fretting fatigue cracking under the form of small cracks at the surface occurs very quickly in the present experiments. Fracture surfaces show that for the same fatigue life, a shot peened specimen presents more crack initiation sites than a ground one.

The observation of the depth of cracks at different fatigue lives and the evolution of residual stresses is in progress.

Discussion

Some discrepancy was observed in fatigue results between the two sets of specimens used. The smaller specimens gave the higher fatigue limit, but no clear reason for such a difference was found.

The flexible beam system showed some limitations in the long term experiments. The contact area often promotes significant wear for this case which leads to geometric perturbation and stress concentration in the specimens. Test results show that fretting fatigue promotes the early nucleation of surface cracks created below the contact. In agreement with the work detailed in reference (4), the fretting fatigue behaviour depends on the ability of such cracks to develop up to a critical size, where the stress intensity factor is high enough to grow the crack to failure. The results clearly show that for the present test conditions the shot peening is effective in preventing 'fretting cracks' growth before they reach the critical size. As mentioned in reference (5), the success of shot peening used as a palliative to improve fretting fatigue, can be attributed to the compressive residual stresses which are able to block small surface cracks.

The improving factor introduced by Chivers and Gordelier (5) is

$$I_f = \sigma_p - \sigma_f/\sigma - \sigma_f$$

where σ is the run-out stress in the absence of fretting, σ_f is the run-out stress with fretting and no palliative, and σ_p the run-out stress with fretting and a palliative treatment. Using this, the effectiveness of the peening treatment in the present test conditions can be quoted between $I_f = 0.63$ and 0.88, which is quite high in comparison to the value, $I_f = 0.45$, found in (5) for similar peening conditions.

The experiments on shot peened and stress relieved specimens tend to prove that the geometric aspect, i.e., the roughness of the surface, is of secondary importance in fretting fatigue resistance.

Conclusions

(1) Fretting fatigue promotes the nucleation of surface cracks below the contact under the influence of important shear stresses; the fatigue limit is then decreased by a factor of three.

(2) The fretting fatigue behaviour depends on the ability of such cracks to develop up to a 'critical' size where the stresses are high enough to grow the crack to failure; it is, therefore, a question of mechanics of structures.

(3) Shot peening, used as a palliative to improve fretting fatigue, is effective due to the compressive residual stresses which stop the surface fretting cracks before they reach a critical size.

Acknowledgements

This work was performed at the Société Fraçnaise de Métallurgie et des Maté riaux with the active participation of C. Bleuzen and T. Groisard (CEAT), P. Merrien and F. Convert (CETIM), J. Foulquier (Aérospatiale), J. C. Le Flour (Renault), M. Pierantoni (Eurocopter), C. Hunter and B. Fournel (IRSID), Y. Le Guernic (MIC), M. Barrallier (ENSAM), and L. Vincent (ECL).

References

(1) WATERHOUSE, R. B. (1990) The effects of surface treatments on the fretting wear of an aluminium alloy (RR58)/Steel (BS 970 080 M40) couple, *Surface engineering*, (Edited by S. A. Meguid), Elsevier Applied Science, pp. 325–334.
(2) FOULQUIER, J. and PETIOT, C. (1993) 'Fretting fatigue behaviour of major helicopter alloys and influence of surface protections, Proceedings of ICAF' 93.
(3) BLEUZEN, C. (1988) Evolution de la susceptibilité au fretting d'alliages aéronautiques, *Conférence 'Le fretting-corrosion'*, Cetim, France.
(4) FOULQUIER, J., JOURNET, B., ALIAGA, D., and PINEAU, A. (1988) Comportement en fretting fatigue d'un acier 3ONCD16 utilisé sur hélicoptère, *Conférence 'Le fretting-corrosion'*, Cetim, France.
(5) CHIVERS, T. C. and GORDELIER, S. C. (1984) Fretting fatigue palliatives: some comparative experiments, *Wear*, **96**, 153–175.

*G. N. Filimonov**

Improving the Fretting Fatigue Resistance of Large Shafts with the use of Surface Cold Working

REFERENCE Filimonov, G. N., **Improving the fretting fatigue resistance of large shafts with the use of surface cold working,** *Fretting Fatigue,* ESIS 18 (Edited by R. B. Waterhouse and C. Lindley) 1994, Mechanical Engineering Publications, London, pp. 483–495.

Introduction

Fretting fatigue processes arise in the contact areas of joined components, if at least one of the components is being cyclically stressed **(1)**. The unfavourable effect of fretting on the fatigue strength is intensified if the tensile strength of the component material increases and also if the components are of larger sizes **(2)–(4)**. A large shaft (or an axle) made from high-strength steel may fail by fatigue in the region of a pressed or shrink-fitted joint (propeller hub, wheel, gear, sleeve, roller bearing) at a nominal stress, which is 5–10 times lower than the endurance limit of a plain shaft of the same size. So the nominal stress in the failure section may account for only 1/20 of the steel's ultimate tensile strength **(5)**.

The beneficial effect of surface cold working on the fatigue behaviour of steel components is well known **(6)–(8)**. An especially high increase in the fatigue strength after cold working can be obtained for components which have stress concentrators or fitted parts **(2)(3)**. This is attributed to the production of surface compressive residual stresses by cold working and to the fact that the residual stresses can be concentrated in the concentrator regions.

Fretting fatigue is an extraordinarily many-sided problem. In this paper an attempt has been made to illustrate only three aspects of this complex problem, based on the experimental results obtained by prolonged tests (though from the standpoint of experimental procedure they are very simple and understandable).

(1) The first part of the paper deals with the extent of fatigue strength reduction under the action of fretting and the possibility of its full recovery (as applied to large shafts with fitted details) by rolling due to the production of compressive stresses in the shaft surface layer.

(2) In the second part, the strengthening effect stability is verified by long-term cyclic loading in air (a widely used medium for components of large machines).

* CRISM, Prometey, St Petersburg, Russia.

(3) In the third part of this paper, an estimation of the joint effect of sea water
    and fretting on steel shaft fatigue resistance is made as well as the efficiency
    of surface cold working by rolling as a means of increasing fatigue resist-
    ance under conditions of fretting in sea water over the time periods, com-
    parable with the operation time of real components (for example, ship's
    shafts).

Experimental procedure

To obtain the results described below, simple testing machines were used to
produce rotating bending fatigue. Loading was by four-point bending.

To provide fretting fatigue conditions, combined specimens (instead of the
usual plain specimens) were applied, Fig. 1. Firstly, a simple plain specimen
with two heads for fastening in the grips of the testing machine was made. This
was then cut into two parts and subsequently jointed with a massive conjoin-
ing sleeve. The relative sizes of the sleeve were selected so as to be similar to
the sizes of a propeller boss, fitted on a ship's shaft. After inserting both specimen halves in the heated sleeve and after subsequent sleeve cooling, this combined specimen was tested in fretting fatigue.

The interference fit (the difference in diameters of sleeve and specimen) was
equal to 0.20–0.23 mm for the 200 mm shaft diameter, and to 0.04–0.05 mm

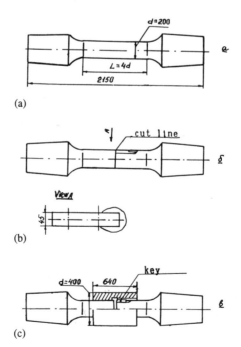

**Fig 1   The sequence of combined specimens production for the fretting fatigue test: (a) plain speci-
men; (b) cut operation; (c) combined specimen**

for the 50 mm shaft diameter. In some instances (when the stress level was more than $\sigma_a = 320$ MN/m^2) the interference fit was increased up to 0.40–0.50 mm for the 200 mm shafts and to 0.07 mm for the 50 mm shafts.

To investigate the efficiency of surface strengthening, the rolling was performed before specimen cutting. In order to study the effect of key fastening in the joint on fretting fatigue properties, keyways were milled in one half of a specimen and/or sleeve. On shafts strengthened by rolling, keyways were milled after rolling.

The large fretting fatigue machines are shown in Fig. 2 (UFMI–200) and Fig. 3 (P–400). The machines make it possible to perform tests of combined specimens with net section diameter of 200 and 400 mm, respectively. Some testing machines of smaller power were used for testing combined specimens with a net section diameter of 5–70 mm.

The results presented below were obtained in the framework of a general research programme in the laboratories of St Petersburg, Odessa, Lviv.

Results

Figure 4 gives fatigue curves in terms of 'stress range amplitude–number of cycles to failure' for combined specimens, made from four different steels:

(a) 0.35 C plain carbon steel in the normalized condition with tempering ($R_{p0.2} > 220$ MN/m^2)
(b) low alloy chromium–nickel steel with molybdenum in the annealed condition ($R_{p0.2} > 320$ MN/m^2)
(c) the same steel hardened and tempered ($R_{p0.2} > 600$ MN/m^2)
(d) high-strength steel with 3 percent nickel, molybdenum and vanadium in the hardened and tempered condition ($R_{p0.2} > 850$ MN/m^2)

The chemical composition of the materials investigated is given in Table 1, the mechanical properties in Table 2. Sleeve couplings were produced from type 45 steel for shafts with $R_{p0.2} > 500$ MN/m^2 and from type 20 Mn steel for shafts with $R_{p0.2} < 400$ MN/m^2 (both in the normalized condition).

Each solid point in Fig. 4 is the result of a fatigue test of a specimen with net section diameter 200 or 180 mm manufactured from an individual forging. The symbol $\bigcirc$ denotes test results of plain specimens in the absence of fretting.

Table 1   Chemical composition of investigated materials

Steel		C	Si	Mn	S	P	Cr	Ni	Mo	V
35	Min	0.32	0.17	0.50	–	–	–	–	–	–
	Max	0.40	0.37	0.80	0.040	0.035	0.25	0.25	–	–
38Cr2Ni2Mo–A	Min	0.33	0.17	0.25	–	–	1.30	1.30	0.20	–
	Max	0.40	0.37	0.50	0.025	0.025	1.70	1.70	0.30	0.05
38CrNi3MoV–A	Min	0.33	0.17	0.25	–	–	1.20	3.00	0.35	0.10
	Max	0.40	0.37	0.50	0.025	0.025	1.50	3.50	0.45	0.18

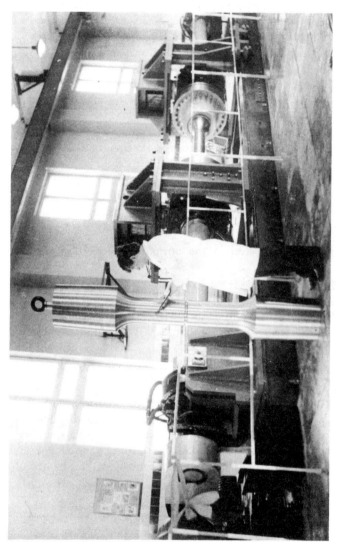

Fig 2   Fatigue machine UFMI–200 for test shafts with net diameter up to 200 mm

Fig 3 Fatigue machine P–400 for test shafts with net diameter up to 400 mm

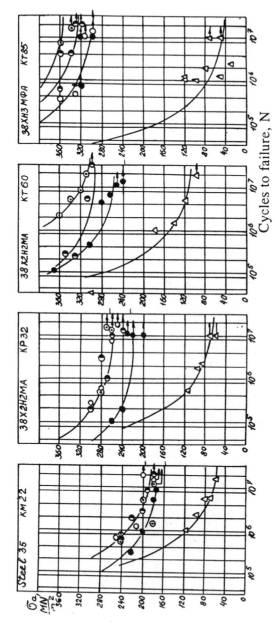

Fig 4   Fatigue test results of shafts with net section diameter 200 mm (steel grade and guaranteed yield strength are given). △ – shafts under fretting fatigue condition without surface strengthening; ○ – plain shafts without surface strengthening; ● – shafts under fretting fatigue condition strengthened by surface rolling with $P = 22$ kN; ⊙ – the same, but the roll loading is increased; ◑ – the same, but there is a key way in the joint

FRETTING RESISTANCE OF LARGE SHAFTS

Table 2 Mechanical properties of investigated materials

Steel	Diameter of specimen	Heat treatment	MN/m^2	$R_{p0.2}$ MN/m^2	$R\,m$ MN/m^2	A %	Z %	KCU J/cm^2
35	200	Normalization	Min	210	530	20	30	30
		with tempering	Max	260	570	30	42	50
35	50	Hardening	Min	350	610	22	54	77
		with tempering	Max	360	620	26	57	125
38Cr2Ni2Mo–A	200	Annealing	Min	320	720	15	25	25
			Max	350	750	25	35	40
38Cr2Ni2Mo–A	200	Hardening	Min	580	790	20	52	130
		with tempering	Max	640	850	23	66	180
38CrNi3MoV–A	200	Hardening	Min	850	1000	13	50	120
	180	with tempering	Max	950	1090	18	64	140

They provide a reference for estimating the effect of fretting and for comparing the efficiency of the applied technological regimes of rolling.

As expected, the plain fatigue strength increases approximately in proportion to tensile strength (or hardness) of steel. So for the type 35 steel, the endurance limit of large-scale specimens is equal to approximately 200 MN/m^2 ($\sim 0.36\ R_m$), and for high-strength steel 320 MN/m^2 ($\sim 0.32\ R_m$).

The test results of combined unstrengthened specimens (denoted by Δ) show that fretting sharply decreases fatigue strength. Due to contact interaction with a fitted sleeve, fatigue strength of large shafts decreases several times over. The higher the shaft steel properties and the larger the shaft diameter, the greater is the fatigue strength reduction.

In the case of very large specimens with sleeves of sufficient rigidity and by prolonged testing under the conditions of joint effect of cyclic loading and fretting during the whole period of testing, the endurance limit decreases approximately to the same level (60–80 MN/m^2), independent of material tensile strength. Moreover, there is a tendency to obtain a lower fretting fatigue limit for shafts made from high-strength steel. This observation is in good agreement with Horger's classic experimental data (4). It should be noted, that such tendencies are typical also for specimens containing very sharp circular notches or cracks.

For mild steel the fatigue strength reduction factor due to fretting was equal to three, and for high-strength steel more than five. It is well known that fretting fatigue strength of shafts joined with fitted sleeves may be significantly increased if a cold worked layer with compressive residual stresses is introduced in the surface, where tangential friction stresses are acting over the contact region. In our case strengthening was produced with shaft cold rolling on a hydraulic two-roll rig. The surface of the shafts before rolling was prepared with fine turning according to the regimes corresponding to turning treatment of actual propeller shafts. Shafts of 200 mm diameter were rolled with hardening rolls of 150 mm diameter, and shafts of 50 mm diameter with 110 mm diameter rolls. In both cases the profile radius was equal to 15 mm. No additional surface treatment was carried out after rolling.

The test results show that after surface plastic deformation of shafts according to optimal technological regimes, fretting fatigue strength of combined specimens made from all the investigated steels is increased to the values which are typical of the fatigue strength for plain specimens without fretting, and even may exceed this level.

In Fig. 4 ● shows test results of combined specimens with shafts made from all investigated steels and rolled by a force of 22 kN, irrespective of steel strength level. It is evident that for all steels such a regime of surface strengthening gave good, but not optimal, results: the fatigue strength did not reach the level which is typical for plain shafts from the corresponding steel.

The calculated value of a hardened layer depth (according to the formula of Heifets $a = \sqrt{(P/2R_{p0.2})}$ decreases if the proof stress of the steel is increased but the rolling force kept constant. With the force $P = 22$ kN the calculated value was equal to approximately 7 mm for shafts from type 35 steel; 5.8 mm for annealed and 4.3 mm for hardened shafts from 38Cr2Ni2Mo–A steel; 3.6 mm for shafts from 38CrNi3MoV–A steel. It should be noted that the depth actually measured greatly depends on the selected measurement procedure. (Incidentally, in practice the hardness of the high-strength 38CrNi3MoV–A steel was not increased after all the rolling regimes.)

In Fig. 4 the symbols ⊙ denote the results of combined specimens after rolling with higher forces $P = 25$, 40, 42, and 35 kN for types of 35, 38Cr2Ni2Mo–A KP32, 38Cr2Ni2Mo–A KT60, and 38CrNi3MoV–A steels, respectively. The fretting fatigue strength of keyless shafts in this case was never less than the plain fatigue strength.

Surface strengthening by rolling also appeared to be effective for keyed shafts, though keyways were milled after rolling (these test results are denoted in Fig. 4 by ◖). The rolling forces were $P = 25$, 40, 50, and 60 kN for types of 35, 38Cr2Ni2Mo–A KP32, 38Cr2Ni2Mo–A KT60, and 38CrNi3MoV–A steels, respectively.

The performed investigations permit the conclusion that, by optimizing the surface cold working, it is possible to eliminate completely the unfavourable effect of fretting on the fatigue resistance of shafts with fitted parts. The degree of increase in fretting fatigue strength as a result of surface cold working is comparable with the extent of fatigue strength reduction of the corresponding steel under the fretting action.

The test results, obtained on specimens of other sizes, confirm the considerations discussed above (Fig 5).

Some investigators, in their researches of plain specimens and specimens with circular notches, have demonstrated that the strengthening effect of the application of surface cold working is maintained and does not vary significantly after prolonged cyclic loading at stress levels lower than the endurance limit of plain unstrengthened specimens. Kudrjavtsev showed that specimens strengthened by rolling had the same fatigue curve after ten years 'keeping' as specimens tested directly after rolling (9).

There are not very many investigations confirming the stability of the strengthening effect under conditions of fretting fatigue, when local fracture of the surface layer may take place. Shkoljnik tested carriage axle models made from medium carbon steel, with net section diameter of 50 mm. He found a two-fold increase in the endurance limit due to surface cold working, and the strengthening effect was maintained up to 500–600 million loading cycles (7).

Figure 6 gives test results of plain (Fig 6(a)) and combined (Fig 6(b)) specimens made from 0.35 percent C carbon steel with a 50 mm net section diameter. The specimens strengthened by rolling did not fail at a loading level of $\sigma_a = \pm 270$ MN/m^2 during fretting fatigue resting of 1.2×10^9 cycles, but without strengthening the failure of similar specimens took place at a loading level of $\sigma_a = \pm 110$ MN/m^2 after only 20 million cycles. It should be noted

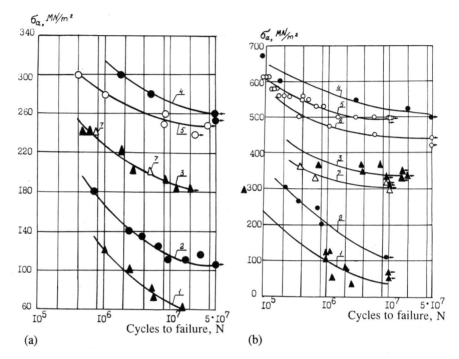

Fig 5 Fatigue test results of shafts from mild carbon steel (a) and high-strength steel (b). 1 – shafts with diameter 200 mm without surface strengthening under fretting conditions; 2 – the same, but shaft diameter – 50 mm; 3 – shafts with diameter 200 mm strengthened by rolling under fretting condition; 4 – the same, but shaft diameter – 50 mm; 5 – plain shafts without surface strengthening, shaft diameter 50 mm by carbon steel, but 8 mm by high-strength steel; 6 – the same, but shaft diameter 40 mm; 7 – the same, but shaft diameter 200 mm.

that the endurance limit of plain specimens of 50 mm diameter was equal to $\sigma_a = \pm 240$ MN/m² before rolling and to $\sigma_a = \pm 320$ MN/m² after rolling. The material yield strength was equal to $R_{p0.2} = 350$ MN/m². Thus surface cold working may be used as an effective way to increase fretting fatigue resistance and the degree of increase does not depend on the duration of cyclic loading if components are operating in air.

Other processes include operating in corrosive environments such as sea water. It is well known that by cyclic loading in sea water, carbon and low alloy steels do not exhibit an endurance limit. This is confirmed by test results in Fig. 6 (curve I). Combined unstrengthened specimens demonstrate a comparatively small additional decrease of fatigue characteristics (curve III). The analysis of the orientation and location of generated cracks suggests that in this case an accelerated failure appears to be not the result of fretting fatigue processes intensified by the presence of a corrosive medium, but rather the result of corrosion fatigue processes, intensified by simultaneous fretting action (rubbing of a protective surface film in the fretting zone, coinciding spatially

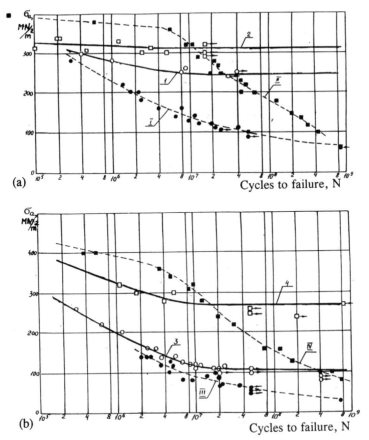

Fig 6 Prolonged cyclic loading effect on surface strengthening efficiency for (a) plain shafts of carbon steel with net section diameter 50 mm; and (b) shafts operated under the conditions of fretting-fatigue (keyless joints). (1) plain shafts with net diameter 50 mm without surface strengthening, tests in air; (I) the same, but tests in sea water; (2) plain shafts strengthened by rolling tests in air; (II) the same, but tests in sea water; (3) combined shafts with net diameter 50 mm without surface strengthening, tests in air; (III) the same, but tests in sea water; (4) combined shafts strengthened by rolling, tests in air; (IV) the same, but tests in sea water.

with the crevice gap zone between shaft and sleeve; additional stress concentration in the surface layer owing to contact friction; intensification of electrochemical corrosion processes such as crevice effect).

At the contact surfaces of combined specimens tested in sea water there are no inclined cracks typical of fretting fatigue. One of the possible reasons for this appears to be the friction force reduction on *contact grounds* because sea water acts as a lubricant and a dividing layer between surfaces.

After strengthening by rolling, the fatigue curves in sea water for plain and combined specimens are not very different (curves II and IV in Fig. 6). The strengthening is effective for test durations of up to 10–20 million loading cycles. Up to this number of cycles both plain and combined specimens demonstrate a higher endurance in sea water than in air.

Experiments in sea water for an insufficient number of cycles have caused some investigators to draw incorrect conclusions about the possibility of entirely eliminating the detrimental effects of corrosion and fretting fatigue with surface cold working. But as can be seen from Fig. 6, by the decrease of the loading level and corresponding increase of number of cycles, the specimens, strengthened by rolling, 'lose' the physical endurance limit and the fatigue curve decreases, reaching 400–600 million cycles. The rate at which fatigue strength decreases as the number of loading cycles increases in sea water for strengthened specimens (both plain and combined) is higher than that for unstrengthened specimens. Curves with and without strengthening approach each other as regards the increase in the number of cycles. Plain specimens in our experiments fractured at the load level $\sigma_a = \pm 100$ MN/m^2 after 400 million cycles and combined specimens at the load level $\sigma_a = \pm 80$ MN/m^2 after 800 million loading cycles in sea water.

Nevertheless, surface cold working may be considered for cyclic loading in corrosive media as an effective means of extending service life: thus, at the nominal level of loading $\sigma_a = \pm 100$ MN/m^2 the lifetime of plain specimens in sea water was increased eightfold by rolling, and the lifetime of combined specimens increased twentyfold. With higher operation loading levels the effect of strengthening is to increase service life (at the load level $\sigma_a = \pm 200$ Mn/m^2 a twentyfold increase in life is shown for plain specimens and a fortyfold increase for combined specimens).

Conclusions

From the information presented the following conclusions can be made.

(1) The fatigue strength of large steel shafts sharply decreases under the action of fretting. The higher the value of the tensile strength the larger the extent of this decrease. Hence the possible use of materials of higher strength is no solution for fretting fatigue.

(2) Using surface cold working (for example, surface rolling) may increase the fretting fatigue strength of large shafts up to a level, which is typical of the plain unrolled shafts from the same material and of the same size. In this way the potential possibilities of high-strength steels may be wholly realized.

(3) In the case of air environment the effectiveness of cold working as regards fatigue strengthening is sufficiently high, even if the number of cycles reaches 10 billion. This refers both to plain and combined specimens, operating under conditions of fretting fatigue.

(4) In the case where steel shafts operate in sea water, the strengthening effect of surface cold working continuously decreases as the number of cycles

increases. The rate of decrease is approximately identical, regardless of whether a plain or a combined shaft is operating in sea water.

(5) The life to failure of cold rolled shafts in sea water is many times longer than that of unstrengthened shafts at all load levels.

(6) If the number of loading cycles does not exceed 10 million, then surface cold worked shafts (plain and combined) have, in sea water, a higher fatigue strength than in air, and this fatigue strength may be higher (by an optimal regime of cold working) than the endurance limit of plain specimens without strengthening.

(7) By increasing the number of cycles up to 10 billion, the fatigue strength of surface cold worked steel shafts in sea water reduces to the fatigue strength level of unstrengthened shafts at a loading endurance of approximately 20–50 millions of cycles.

References

(1) WATERHOUSE, R. B. (Editor) (1981) *Fretting fatigue*, Applied Science, London.
(2) KUDRJAVTSEV J. V., NAUMTCHENKOV, N. B., and SSAVVINA, N. M. (1971) *Fatigue of large machine components*, Mashinostrojenije, Moscow.
(3) FILIMONOV G. N., BALATSKIJ, L. T. (1973) *Fretting in ship machinery joints*, Sudostrojenije, Leningrad.
(4) HORGER O. J. (1956) Fatigue of large shafts by fretting corrosion, *Proceedings of International Conference on Fatigue of Metals*, Mechanical Engineering Publications, London.
(5) HEYWOOD R. B. (1962) *Designing against fatigue*, Chapman and Hall, London.
(6) KUDRJAVTSEV J. V. (1951) *Internal stresses as a strength reserve in machine building*, Mashgiz, Moscow.
(7) SHKOLNIK, L. M. and SHAKHOV, V. I. (1964) *Surface rolling technology and rigs for strengthening and finishing of components*, Mashinostrojenije, Moscow.
(8) BRASLAVSKIJ, V. M. (1975) *Rolling technology of large components*, Mashinostrojenije, Moscow.
(9) KUDRJAVTSEV, J. V. and SSAVVINA, N. M. (1965) Influence of ten-year keeping on fatigue strength of components with residual stresses, *Improvement of fatigue life of machine components by surface cold working*, Mashinostrojenije, Moscow.

C. Petiot*, J. Foulquier†, B. Journet†, and L. Vincent*

Fretting Fatigue Behaviour of a Chromium, Vanadium, Molybdenum Steel Influence of Surface Protection

REFERENCE Petiot, C., Foulquier, J., Journet, B., and Vincent, L., **Fretting fatigue behaviour of a chromium, vanadium, molybdenum steel – influence of surface protection**, *Fretting Fatigue*, ESIS 18 (Edited by R. B. Waterhouse and T. C. Lindley) 1994, Mechanical Engineering Publications, London, pp. 497–512.

ABSTRACT The fretting fatigue behaviour of a 3Cr–1MoV (French AFNOR standard 32CDV13) steel in contact with a roll bearing steel is determined using fretting fatigue maps and S–N curves. This approach is used to evaluate and understand the influence of two currently used surface treatment techniques: shot-peening and cadmium coating.

Notation

Fc	Clamping force (N)
Ft	Frictional force (tangential to the contact area) (kg)
L	Fatigue load linked to S_{max} (kN)
N	Number of cycles
P_0	Maximum Hertz pressure (MPa)
R	Load ratio $= L_{min}/L_{max}$
Ra	Surface roughness (μm)
S	Oscillatory stress in the specimen (MPa)
S_{max}	Maximum stress in the specimen (MPa)
RCFM	Running Condition Fretting Map
MRFM	Material Response Fretting Map
CI	Crack Initiation
PD	Particle Detachment

Introduction

Surface damage induced by small amplitude oscillatory displacements between metal contacting components is termed fretting. This phenomenon often arises from cyclic stressing of one or both components and is then known as fretting fatigue. According to the contact geometry and the in-service loading, fretting

* Ecole Centrale de Lyon, Department of Materials, Mechanical, and Physical Engineering, URA CNRS 447, 36 avenue Guy de Collongue, BP163, 69131 Ecully Cedex, France.

† Aerospatiale, Louis Blériot Joint Research Centre, Materials Department, 12 rue Pasteur, BP76, 92152 Suresnes Cedex, France.

fatigue can result in either material system clamping or material wear and free motion or early fatigue crack initiation and propagation until failure, thereby drastically reducing the service life of the components.

During the past ten years, many studies have focussed on the link between operating parameters and the damage mechanisms under fretting wear conditions (clamping force, imposed displacement without cyclic stressing). A methodology was established by Vingsbo et al. (1) and later by Vincent et al. (2)(3) based on fretting maps (test parameters diagrams). It is possible to distinguish the Running Condition Fretting Map (RCFM), that is the accommodation of displacements (stick/gross slip/partial slip) and the Material Response Fretting Map (MRFM), i.e., the damage mechanism in the material (no damage/ particle detachment/crack initiation).

In helicopters, because of the large number of components in contact and the large number of fatigue cycles, fretting fatigue resistance of materials is a major concern for designers. In order to minimize the effect of fretting on the service life of components, surface treatment techniques are currently applied to modify the surface properties of materials and allow promising development. However, designers still need data and guidelines to determine the best surface protection for a given application. For this purpose, fretting fatigue maps must be established.

Objectives

The Aerospatiale Louis Blériot research centre, in cooperation with Eurocopter France (Aerospatiale helicopter division) and Ecole Centrale de Lyon, is currently carrying out research work on the fretting fatigue behaviour of a 3 Cr–1MoV steel (French AFNOR standard 32CDV13 steel) and the influence of surface protections. This chromium, vanadium, molybdenum steel is used for manufacturing rotating parts of helicopters, such as highly loaded gears, which very often encounter fretting fatigue problems.

The objective of the present paper is to first set up a methodology to assess the would-be operating damage mechanisms with respect to service parameters and, second to assess the benefits brought up by surface protections through this methodology. The long-term goal of this study is the establishment of a fretting guide for designers (data and guidelines for surface protection choice) via the determination of criteria for predicting the different damage.

The study proceeds, as follows:

(1) Unprotected 3Cr–1MoV steel is focused on fretting fatigue conditions simulated in the laboratory and a reliable testing procedure and tools are defined for evaluating fretting fatigue behaviour of materials. Fretting fatigue regimes are identified, that is, the accommodation of displacements and damage mechanisms with the operating parameters (similar to the fretting map approach established under fretting wear conditions).

(2) This procedure is used to determine the role of different surface treatment techniques and understand their response to operating parameters (compared to unprotected steel).

Material

3Cr–1MoV steel was supplied by Aubert and Duval as round bar ($\phi = 100$ mm). Its chemical composition is given in Table 1. It was quenched (950°C) and tempered (635°C) to the mechanical properties given in Table 2.

Experimental technique

A specific set-up is used to simulate fretting fatigue conditions in the laboratory. Two cylindrical fretting pads (diameter 10 mm) are clamped against both surfaces of a flat uniaxial fatigue specimen tested under constant amplitude loading ($R = 0.1$) at a frequency of 20 Hz (Fig. 1 and 2).

The contact geometry is a cylinder/plane type. Therefore, the pressure distribution is given by the Hertz theory of contact (4). The pads are made of 100C6 roll bearing steel (hardness = 62 HRC) to simulate the most currently encountered contact configuration in helicopters. Specimen and pads are machined with a surface finish with an Ra value of 0.45 and 0.1 μm respectively. The imposed oscillatory motions between the pads and the specimen are linked to the imposed oscillatory fatigue stress S in the specimen. For a maximum stress $S_{max} = 500$ MPa, the semi amplitude is 55 μm.

The flexible beams are equipped with strain gauges in order to measure the clamping force Fc between pads and specimen and the friction force Ft (tangential to the contact area) related to the displacements accommodation.

The test parameters are

– Fc clamping force: 40N < Fc < 140N
                               388 MPa < Po < 726MPa
                               (Po = maximum Hertz pressure)
– S_{max} maximum stress in the specimen.

The variations of Ft are recorded during each fatigue cycle and plotted as a function of S (friction loops).

Table 1   Chemical composition of 3Cr–1MoV steel (Aubert and Duval 32CDV13)

C	Si	Mn	S	P	Ni	Cr	Mo	V	Fe
0.333	0.28	0.55	<0.002	0.008	0.05	2.93	0.83	0.3	Compl.

Table 2   Mechanical properties of 3Cr–1MoV steel (Aerospatiale)

3Cr–1MoV	Yield strength (MPa)	Tensile strength (MPa)	Young's Modulus (GPa)	El (%)	RA (%)	Hardness (Hv 50g)
	980	1140	215	16	70	360

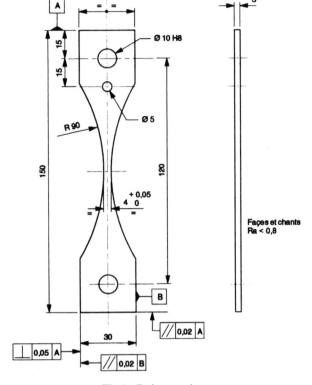

Façes et chants
Ra < 0,8

Fig 1   Fatigue specimen

Fretting fatigue behaviour of unprotected 3Cr–1MoV steel

Fretting fatigue maps

Focussing on unprotected 3Cr–1MoV steel, the fretting fatigue regimes (accommodation of displacements) and damage mechanisms are identified with operating parameters (Fc, S_{max}) by varying the clamping force for each S_{max} value investigated.

The friction loops (Ft pad deformation, S) are used for assessing the fretting fatigue regime taking place. Three regimes are evidenced by visual inspection:

- *Stick regime*   This is associated with friction loops which keep a non-evolutive, relatively closed shape, from the first fretting fatigue cycle until the end of the test (see Fig. 3).
- *Gross slip regime*   This is associated with loops which show at a trapezoidal shape all along the test, even though the Ft amplitude is strongly modified (see Fig. 4).
- *Partial slip mixed-mode regime*   This is associated with loops which show an elliptic shape, usually obtained after a few hundred cycles of slip (see Fig. 5).

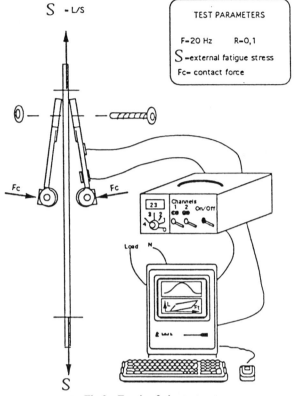

Fig 2   Fretting fatigue set-up

A RCFM can then be drawn in a $Fc-S_{max}$ diagram which gives the limits of the three fretting fatigue regimes. Such a map is similar to the RCFM plotted by Vincent *et al* **(2)(3)** under fretting wear conditions.

For each test carried out, the damage mechanism is determined via optical observations of the contact area and/or cross-sections during interrupted fretting fatigue tests to look for cracks.

It appears that each of the above-mentioned fretting fatigue regimes is related to a particular damage mechanism.

- The *stick regime* does not create any damage in the specimen before at least 10^7 cycles. This means that the combination of S and contact induced stresses (related to Ft) is not sufficient to create any damage in the specimen (crack initiation or plastic deformation), as shown on Fig. 6. In this regime, the fretting fatigue life exceeds 10^7 cycles.
- The *gross slip regime* results in *particle detachment* as usually observed by Vincent *et al* **(2)(3)** under fretting wear conditions (see Fig. 6).
- The *mixed-mode regime* results in fatigue crack initiation (see Fig. 6) after a number of cycles (less than 30 percent of the fatigue life) depending on the applied stresses (S_{max} + contact induced stresses). However, in all cases, this

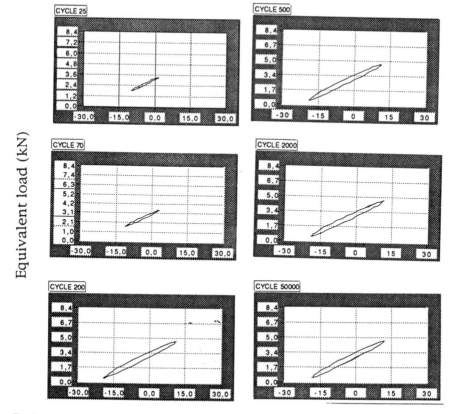

Pad deformation (μDef) proportional to tangential contact force

Fig 3 Evolution of friction loops characteristic of the stick regime with $N(S_{max} = 400MPa$ and $Fc = 100N)$

occurs before 10^7 cycles. A crack is considered a fatigue crack, if it is deeper than 10 μm.

Therefore, a MRFM can be drawn in a Fc–S_{max} diagram, giving the limits between the different damage mechanisms taking place in the specimen after a maximum number of 10^7 cycles: no damage; particle detachment; crack initiation. The MRFM (plotted after 10^7 cycles) and the RCFM are identical. They are plotted on the same diagram on Fig. 7. Such diagrams associate both the kinematics of contact (fretting fatigue regime) and the corresponding damage mechanism with the operating parameters.

S–N curves

Because of the large number of fatigue cycles which helicopter components must endure, the influence of fretting on the fatigue resistance of metals is a

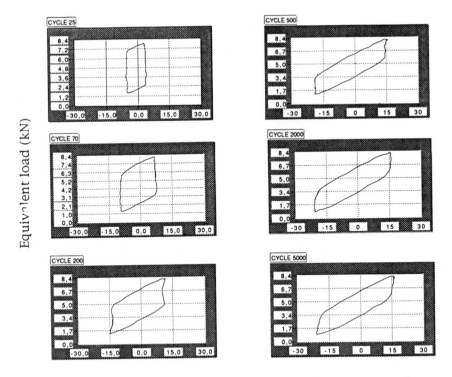

Pad deformation (μDef) proportional to tangential contact force

Fig 4 Evolution of friction loops characteristic of the gross slip regime with $N(S_{max} = 700$ MPa and $Fc = 80$N)

key criterion for designers. As far as the fatigue resistance of an alloy is concerned, the 'partial slip–crack initiation' regime is the most dangerous. In this regime, the loss of fatigue resistance of 3Cr–1MoV steel is evaluated by comparing the fatigue and fretting fatigue S–N curves (see Fig. 8).

In the crack initiation regime the contact force has no effect on the fretting fatigue lifetime for a given S_{max} value. This was evidenced on 3Cr–1MoV specimens having a lower section than the specimens described in the paper, under the same test conditions (see (**6**)). It, therefore, can be concluded, that the fretting fatigue life is related to the ability of S_{max} to let the fretting initiated crack propagate ($S_{max} >$ fretting fatigue limit) or not ($S_{max} <$ fretting fatigue limit). The fretting fatigue limit is, therefore, linked to a crack propagation threshold condition (**5**).

Also in the crack initiation regime, the reduction factor on the fatigue limit (no failure after 10^7 cycles) due to fretting is 2.8 (fatigue limit = 975 MPa, fretting fatigue limit = 350 MPa).

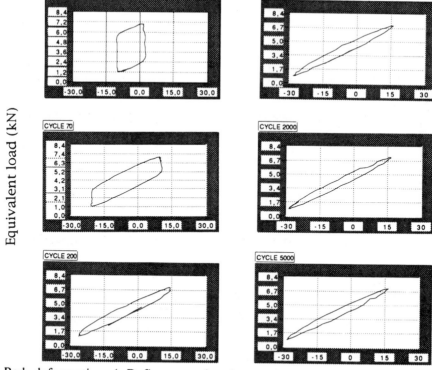

Pad deformation (μDef) proportional to tangential contact force

Fig 5 Evolution of friction loops characteristic of the partial slip regime with $N(S_{max} = 600\text{MPa}$ and $Fc = 100\text{N})$

Synthesis

Based on a reproducible laboratory simulation, fretting fatigue maps and S–N curves are complementary tools for evaluating the fretting fatigue behaviour of contacting materials and, therefore, ranking materials or testing surface pro-

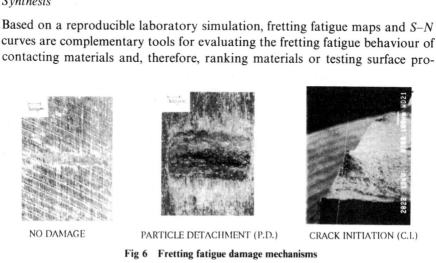

NO DAMAGE   PARTICLE DETACHMENT (P.D.)   CRACK INITIATION (C.I.)

Fig 6 Fretting fatigue damage mechanisms

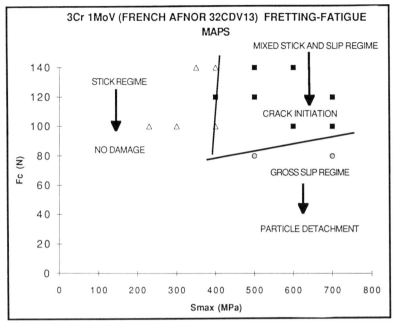

Fig 7   Comparison between RCFM and the MRFM of the unprotected 3Cr–1MoV steel

tections. The fretting maps provide information on the susceptibility of the alloy to fretting crack initiation. Then, the fretting fatigue $S–N$ curve provides an evaluation of the severity of fretting on its fatigue resistance, by determining conditions of propagation and arrest of fretting cracks. Moreover, the fretting fatigue maps form the basis of the determination of a criterion for each damage mechanism (crack initiation and particle detachment). They make a link between:

(a) the operating parameters (Fc, Ft, S_{max});
(b) the kinematics of contact (accommodation of displacements);
(c) the damage mechanisms.

(a) and (b) enable the calculation of the stresses and deformations in the specimen. (c) suggests a suitable criterion for each mechanism: crack initiation is related to overstress and particle detachment to overstrain.

Influence of surface treatment techniques

Two currently used surface treatment (ST) techniques (shot-peening and cadmium coating) are investigated using the methodology described above. The conclusions serve as reference cases for evaluation of emerging techniques. The effect of these protections on the fretting fatigue behaviour of the steel are tested by fretting fatigue maps and the fretting fatigue $S–N$ curve. The results

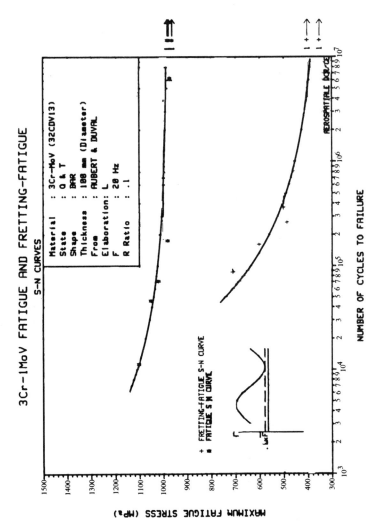

Fig 8 Comparison between fatigue and fretting fatigue *S–N* curves of a 3Cr–2MoV steel

are compared to those of the unprotected steel. The maps illustrate the effect of the ST technique on fretting crack initiation in the steel. As long as the fatigue life of the substrate is concerned, an efficient ST technique should reduce the 'partial slip–crack initiation' area. The S–N curve illustrates its influence on the propagation of fretting initiated cracks.

Shot-peening

In the case of shot-peened steel, the link between friction loop shape evolutions and fretting fatigue regimes established for unprotected steel is still valid (see Fig. 11). Therefore, the 3Cr–1MoV and the shot-peened 3Cr–1MoV RCFM are identical. The stick and gross slip regimes for shot-peened steel correspond to no damage and particle detachment, respectively, all along the test. In the partial slip regime, crack initiation in the shot-peened specimen occurs at nearly the same number of cycles as in the unprotected steel. Because of the compressive residual stresses in the specimen (see Figs 9 and 10), the propagation of fretting initiated cracks and failure of the specimen is shifted to high S_{max} values, when compared to unprotected steel.

The non-propagation of fretting cracks in the shot-peened steel ($S_{max} <$ fretting fatigue limit) is associated with an important formation of particles and wear of the surface after about 10^6 cycles. This was not evidenced on unprotected 3Cr–1MoV steel.

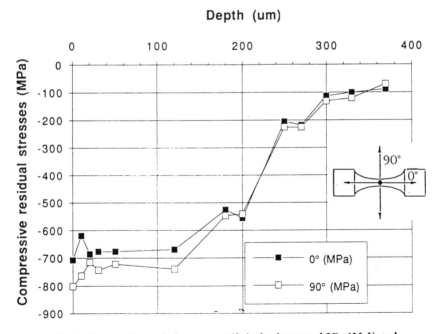

Fig 9   Compressive residual stresses profile in the shot-peened 3Cr–1MoV steel

FATIGUE RELAXATION OF COMPRESSIVE RESIDUAL STRESSES

IN SHOT-PEENED 3Cr-1MoV STEEL

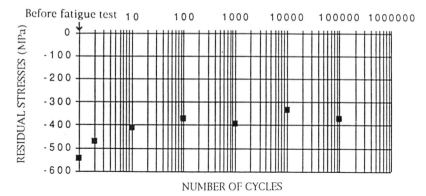

Fig 10   Compressive residual stresses relaxation in shot-peened 3Cr–1MoV steel for S_{max} = 800MPa in pure fatigue loading

The damage mechanism can be thought of as a battle between crack initiation and particle detachment. If crack propagation is possible (S_{max} > fretting fatigue limit), particle detachment is very limited and does not modify the kinematic of contact. The damage mechanism then leads to fatigue failure of the specimen. If, on the contrary, crack propagation is not possible (S_{max} < fretting fatigue limit), particle detachment increases with the number of cycles, modifying the kinematic of contact and leading to wear of the surface and elimination of the cracks. This second phenomenon is observed only on shot-

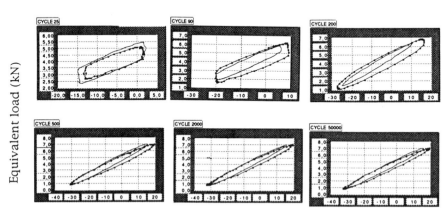

Pad deformation (μDef) proportional to tangential contact force

Fig 11   Comparison of the friction loops between shot-peened and unprotected 3Cr–1MoV steel for S_{max} = 600MPa and Fc = 140N, both curves in partial slip regime. Solid lines - unprotected steel; dotted lines - shot-peened steel

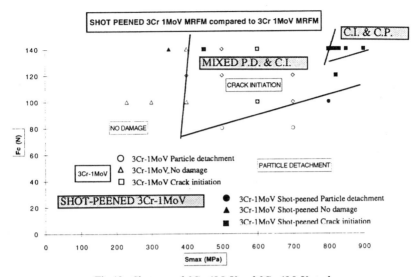

Fig 12   Shot-peened 3Cr–1MoV and 3Cr–1MoV steel

peened steel because its fretting fatigue limit (800 MPa) (see Fig. 15) is higher than that of unprotected steel (350 MPa). Conditions of crack arrest ($S_{max} <$ fretting fatigue limit) result in larger relative displacements which are more favourable to particle detachment. The surface roughness ($Ra = 2.2$ μm) and hardness of shot-peened specimens can also be a factor that favours the formation of particles.

Therefore, in the case of shot-peened steel, the 'conventional' crack initiation area of the MRFM can be divided into a crack initiation and wear area, and a crack initiation and propagation area as shown in Fig. 12. The fretting fatigue lifetime in the second area is given by the S–N curve (see Fig. 15). The shot-peened steel fretting fatigue limit (800 MPa) is still lower than the fatigue limit (975 MPa).

Cadmium coating

The electro-deposited cadmium coating is 7 μm thick.

For test parameters within the 'gross slip–particle detachment' area of the unprotected 3Cr–1MoV steel map, the presence of Cd coating does not modify the accommodation regime and the damage mechanism. The coating is rapidly eliminated from the contact area (after about 2000 cycles).

For test parameters within the 'partial slip–crack initiation' area of the unprotected steel map, the presence of Cd coating slightly delays the stabilization of the friction loops to an elliptic shape, typical of the partial slip regime (2000–5000 cycles) (see Fig. 13). During the first 2000–5000 cycles, the gross slip regime is established and the coating is progressively eliminated from the

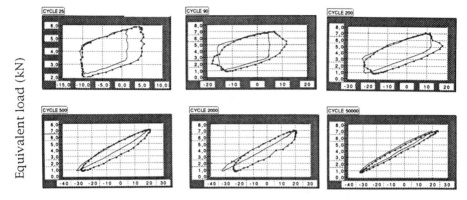

Pad deformation (μDef) proportional to tangential contact force

Fig 13 Comparison of friction loops between cadmium coated and unprotected 3Cr–1MoV steel for S_{max} = 600MPa and Fc = 140N, both curves in partial slip regime. Solid lines – uncoated steel; dotted lines – cadmium coated steel.

contact area. Afterwards, the fretting fatigue regime is similar to that of the unprotected steel. The MRFM is also similar to that of the unprotected steel. However, the damage mechanism seems to progress in three steps. The first one is the coating elimination from the contact area after a few thousand cycles and the formation of folds at the border of the contact area (see Fig. 14). The second and third steps are, respectively, crack initiation in the uncoated contact area and its propagation.

Within the crack initiation regime, only slight differences on the fretting fatigue lifetime are observed between the unprotected steel and the coated steel. Almost no difference is observed in the fretting fatigue limit, as shown in Fig. 15.

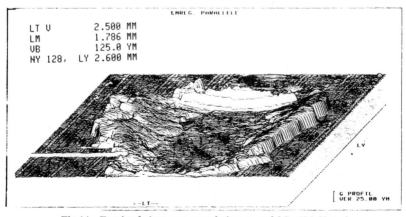

Fig 14 Fretting fatigue scar on cadmium coated 3Cr–1MoV steel

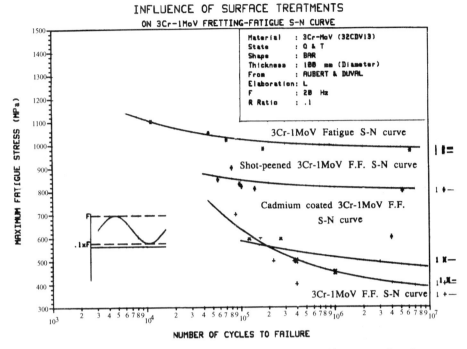

Fig 15   Comparison of protected 3Cr–1MoV fretting *S–N* curves with unprotected steel ones

Conclusions

A methodology based on fretting fatigue maps (running condition fretting map and material response fretting map) and *S–N* curves, is described in order to characterize the fretting fatigue behaviour of materials.

The results obtained on the 3Cr–1MoV steel show three equivalent areas in the running condition fretting map and in the material response fretting map which are the:

– stick–no damage area (life $> 10^7$ cycles)
– partial slip–crack initiation area (life $< 10^7$ cycles).
– gross slip–particle detachment area (life $> 10^7$ cycles)

This methodology is used to test the influence of two currently used surface treatment techniques on the fretting fatigue behaviour of a 3Cr–1MoV steel.

The main effects of *shot-peening* are:

– no influence on the kinematics of contact (running condition fretting map),
– no influence on the damage mechanisms (material response fretting map, especially crack initiation),
– crack propagation shifted to higher S_{max} values (compressive residual stresses).

The main effects of *cadmium coating* are:

- the protection is eliminated from the contact after a few thousand cycles (2000–5000),
- therefore, it has no influence on the kinematics of contact (running condition fretting map), and damage mechanisms (material response fretting map),
- crack propagation is similar to that of unprotected steel.

Fretting fatigue maps will be the basis of the determination of a suitable criterion for each damage mechanism (crack initiation and particle detachment).

Acknowledgements

The authors would like to thank Mr Pierantoni, from Eurocopter France, for supplying the 3Cr–MoV bar and surface treatments.

References

(1) VINGSBO, O. and SODERBERG, D. (1988) On fretting maps, *Wear*, **126**, 131–147.
(2) VINCENT, L., BERTHIER, Y., AND GODET, M. (1992) Testing methods in fretting fatigue: a critical appraisal, *Standardization of Fretting fatigue test methods and equipment, ASTM STP 1159*, (Edited by M. Helmi Attia and R. B. Waterhouse) Philadelphia, pp. 33–48.
(3) ZHOU, Z. R. (1992) Fissuration induite en petits debattements: application au cas d'alliages d'aluminium aéronautiques, PhD Thesis, Ecole Centrale de Lyon, France.
(4) JOHNSON, K. L. (1985) *Contact mechanics*, Cambridge University Press.
(5) FOULQUIER, J. (1991) Comportement en fretting fatigue d'un acier 30NCD16, rapport Aérospatiale n° DCR/M–60119/F–91.
(6) PETIOT, C. (1993) Comportement en fretting fatigue de l'acier 32CDV13–Influence des traitements de surface, rapport Aérospatiale n° DCR/M–60811/1–93.

K. Sato* and S. Kodama†

Effect of TiN Coating by the CVD and PVD Processes on Fretting Fatigue Characteristics in Steel

REFERENCE Sato, K. and Kodama, S., **Effect of TiN coating by the CVD and PVD processes on fretting fatigue characteristics in steel,** *Fretting Fatigue,* ESIS 18 (Edited by R. B. Waterhouse and T. C. Lindley) 1994, Mechanical Engineering Publications, London, pp. 513–526.

ABSTRACT A ceramics coating film of titanium nitride (TiN) is applied to improve the fretting fatigue characteristics of a carbon steel (S45C) and a high speed tool steel (SKH56). The effects of two coating methods, that is the chemical vapour deposition (CVD) and physical vapour deposition (PVD) processes, and of the coated TiN thickness of 2, 6, and 10 μm, on fretting S–N properties are examined. The PVD method has advantages over the CVD method and the thicker film is more effective. Such results are discussed from fretting wear observations, X-ray residual stress measurements, and finite element analysis.

Introduction

Since fretting is a surface degradation process due to mechanical and chemical attack, surface treatments are effective strategies against fretting problems; i.e., fretting wear, fretting corrosion, and fretting fatigue. Several papers have presented methods for improving fretting characteristics by surface treatments.

Gordelier and Chivers (1)(2) presented a review of palliative for fretting fatigue and carried out experiments. They pointed out the effect of surface treatments and coatings. Gabel and Bethke (3) determined the optimum fretting fatigue coating resistance of a number of coating materials for titanium specimens. Hoeppner and Gates (4) reported some approaches for the alleviation or elimination of fretting fatigue in the design of engineering structures and mechanical components. Taylor and Waterhouse (5) tried molybdenum coating by spraying in order to examine the beneficial effect of hard metal coating on combating fretting fatigue. Ohmae, Nakai, and Tsukizoe (6) studied the effect of ion plated gold, silver, and boron carbide films on fretting wear. Isogai (7) studied the effect of electroplating on fretting fatigue behaviour. The effect of shot peening on fretting fatigue has been investigated by Waterhouse and Saunders (8), Waterhouse, Noble, and Leadbeater (9), and Mutoh and Tanaka (10). Shot peening is a useful surface treatment against fretting fatigue because of the compressive residual stress in the surface layer.

In the present paper, the effect of TiN coating on fretting fatigue characteristics is studied, because high quality coating technologies have been widely applied to metal and non-metal products as a part of surface treatments. There

* Department of Mechanical Engineering, Chiba University, 1–33 Yayoi-cho, Inage-ku, Chiba, 263 Japan.
† Department of Precision Engineering, Tokyo Metropolitan University, 1–1 Minami-Ohsawa, Hachiouji-city, Tokyo, 192 – 03 Japan.

is considerable interest in the application of ceramic coatings, such as TiC, TiN, SiC, and their composites.

Materials and test methods

TiN was selected because of its advantageous characteristics as a coating material for steel specimens; namely, its hardness is not too high but nevertheless it has excellent wear resistance, and its Young's modulus is not very different from that of the substrate. Also it is not difficult to apply this coating to engineering materials. Table 1 shows the mechanical and physical properties of TiN. In this study, the TiN film was applied by the CVD (chemical vapour deposition) and PVD (physical vapour deposition) processes on two kinds of steel.

Two types of fretting fatigue tests were carried out to determine S–N properties. Rotating bending tests were performed on carbon steel specimens (JIS S45C) having 0.45 weight percent carbon, and plane bending tests were carried out on sintered high speed tool steel specimens (modified JIS SKH56). Tables 2 and 3 show the chemical compositions of the S45C steel and the SKH56 steel and their mechanical properties, respectively. Schematic diagrams of the fretting fatigue tests apparatus used are shown in Figs 1 and 2. In both rotating and bending fretting fatigue tests, the specimens are compressed by a pair of fretting pads, so that fretting occurs at the contact edge. The fretting pads used were the same materials as the specimens, but without the TiN film. The nominal contact pressure of 40 and 100 MPa was applied for the S45C and SKH56 specimens, respectively.

The TiN coating processes used in this study were as follows: in the PVD method, at a temperature of 773 K, the heating time was 2 h per 2 μm; in the CVD method, at a temperature of 1098K, the heating time was 2.5 h per 2 μm.

Table 1   Comparison of physical and mechanical properties between TiN and steel

Materials	TiN (Titanium-nitride)	Steel
Crystal structure	FCC	BCC
Density (g/cm^3)	5.44	7.87
Linear thermal coefficient	9.35	11.2
of expansion ($\times 10^{-6}$)	9.35	11.2
Melting point (K)	3200	1750
Young's modulus (GPa)	250	206
Poisson's ratio	0.19	0.30

Table 2   Chemical composition of the S45C and modified SKH56 specimens Composition (%wt)

Elements	C	Si	Mn	P	S	Cr	Mo	W	V	Co
1S45C	0.43	0.22	0.67	0.011	0.016	–	–	–	–	–
SKH56	1.0	–	–	–	–	4.2	6.5	5.5	1.6	8.0

Table 3 Mechanical properties of the specimens

Materials	S45C					SKH56				
Method of coating	CVD	CVD	PVD	PVD	PVD	CVD	CVD	PVD	PVD	PVD
Thickness of TiN film (μm)	2	6	2	6	2	6	10	2	6	10
Ultimate tensile stress (MPa)	477	480	697	-	1450	>1530	1320	804	798	796
Elongation (%)	38.8	31.9	9.6	-	-	-	-	-	-	-
Reduction in area (%)	54.5	46.3	30.6	-	25.9	17.1	17.9	3.3	-	5.4
Vickers' hardness number	185	186	278	251	335	414	432	949	1232	1325

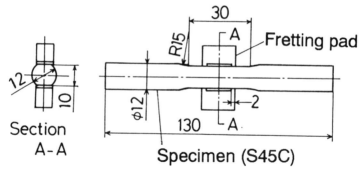

Fig 1 Shape and dimensions of rotating bending fretting fatigue tests with S45C specimens. (All
       dimensions in mm)

By these processes, TiN films of thicknesses 2 and 6 μm were coated on the
S45C specimens, and films of thicknesses 2, 6, and 10 μm were coated on the
SKH56 specimens.

Results and discussions

Rotating bending fretting fatigue tests with carbon steel S45C

The $S–N$ properties of the S45C specimens obtained from rotating bending
fretting fatigue tests are shown in Fig. 3. The $S–N$ properties were improved
by the TiN coating. The TiN coatings of 6 μm displayed more advantage on
the $S–N$ properties than those of 2 μm for both CVD and PVD methods. The
thicker TiN film has a beneficial effect on the $S–N$ properties because of its
wear resistance and shimming effect caused by the release of the stress concen-

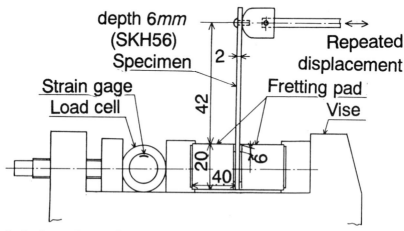

Fig 2 Specimen and test method of plane bending fretting fatigue tests with SKH56 specimens. (All
       dimensions mm)

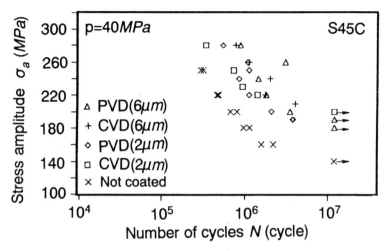

Fig 3   *S–N* properties of the S45C specimens

tration at the contact edge of the fretting pad. It is noted that the reduced specimen hardness due to heating in the coating process did not affect the *S–N* properties (Fig. 4). The reduced specimen hardness in the CVD process resulted in a decrease in notch sensitivity. A similar result, i.e., fretting strength is not affected by specimen hardness, has been obtained from tests with press-fitted specimens of S45C **(11)**.

Plane bending fretting fatigue tests with high speed tool steel SKH56

Figure 5 shows the S–N results of the SKH56 specimens obtained from the plane bending fretting fatigue tests. The *S–N* results for the PVD method are

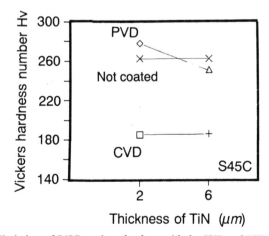

Fig 4   **Variations of S45C specimen hardness with the CVD and PVD processes**

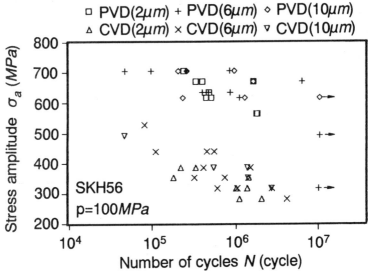

Fig 5 *S–N* properties of the SKH56 specimens

clearly superior to those for the CVD method. The advantage of thicker
coating films, such as 6 and 10 μm, can be also seen.

The variation of specimen hardness with the coating process is shown in
Fig. 6. The hardness increased with the thickness of the TiN film in the PVD
method, but was noticeably much lower in the CVD method, only increasing
slightly with thickness. The fretting fatigue properties of this material correlated with the specimen hardness. The poor fretting fatigue properties of the

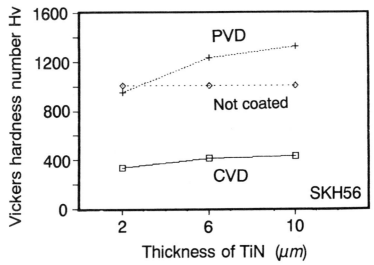

Fig 6 Variations of SKH56 specimen hardness with the CVD and PVD processes

specimens coated by the CVD method resulted from heat effects in the coating process. In practice, therefore, the CVD method is not conventionally used for high speed tool steel.

Wear resistance of TiN coated films

The improvement in wear resistance due to the TiN coating was only slight for the S45C specimens, especially for the 2 μm coating. This was due to the low hardness of the S45C substrate (Hv = 180–280). However, metal transfer from the fretting pad onto the specimen was clearly observed. This fact shows that the TiN film has a better wear resistance than the fretting pad. Figure 7 shows a worn surface for the fretted S45C specimen. Figure 7(a) is a photograph taken of the as-worn specimen surface, and Fig. 7(b) shows the specimen surface after it has been lightly polished with emery paper. The transferred metal observed as wear products in Fig. 7(a) were removed by the polishing, Fig. 7(b). Therefore, only severe wear that was observed on the specimen surface before light polishing is apparent.

By examining the specimen surface with a microscope, it was seen that the fretting wear of the coated specimen occurred after the microcracking and/or peeling of the TiN film (Fig. 8), due to the stress concentration around the oxidized transferred metal (Fig. 9). In Fig. 9, it can be seen that microcracks spread radially from the oxide, this was identified as ferrous oxide by electron probe microscope analysis.

The excellent wear resistance of the SKH56 substrate is due to its high degree of hardness. In order to examine the wear resistance of the TiN coated specimens in more detail, ball-on-flat wear tests were carried out under fretting wear conditions. The tests were conducted using the vibrator for the plane bending fretting tests. The steel ball, 10 mm in diameter, was pressed onto the specimen. The fretting was carried out under a constant relative slip amplitude of 20 μm and maximum pressure of 1300 MPa (Fig. 10).

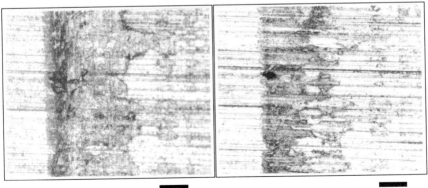

As worn   0.1mm   Polished surface   0.1mm

Fig 7   Appearance of metal transfer from the fretting pad onto the TiN film of specimen

50 μm

PVD, 2 μm, σ =220MPa

Fig 8   Microcracking in the TiN film

These test results show that the thicker TiN film is more effective in preventing fretting wear. Figure 11(a) shows the wear scar for an uncoated SKH56 specimen, and Figs. 11(b)–11(d) show the wear scars for the SKH56 specimen coated by the PVD method. Figure 11(b) indicates the appearance before polishing. Figures 11(c) and 11(d) illustrate the worn surface after light polishing to remove loose wear debris. From Figs. 11(a), (c), and (d) it can be seen that

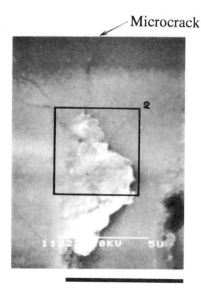

5μm

The box in this figure shows the X-ray irradiated area for No.2 test.

Fig 9   A view of microcracks spread radially from the oxide

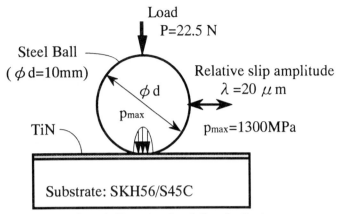

Fig 10   Schematic illustrations for a ball-on-flat fretting wear test

the TiN coating is effective in reducing fretting wear, and the effectiveness increases with thickness of the TiN film.

X-ray residual stress analysis

The residual stress in the TiN film was measured by the X-ray diffraction method using to the conditions listed in Table 4. A compressive residual stress

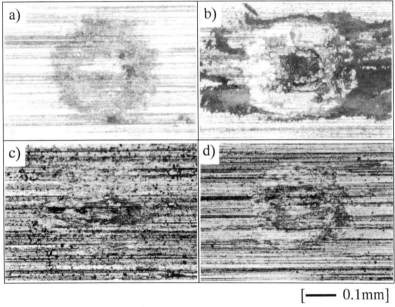

[——— 0.1mm]

a) un coated   b) PVD, $2\,\mu$m, as worn
c) PVD, $6\,\mu$m, polished   d) PVD, $2\,\mu$m, polished

Fig 11   Appearance of fretting wear on the SKH56 specimen

Table 4   X-ray diffraction conditions for residual stress measurements

Items	Substrate	TiN
Characteristic X-ray	Cr–Kα	Co–Kα
Incident angle (degrees)	0, 15, 30, 45	0, 10, 20, 30, 40
Diffraction plane	α-Fe(211)	TiN(420)
Voltage (kV)	30	30
Current (mA)	9	10
Irradiated area (mm^2)	0.8 × 6	5 × 4
Preset time (sec)	20	6
Counter	Scintillation counter	

of 1.3 and 2.0 GPa was obtained from the TiN films of 6 and 10 μm in thickness, respectively, which were coated on the SKH56 specimens by the CVD method. Unfortunately, it was impossible to measure the residual stress in all the PVD films because of profile broadening by ion embedding, and in the CVD film of 2 μm thickness because of penetration of X-ray. However, in the case of a thin film the residual stress of substrate could be measured through the film. The residual stresses in the S45C substrate were -20 MPa in the CVD method and -230 MPa in the PVD method, which were rather small compared with the residual stresses in the films.

It is considered that the residual stress in the coating results from mis-match in the thermal expansion coefficient between TiN and substrate. That is, residual stress is estimated by the following equation

$$\sigma_{TiN} = (\alpha_{TiN} - \alpha_{Fe})\Delta T E_{TiN}/(1 - \nu_{TiN}) \tag{1}$$

$$\sigma_{Fe} = (\alpha_{Fe} - \alpha_{TiN})\Delta T E_{Fe}/(1 - \nu_{Fe}) \tag{2}$$

By using values of thermal coefficient α, Young's modulus E, and Poisson's ratio ν for the TiN and substrate (Fe), and temperature change ΔT, the residual stresses for σ_{TiN} and σ_{Fe} can be calculated. The following values were obtained.

$\sigma_{TiN} = -470$ MPa, $\sigma_{Fe} = 450$ MPa in the CVD process

$\sigma_{TiN} = -290$ MPa, $\sigma_{Fe} = 270$ MPa in the PVD process.

The calculated residual stress was comprised of compression in the TiN film and tension in the substrate for both CVD and PVD processes. In the PVD process, additional compressive stresses were produced in both film and substrate by ions embedding during the coating process. The compressive stress by ions embedding was in the order of giga-Pascal (GPa) in WC $-$ 10%Co alloys (12). By adding this compressive stress to the above calculated stress, it is shifted to a higher compressive value, and the estimated residual stress is close to the value measured by the X-ray method. Both residual stresses in the film and substrate, however, did not contribute so much to the fatigue life, since they existed at very shallow surface depth and were relaxed by applied cyclic stress. The extent of fatigue life was only during the period for stress relaxation. It is known that shot peening is effective in improving the fretting fatigue characteristics because in this case the residual stress layer is much deeper than in the case of thin coating films. Therefore, combining shot peening with TiN coating will probably be very effective against fretting fatigue.

Finite element analysis on the film thickness effect

The stress distributions in the contact surface and subsurface were calculated by using an elastic finite element analysis as a contact problem. Figure 12 shows the finite element meshes and boundary conditions used. To simplify the calculations the analysis was done for push–pull loading, instead of bending, for the S45C specimens with and without the TiN film. The finite element model for the film thicknesses of 10 and 60 μm was used to enhance the effect of film thickness. The axial stress σ_a of $+200$ and -200 MPa and the nominal contact pressure of 40 MPa were applied. The coefficient of friction of 0.67 was used.

Figures 13(a)–13(c) show Mises' equivalent stress contours at the contact edge for the specimen models with and without the TiN film. There was not

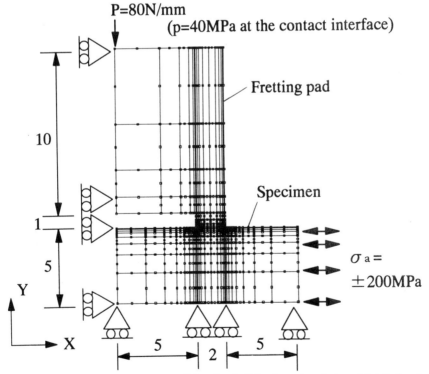

Fig 12   **Mesh divisions and boundary conditions for the finite element analysis, and detail of contact region**

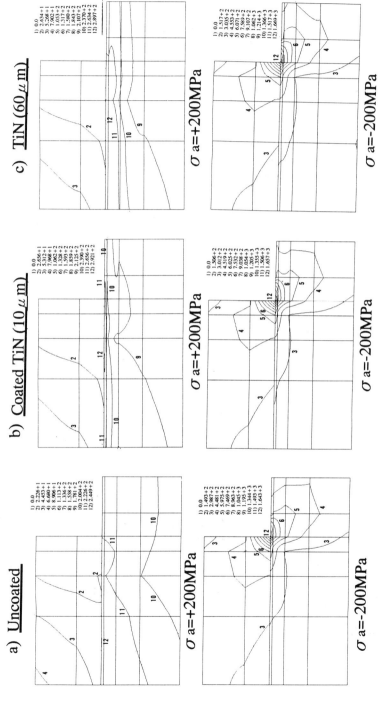

Fig 13 Mises' equivalent stress contours: (a) uncoated specimen, (b) coated specimen of 10 μm TiN thickness, and (c) coated specimen of 60 μm TiN thickness

much difference in the stress contours in specimens with and without the TiN film. In the assessment of fretting fatigue characteristics, however, it is important to evaluate the value of stress amplitude between maximum and minimum applied stresses, because stress concentration at the contact edge is changed by the loading direction (13).

The equivalent stress amplitudes of Tresca and Mises were calculated by using the following formulae respectively.

$$\sigma_{\text{Tresca, a}} = (\sigma_{x, a}^2 - 2\sigma_{x, a}\sigma_{y, a} + \sigma_{y, a}^2 + 4\tau_{xy, a}^2)^{1/2} \tag{3}$$

$$\sigma_{\text{Mises, a}} = (\sigma_{x, a}^2 - \sigma_{x, a}\sigma_{y, a} + \sigma_{y, a}^2 + 3\tau_{xy, a}^2)^{1/2} \tag{4}$$

where

$$\sigma_{x, a} = (\sigma_{xt} - \sigma_{xc})/2 \tag{5}$$

$$\sigma_{y, a} = (\sigma_{yc} - \sigma_{yt})/2 \tag{6}$$

$$\tau_{xy, a} = (\tau_{xyt} - \tau_{xyc})/2 \tag{7}$$

where σ_x, σ_y and τ_{xy} show direct and shear stresses, and subscripts t and c correspond to tensile and compressive applied stress.

The results are summarized in Table 5. The stresses σ_x, σ_y and τ_{xy} are the values at the point of substrate beneath the contact edge where the fatigue crack initiates. The equivalent stress amplitudes of $\sigma_{\text{Tresca,a}}$ and $\sigma_{\text{Mises,a}}$ decreased with increasing TiN film thickness due to the shimming effect of the TiN film. This shows that the thicker TiN film inhibits crack initiation.

Conclusions

(1) The TiN film coating is effective in improving fretting fatigue characteristics. The thicker films are more effective because of their wear resistance.

Table 5 Equivalent stress amplitude values: TiN coated versus uncoated specimens

Stresses	TiN uncoated	TiN coated 10 μm	TiN coated 60 μm
Under $\sigma_a = +200$MPa			
σ_x	255	280	265
τ_y	0	0	0
τ_{xy}	0	0	0
Under $\sigma_a = -200$MPa			
σ_x	−1990	−1500	−510
σ_y	−1423	−1000	−730
τ_{xy}	580	500	300
Equivalent stress amplitude			
$\sigma_{\text{Tresca, a}}$	1923	1477	808
$\sigma_{\text{Mises, a}}$	1678	1294	700

(2) The PVD method has the advantage over the CVD method because the latter decreases the hardness of the substrate in its coating process. It should be noted that the specimen hardness strongly influences the fretting fatigue characteristics.

(3) The advantage of the thicker TiN films was supported by the observation of micro-cracking and wear processes and also by finite element analysis.

(4) The residual stress in the TiN film processed by the PVD method was compressive, but the improvement of the fretting fatigue life was rather small.

(5) The effectiveness of the thicker TiN film was confirmed by calculation of the equivalent stress amplitude based on finite element analysis.

References

(1) GORDELIER, S. C. and CHIVERS, T. C. (1979) A literature review of palliative for fretting fatigue, *Wear*, **56**, 177–190.

(2) CHIVERS, T. C. and GORDELIER, S. C. (1984) Fretting fatigue palliatives: some comparative experiments, *Wear*, **96**, 153–175.

(3) GABEL, M. K. and BETHKE, J. J. (1978) Coating for fretting prevention, *Wear*, **46**, 81–96

(4) HOEPPNER, D. W. and GATES, F. L. (1981) Fretting fatigue considerations in engineering design, *Wear*, **70**, 155–164.

(5) TAYLOR, D. E. and WATERHOUSE, R. B. (1972) Sprayed molybdenum coatings as a protection against fretting fatigue, *Wear*, **20**, 401–407.

(6) OHMAE, N., NAKAI, T., and TSUKIZOE, T. (1974) Prevention of fretting by ion plated film, *Wear*, **30**, 299–309.

(7) ISOGAI, K. (1983) A study on fretting fatigue–the effect of electroplating on the fretting fatigue behaviour of carbon steel, Research and Investigation Bureau of the Institute for Sea Training, Report No. 59, 53–114.

(8) WATERHOUSE, R. B. and SAUNDERS, D. A. (1979) The effect of shot peening on the fretting fatigue behaviour of an austenitic stainless steel and a mild steel, *Wear*, **53**, 381–386.

(9) WATERHOUSE, R. B., NOBLE, B., and LEADBREATER, G. (1983) The effect of shot peening on the fretting-fatigue strength of aged-hardened aluminum alloy (2024A) and an austenitic stainless steel (En58A) *J. mech. Working Technol.*, **8**, 147–153

(10) MUTOH, Y. and TANAKA, K. (1988) Fretting fatigue in several steels and a cast iron, *Wear*, **125**, 175–191.

(11) SATO, K., FUJII, H., AKIYAMA, H., and KODAMA, S. (1986) Fatigue strength and crack propagation behaviour of press-fitted axle of carbon steel, *J. Soc. Mater. Sci. Japan*, **35**, 407–413.

(12) SUZUKI, H., MATSUBARA, H., MATSUO, A., and SHIBUKI, K. (1985) Residual compressive stress in Ti(C,N) layer deposited an cemented carbide by ion plating (PVD) process, *J. Japan Inst. Metals*, **14**, 773–778.

(13) SATO, K. (1992) Determination and control of contact pressure distribution in fretting fatigue, *Standardization of fretting fatigue test methods and equipment, ASTM STP 1159*, ASTM, Philadelphia, pp. 85–100.

Index